# ERGONOMICS IN ASIA: DEVELOPMENT, OPPORTUNITIES, AND CHALLENGES

SELECTED PAPERS OF THE 2ND EAST ASIAN ERGONOMICS FEDERATION SYMPOSIUM (EAEFS 2011), HSINCHU, TAIWAN, 4–8 OCTOBER 2011

# Ergonomics In Asia: Development, Opportunities, and Challenges


*Editors*

Yuh-Chuan Shih

*Department of Logistics Management, National Defense University, Taipei, Taiwan*

Sheau-Farn Max Liang

*Department of Industrial Engineering and Management, National Taipei University of Technology, Taipei, Taiwan*

*Co-Editors*

Yu-Hsing Huang

*Department of Industrial Management, National Pingtung University of Science and Technology, Pingtung, Taiwan*

Yu-Cheng Lin

*Department of Industrial Engineering and Management, Overseas Chinese University Taichung, Taiwan*

Chih-Long Lin

*Department of Crafts and Design, National Taiwan University of Arts, Taipei, Taiwan*



CRC Press
Taylor & Francis Group
Boca Raton London New York

CRC Press is an imprint of the
Taylor & Francis Group, an **informa** business
A BALKEMA BOOK

Typeset by Vikatan Publishing Solutions (P) Ltd., Chennai, India

CRC Press
Taylor & Francis Group
6000 Broken Sound Parkway NW, Suite 300
Boca Raton, FL 33487-2742

First issued in hardback 2019



ISBN-13: 978-0-415-68414-9 (hbk)
ISBN-13: 978-0-203-11999-0 (ebk)



**Visit the Taylor & Francis Web site at**
**http://www.taylorandfrancis.com**

**and the CRC Press Web site at**
**http://www.crcpress.com**

*Ergonomics in Asia – Shih & Liang (eds)*
*© 2012 Taylor & Francis Group, London, ISBN 978-0-415-68414-9*

# Table of contents

*Ergonomics in Asia – Shih & Liang (eds)*
*© 2012 Taylor & Francis Group, London, ISBN 978-0-415-68414-9*

# Preface

This volume contains the 48 selected papers from 13 countries that presented at the Second East Asian Ergonomics Federation Symposium (EAEFS), held from October 4 to 8, 2011 at the National Tsing Hua University in Taiwan. The Second EAEFS, endorsed by the International Ergonomics Association (IEA), aims to consolidate the established ties within East Asia and to extend the cooperation among ergonomics researchers, professionals, and practitioners in the region to other regions in Asia and the world.

The papers contributing to this book address the latest research and application in accordance with the theme of the symposium, "Ergonomics in Asia: Development, Opportunities and Challenges," and cover the following areas: Virtual Environments & Design, Aging & Work Ability, Usability & Interface, Biomechanics & Anthropometry, and Occupational Safety & Health.

We are most grateful to the following sponsors of the conference:
National Science Council
Ministry of Education
Institute of Occupational Safety and Health
Ergonomics Society of Taiwan
National Tsing Hua University
Feng Chia University

and to the cooperating organizations (in alphabetical order):
Chinese Ergonomics Society
Ergonomics Society of Indonesia
Ergonomics Society of Korea
Ergonomics Society of Singapore
Ergonomics Society of Thailand
Hong Kong Ergonomics Society
Human Factors and Ergonomics Society of Malaysia
Indian Society of Ergonomics
Israel Ergonomics Association
Japan Ergonomics Society
Philippines Ergonomics Society
Society for Occupational Safety, Health & Ergonomics, Japan
Turkish Ergonomics Society

We wish to thank the following Committee members who diligently contributed to the success of the symposium and to the direction of the content of this book. The symposium Committee members include.

*Ergonomics in Asia – Shih & Liang (eds)*
*© 2012 Taylor & Francis Group, London, ISBN 978-0-415-68414-9*

# Organizational committee

## ORGANIZER

Eric Min-yang Wang
*National Tsing Hua University*
*Vice President and Secretary General*
*International Ergonomics Association*

## CONFERENCE CHAIRS

Wen-Ko Chiou
*Chang Gung University*
*President*
*Ergonomics Society of Taiwan*

Chiu-Hsiang Joe Lin
*National Taiwan University of Science and Technology*
*Immediate Past President*
*Ergonomics Society of Taiwan*

*General Chair*

Kuo-Hao Tang
*Feng Chia University*

*General Co-chair*

Di-Yi Lin
*I-Shou University*

*Secretariats*

Chris Kuo-Wei Su, *National Kaohsiung First University of Science and Technology*
Yu-Hsing Huang, *National Pingtung University of Science & Technology*

*Scientific Program*

Yuh-Chuan Shih, *National Defense University*
Sheau-Farn Max Liang, *National Taipei University of Technology*
Chih-Long Lin, *National Taiwan University of Arts*

*Promotion, Public Relationship, and Exhibition*

Yu-Cheng Lin, *Overseas Chinese University*
Hsin-Chieh Wu, *Chaoyang University of Technology*
Min-Chi Chiu, *School of Occupational Therapy, Chung Shan Medical University*

*Hotel, Accommodation, Transportation, Registration and Information*

Chung-San Yu, *National Tsing Hua University*
Yu-Hung Chi, *National Tsing Hua University*
Rui-Fen Lin, *National Tsing Hua University*

*Scenic and Technical Tours*

Ying-Lien Li, *Chaoyang University of Technology*
Po-Hung Lin, *Huafan University*

*Ergonomics in Asia – Shih & Liang (eds)*
*© 2012 Taylor & Francis Group, London, ISBN 978-0-415-68414-9*

# International advisory committee

## CHAIR

| | |
|---|---|
| Sheng Wang | President of Chinese Ergonomics Society<br>Professor<br>Department of Occupational and Environmental Health<br>Peking University Health Science Center<br>China |

## VICE-CHAIR

| | |
|---|---|
| Chetwyn Chan | President of Hong Kong Ergonomics Society<br>Chair Professor of Rehabilitation Sciences<br>Applied Cognitive Neuroscience Laboratory<br>The Hong Kong Polytechnic University<br>Hong Kong Special Administration Zone<br>China |

## MEMBERS

| | |
|---|---|
| Chalermchai Chaikittiporn | President of Ergonomics Society of Thailand<br>Associated Professor<br>Department of Occupational Health and Safety<br>Faculty of Public Health, Mahidol University<br>Thailand |
| Chia-Fen Chi | Dean of International Affairs<br>Professor<br>Department of Industrial Management<br>National Taiwan University of Science and Technology<br>Taiwan, ROC |
| Sheue-Ling Hwang | Professor<br>Department of Industrial Engineering & Engineering Management<br>University of National Tsing Hua University<br>Taiwan, ROC |
| Hardianto Iridiastadi | President of Ergonomics Society of Indonesia<br>Professor<br>Faculty of Industrial Technology (FTI)<br>Institute Technology Bandung<br>Indonesia |
| Eui-Seung Jung | President of Ergonomics Society of Korea<br>Professor<br>School of Industrial Management Engineering<br>Korea University<br>Korea |

| | |
|---|---|
| Halimahtun M. Khalid | President of Human Factors and Ergonomics Society of Malaysia<br>Damai Sciences Sdn Bhd<br>Malaysia |
| Masaharu Kumashiro | Professor<br>Department of Ergonomics<br>University of Occupational and Environmental Health<br>Japan |
| Tsai-Chang Lee | Professor<br>Department of Business Administration<br>Kun Shan University<br>Taiwan, ROC |
| Ray Yair Lifshitz | President of Israel Ergonomics Association<br>Research Center for Ergonomics<br>Industrial Engineering and Management<br>Technion—Israel Institute of Technology<br>Israel |
| Kee Yong Lim | Managing Director & Chief Consultant<br>Human Centered Analysis & Design Pte Ltd<br>Singapore |
| Chiu-Hsiang Joe Lin | Professor<br>Department of Industrial Management<br>National Taiwan University of Science and Technology<br>Taiwan, ROC |
| Adnyana Manuaba | Professor<br>Department of Physiology<br>Faculty of Medicine<br>University of Udayana<br>Indonesia |
| Ahmet Fahri Özok | President of Turkish Ergonomics Society<br>Professor<br>Department of Industrial Engineering<br>Istanbul Kültür University<br>Turkey |
| Gaur Gopal Ray | President of Indian Society of Ergonomics<br>Professor<br>Industrial Design Centre<br>Indian Institute of Technology<br>Bombay, India |
| Susumu Saito | President of Japan Ergonomics Society<br>The Institute for Science of Labour<br>Japan |
| Rosemary Seva | President of Philippines Ergonomics Society<br>Professor<br>Department of Industrial Engineering<br>De La Salle University<br>Philippines |
| Evelyn Guat-Lin Tan | Professor<br>School of Housing, Building & Planning<br>University Sains<br>Malaysia |

| | |
|---|---|
| Kuo-Hao Tang | Professor<br>Department of Industrial Engineering & Systems Management<br>Feng Chia University<br>Taiwan, ROC |
| Mao-Jiun Wang | Professor<br>Department of Industrial Engineering & Engineering Management<br>National Tsing Hua University<br>Taiwan, ROC |

Yuh-Chuan Shih, *National Defense University*
Sheau-Farn Max Liang, *National Taipei University of Technology*

# *Part I: Virtual environments and design*

*Ergonomics in Asia – Shih & Liang (eds)*
 *ISBN 978-0-415-68414-9*

# Can virtuality become a professional design medium? A study of architectural information in virtual reality

Kaihsiang Liang
*Graduate Institute of Architecture in National Chiao Tung University, Hsinchu, Taiwan*

ABSTRACT: This research investigates the possibility of a virtual reality interface as a presentational medium in architectural design and the design process. The key point is to understand the effect on virtual reality presentation. There is analysis and discussion of whether a virtual reality interface is suitable as a presentational medium for architectural design or not. The conclusions are presented as two aspects of virtual reality as a presentational medium in architectural design. First, this research points out that virtual reality is suitable for architectural presentation for a design which has already been finished. Second, a virtual reality interface does not need to be applied to professional architectural presentation.

*Keywords*: virtual reality, design media, architectural media, human–computer interaction

## 1 INTRODUCTION

Architecture is the domain focusing on how to create space and how to present space. Because of the need to use different media, the final work will differ from its earlier stages (Lim 2003). Architectural media have two functions: as design media and presentation media. Design media are an aid to the designing process, and presentation media are for presenting the design itself (Schon & Wiggins 1992, Liu 1996, Liang 2010). The former is, for example, a plan drawing; the latter may be a perspective drawing. Concerning users who use the media, design media are for architectural experts; presentation media are not only for architectural experts but also for the general public who do not have an architectural design background. Therefore, design media need to include professional architectural information, but presentation media do not. This study analyzed design media and identified three main factors, which are "identify objects", "judge scale", and "recall memory" (see Table 1).

Table 1. The details of three main factors of design media's professional architectural information.

| Main factors of design media's professional architectural information | Details of main factors |
|---|---|
| Identify objects | To test whether people can clearly identify the geometric objects in design media; the features of objects include their function and geometry. |
| Judge scale | To test whether people can accurately judge the scales of objects and space in design media, including the length, width, and height. |
| Recall memory | To test whether people can accurately recall the location of objects in design media. If so, it shows the professional architectural information is being accurately conveyed to viewers, and can arouse people's attention to store information in the viewer's memory. |

The development of architectural presentation media focuses on the quality of immersion and interaction between people and space, such as virtual reality (Roussou 2001). The difference between virtual reality and other media is the ability for a human to feel and experience the design under construction. People can transmit their experiences by intuitive ways in virtual reality (Schubert & Friedmann 1998, Wu 2009). Therefore, virtual reality is available to be a presentation media.

The research problem is to investigate whether virtual reality can become a professional design medium in architectural design. Professional design media include plan drawings, elevation drawings, section drawings and perspective drawings, which can all show the information of function, geometry and spatial scale; presentational media are for example interior spatial simulations and exterior spatial simulations, which can all easily show the visualization of the architectural design. The key point is to understand the effect on virtual reality presenting, including use of objective questions and subjective questionnaires. I analyze and discuss whether virtual reality is suitable as a design medium in architecture or not. From past research, people know virtual reality is a good presentation media, but is virtual reality a valuable design media?

## 2 METHODOLOGY AND STEPS

This study uses one experiment and one questionnaire to analyze whether virtual reality can show designers the adequacy of architectural information as a professional design medium. This experimental design is using single-factor, randomized block and within-subject effects in VR-CAVE. Concerning the architectural information, there are three factors to objective experiment, which are "identify objects", "judge scale", and "recall memory." In addition to objective spatial evaluation, I use the Likert scale to evaluate the subjective experience of subjects, including users' preferences, the sense of three-dimensional space, and the sense of comfort.

### 2.1 *Experience design*

This experimental design is using single-factor, randomized block and within-subject effects. The subjects are randomly assigned to receive two tests. Digital three-dimensional modeling software, 3ds Max, is used to create a digital three-dimensional scene and to set up three cameras to render the images. There are two kinds of images, drawings in virtual reality and drawings in non-virtual reality.

I did not only record the data in the experiment, but at the same time I also used a timer to record the time subjects spent in answering questions. At the end of the experiment, there is a questionnaire to fill in for the subjects' subjective feelings. The results of analyzing the questionnaires can be compared with the experimental results in order to obtain more information.

Concerning the dependent variables, there are "identify objects", "judge scale", and "recall memory." "Identify objects" is about how many seconds the subject searches the chosen objects. "Judge scale" is about how the subject answers some questions about judging the scale of the same object in two different presentations, drawings in virtual reality and drawings in non-virtual reality. The operative criterion is the ratio of the scale of the subject's judgment compared to the correct scale I actually measure, and a smaller ratio means more accurate judgment. "Recall memory" is how many seconds the subject takes to answer the question about the previous image.

In addition to the above objective spatial evaluation, I also evaluate the subjective experience of subjects, including users' preferences, the sense of three-dimensional space, and the sense of comfort. I use the Likert scale to evaluate these four items, for example 1 means that the subject likes it very much, 2 means that the subject likes it, 3 means that subject dislikes it, 4 means that subject dislikes it very much.

### 2.2 *Experimental system*

#### 2.2.1 *Experimental equipment*

The experimental equipment, passive VR-CAVE, is designed by the Industrial Technology Research Institute in Taiwan, and it principally includes six rear projectors, seven computers and three screens. The experimental system consists of PC clusters, including six display client computers and one control server computer. The output equipment is composed of six projectors and three planar projection screens in which each side angle is 120 degrees. In the operational mode the server computer controls the screens of six client computers, and the contents of three projectors are simultaneously updated in three projector screens by the server computer.

#### 2.2.2 *Software*

Six client computers link to the control server through the TCP/IP communication. The control server makes the client computers operate synchronously, and also dominates the display projectors. It transmits data to six client computers every 10 ms, and the data are benchmarks for the client computers to display the six contents of six projection screens.

What each side of the projection screen shows is the content of two superimposed images through two kinds of polarizing filters, which are horizontal and vertical. When viewers wear glasses with polarized filters, each of their eyes can receive the corresponding image I set up.

The drawings in virtual reality are rendered by 3ds Max. I create a scene, a museum to be the material of an architectural design in the experiment. Then I set up three virtual cameras of which the right and left cameras simulate the two views of human eyes and the middle one is set for the use of drawings in non-virtual reality (see Figure 1).

#### 2.2.3 *Experimental steps*

First of all, subjects are told about the experimental purposes, processes and notes. Second, I display an image as an example to make the subjects warm up, and answer subjects' questions about the example. Then, I confirm that subjects are ready and have no questions. Finally, I display the images which are randomly generated by a computer. There are twenty images of every experiment, including ten drawings in virtual reality and ten drawings in non-virtual reality. There are three steps for questioning subjects (see Figure 2).

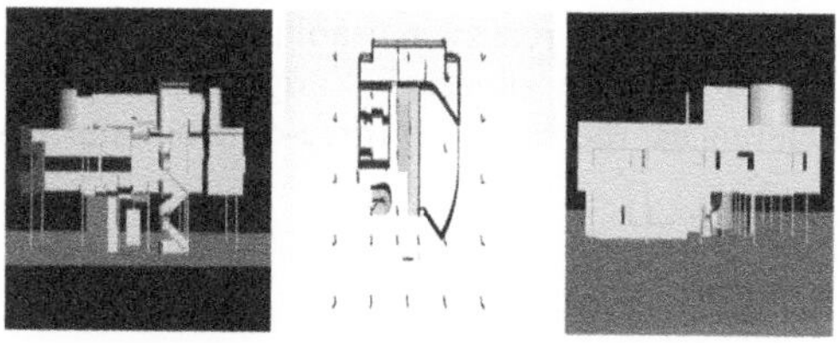

Figure 1. Examples of experimental images.

Figure 2. An example of a subject in the experiment.

*Identify objects*
"Identify objects" is to test whether people can clearly identify the geometric objects in the design media, and the features of objects including their function and geometry. The basis of "identify objects" is the length of the time taken for subjects to find the objects, and the shorter the time shows the object in the drawing is clearer. Therefore, the first step is to measure how many seconds it takes to correctly search the chosen objective.

*Judge scale*
"Judge scale" is to test whether people can accurately judge the scales of objects and space in design media, including the length, width, and height. The basis of "judge scale" is the ratio of the length from the subject; I used the estimated length of the object and the measurement of the actual length; if the ratio is close to 100%, this shows people are more able to accurately determine the scales. Therefore, the second step is to measure the scale by which subjects judge the chosen object of the same scene.

*Recall memory*
"Recall memory" is to test whether people can accurately recall the location of objects in design media. If so, it shows the professional architectural information is accurately conveyed to viewers, and it can arouse people's attention to store information in the viewer's memory. The basis of "recall memory" is the ratio of correct answers to all questions; if the ratio is close to 100%, this shows people are more able to accurately recall their memory and store more information. The third step is to measure how many details subjects can remember. I display a whole black image among every image to test subjects' understanding of the previous image. Then, I analyze the numbers of correct answers to understand the "recall memory."

## 3 RESULTS

The research problem is to investigate whether virtual reality can become a professional design medium in architectural design. The key point is to understand the effect on virtual reality presenting, including objective questions and subjective questionnaires, which can answer the problem, is virtual reality a suitable design media? Based on the experiment with single-factor and within-subject effects, this study has verified the difference between virtual reality and non-virtual reality in usability criteria and users' subjective experiences. The former includes "identify objects," "judge scale", and "recall memory;" the latter includes users' preferences, the sense of three-dimensional space, and the sense of comfort.

### 3.1 *Identify objects*

In terms of the factor "identify objects," the experimental data show that the time subjects search objects in virtual reality is quicker than it is in a non-virtual reality environment. The result shows that people can clearly identify the geometric objects in virtual reality, including the objects' function and geometry. Therefore, in terms of the factor "identify objects," virtual reality can be a professional design medium.

It is very important that viewers can clearly and quickly search objects for architectural design media, because there are usually a lot of objects of an architectural design in one architectural drawing, and every object has its different shape, size, and function. Architects design architectures and understand other cases with architectural drawings. If virtual reality can make people identify objects in VR fast, virtual reality can also reduce the time taken for architects to read architectural drawings. It is helpful for when the designer is not only one architect but a design team. Because the architectural objects in virtual reality affect the architect's design thinking more clearly and quickly than in non-virtual reality, virtual reality is more suitable as a professional design media than non-virtual reality.

### 3.2 *Judge scale*

In terms of the factor "judge scale," the experimental data show that the scale of subjects' judgment in virtual reality is no more correct than the scale in a non-virtual reality environment. Experimental result shows the lengths in virtual reality and non-virtual reality are almost the same. Although they are not judging the scales of objects and space more accurately in virtual reality, neither do people judge the scales more inaccurately. Therefore, in terms of the factor "judge scale," virtual reality can be a professional design medium, because its value of scale is the same as other professional design media.

According to the above, virtual reality and non-virtual reality do not affect architectural designers' judgment of the spatial scale. "Size" is one of the most important elements in professional architectural design media. Virtual reality makes people experience the three-dimensional space by parallax. Parallax is where people's two eyes see different images at the same time. Because of our power of processing 3D, the scale of viewers' judging in virtual reality is the same as in non-virtual reality. In terms of the factor "judge scale," virtual reality can be a professional design medium.

### 3.3 *Recall memory*

In terms of the factor "recall memory", the key point is the time taken for subjects to recall the previous image I showed and correctly answer an experimental question. The experimental data show that the time for subjects to recall in a virtual reality environment is quicker than it is in a non-virtual reality environment. The experimental result shows that "recall memory" in virtual reality is superior to non-virtual reality. Therefore, people can accurately recall the location of objects in design media. It shows that professional architectural information is accurately conveyed to viewers, and it can arouse people's attention to store information in their visual memory. In terms of the factor "recall memory," virtual reality can be a professional design medium.

"Recall memory" in virtual reality affects viewers more than it does in non-virtual reality. This also means that architects' design thinking will be more active and a more efficient use of working memory with virtual reality as a design medium. Experimental results show that in using virtual reality as a professional design medium, architects can more easily integrate complex information and a large number of plans, sections, and other professional architectural drawings.

### 3.4 *Subjective experience*

Comparing with the drawings in non-virtual reality, the results showed that subjects prefer to watch the drawings in virtual reality for the presentation of architectural design. After analyzing the data in the questionnaire, the reason is found to be because the viewers can quickly understand the architectural design, including the scales and locations of architectural elements. Therefore, viewers prefer to read virtual reality as a professional design medium. Concerning the sense of three-dimensional space, the experimental results show that subjects have a stronger sense of three-dimensional space in virtual reality than in non-virtual reality. While non-virtual reality can also provide three-dimensional sense by monocular cues, virtual reality not only has monocular cues but also binocular cues. Therefore, virtual reality offers viewers more sense of the three-dimensional. The above subjective experiences in virtual reality are better than in non-virtual reality. However, concerning the sense of comfort, when presenting architectural space in virtual reality for more than 15 minutes, the subjects' sense of comfort will be lower than in non-virtual reality.

This result shows the limitation of virtual reality as a professional design medium, which means virtual reality is not suitable to watch for a long time. Architects must pay more attention to watching and to understanding design media because they need a high level of abstraction and affordability (2004). Therefore, compared with the commercial movie played in 2009, Avatar, people need to know all the details and layout when they watch architectural

Table 2. The comparison between virtual reality and non-virtual reality.

| | Main factors of professional design media | Comparison between virtual reality and non-virtual reality |
|---|---|---|
| Objective experience | Identify objects | Better |
| | Judge scale | The same |
| | Recall memory | Better |
| Subjective experience | The sense of preference | Better |
| | The sense of three-dimensional space | Better |
| | The sense of comfort | Worse |

design. That's why a VR commercial movie can be played for about 3 hours, but VR design media are not suitable to be watched for over 15 minutes. On the other hand, watching the drawings in virtual reality must be in a lighting-controlled environment, such as VR-CAVE, which is another shortcoming if virtual reality is used as a professional design medium.

## 4 CONCLUSION

The research problem is to investigate whether virtual reality can become a professional design medium in architectural design. The conclusion is that virtual reality is suitable for a design medium in architecture. After analyzing the experimental data and questionnaire replies, the result shows virtual reality is not only a good presentation media, but also a suitable design media. Most objective factors in the experiment and subjective factors in the questionnaire support this conclusion. The former includes the main three factors of design media, "identify objects", "judge scale", and "recall memory;" the latter includes the sense of preference, and the sense of three-dimensional space.

Virtual reality can demonstrate clearly the identity of geometric objects, including the objects' function and geometry. Virtual reality as a design medium is more useful than non-virtual reality, especially when the designers are a team. Also, virtual reality affects the architect's design thinking more clearly and quickly than non-virtual reality. Architects' design thinking will be more active and a more efficient use of working memory with virtual reality. Concerning the size, virtual reality shows designers the same lengths. Recalling the location of objects in virtual reality is also quicker than in non-virtual reality. Professional architectural information can be accurately conveyed to viewers with virtual reality, and virtual reality can also arouse designers' attention to store information in the viewers' memory. Therefore, architects can more easily integrate and read complex information and a large number of professional architectural drawings with virtual reality. Concerning subjective feelings, virtual reality is much better than non-virtual reality in designers' preferences as is the sense of three-dimensional space. However, because it is not better in the sense of comfort, virtual reality is not suitable to present for more than 15 minutes.

This study provides the guidelines for virtual reality as a professional design medium. Because of limited equipment, there are issues for future research, such as real-time system and augmented reality.

## REFERENCES

Carroll, F., Smyth, M. & Dryden, L. (2004). Visual-Narrative and Virtual Reality, *The International Association of Visual Literacy,* IVLA 2004, Johannesburg, South Africa.

Kalay, Y.E. (2004). *Architecture's new media:principles, theories, and methods ofcomputer-aided design,* MIT Press.

Liang, K.H. (2010). Future architecture and past architecture in virtual reality. In D.-Y.M. Lin and H.-C. Chen (eds), *Ergonomics for All*, CRC/Balkema, pp. 421–425, an imprint of Taylor & Francis: Leiden, The Netherlands.
Lim, C.K. (2003). An insight into the freedom of using a pen: pen-based system and pen-and-paper, *In Proceedings of the 22nd Conference on Association for Computer Aided Design in Architecture 2003,* pp. 382–391, Indianapolis, Indiana.
Liu, Y.T. (1996). Understanding Architecture in the Computer Era. pp. 127–153, Taiwan.
Mitchell, W.J. & McCullough, M. (1995). *Digital Design Media*, New York: John Wiley & Sons.
Roussou, M. (2001). The interplay between form, story and history: The use of narrative in cultural and educational VR, In: Balet, O., Subsol, G., Torguet, P. (Eds.): *Virtual Storytelling: Using Virtual Reality Technologies for Storytelling*. Springer-Verlag, Berlin, 181–190.
Schon, D.A. & Wiggins, G. (1992). Kinds of seeing and their function in designing. *Design Studies*, Vol. 13, No. 2, pp. 135–156.
Schubert, T. & Friedmann, F. 1998. Embodied presence in virtual environments, In Ray Paton and Irene Neilson (ed.), *Visual representations and interpretations*. London: Springer–Verlag, pp. 268–278.
Szalapaj, P. (2005). *Contemporary Architecture and the Digital Design Process*, Architectural Press
Wu, Y.L. (2009). Some Phenomena in The Spatial Representation of Virtual Reality, *Proceedings of the 14th International Conference on Computer Aided Architectural Design Research in Asia*, pp. 143–152.

*Ergonomics in Asia – Shih & Liang (eds)*
*© 2012 Taylor & Francis Group, London, ISBN 978-0-415-68414-9*

# An evaluation from presence perspective of customer experiences in virtual environments

Chun Y. Lee, Chinmei Chou & Tien L. Sun
*Department of IEM, Yuan Ze University, Chung-Li, Taiwan*

ABSTRACT: In real environments, corporations often take the overall environmental atmosphere into consideration, hoping that customers would feel the spirit they've created. But we must also concern ourselves with the question of whether virtual environments could offer a similar atmosphere. Thus, this study aims to discuss customer experiences, researching the presence of virtual environments in both active and passive patterns. Owing to the concern of the security and seriousness of nuclear power plants, this study will be based on the perspective of customer experiences, so as to understand the presence of customer experiences in a virtual environment. Thirty-two participants were asked to complete a presence questionnaire (Witmer and Singer, 1998). There is found to be no significant difference between two kinds of customer experiences. The results also show that four factors are not significantly different.

*Keywords*: customer experience, presence, virtual reality

## 1 INTRODUCTION

Customer experience has already been called the initiator of a business "storm" in the next generation. In the 1980s, quality was the emphasis, and in the 1990s, it had turned into brand names; now customer experience is the dominant focus (Simon, 2004). Considering global competitiveness, a large degree of customization and individualization has already become an unstoppable trend, but in an age full of economic uncertainty and austerity, understanding long-term operational needs in advance could probably create another zenith for corporations.

Customer experiences, namely interaction between corporations and customers, fuse physical function, sensory stimulation, and emotional inducement. These interactions allow direct measuring of customers' expectations (Beyond Philosophy, 2011). Owing to concerns about budgets and security, the promotion of company ethos, or brands often fail to offer actual products or services for customer experience.

Thus, 3D virtual reality is used to construct a virtual environment, hoping to help customers experience the spirit of a given corporation via 3D sensory stimulations. In addition, considering the fact that demonstrating actual products requires a large space, if virtual reality is constructed in a real environment, the core value of interactive experience in virtual reality will be enshrined.

In real environments, corporations tend to consider the atmosphere of the overall environment, hoping customers can feel the spirit they've created; the situation for virtual environments is the same. But we would like to discuss whether virtual environments could offer the same atmosphere. Gibson (1966) utilized the perspectives of ecology to illustrate the mechanics of how human beings perceive external environment, and further proposed the term "presence," defining it as the concept of the mental perception of the environment. Other than the challenge of constructing the environment, this study aims to discuss whether a virtual environment could offer a corresponding presence similar to the real environment.

## 2 LITERATURE REVIEW

### 2.1 *Customer experiences*

Pine et al. (1998) divided customer experiences into two perspectives. The first one is customer participation, and the second one is cohesiveness, as shown in Figure 1. In customer participation, active and passive approaches are involved. During active participation, customers play key roles in the environment, and interact with it; while in passive participation, customers are not influential, unable to interact with the environment. Hence, if customers' experiences are passive, it is more difficult for them to experience the procedure. On the contrary, if the experiences are active, customers will greatly influence the experience procedure.

The second perspective of customer experience is cohesiveness, and this can further be divided into absorption and immersion. Absorption indicates that customers are not directly immersed in the ongoing events in the environment, and immersion shows that customers are fully involved in the overall environment, sounds, and scents. Taking the control room in a nuclear power plant as an example, absorption experiences show that visitors could only observe the operator from the outside; while immersion experiences allow visitors to cross the warning line and stand beside the operator and observe.

In addition, as indicated in Figure 1, based on these two perspectives, Pine et al. subdivided them into four categories of customer experiences. One is "Entertainment experiences" that customers are passively involved in the service environment during the service. For the control room in a nuclear power plant, customers will observe the operation from outside the room.

The second one is "Educational experiences," in which customers are actively involved in the events and environment full-heartedly. This kind of experience allows customers to touch and operate the simulation buttons freely, but leaves them unable to operate with operators.

The third kind is "Esthetic experience," which allows customers to become immersed in the environment, but they are merely passive participators. In a nuclear power plant, customers can enter the operation area and observe the operation procedures, but they cannot touch any control button.

The last category is "Escapist experiences." Customers are actively involved in the procedure, and are immersed in events and environment; namely, in the control room, customers can enter operation area and touch the control buttons.

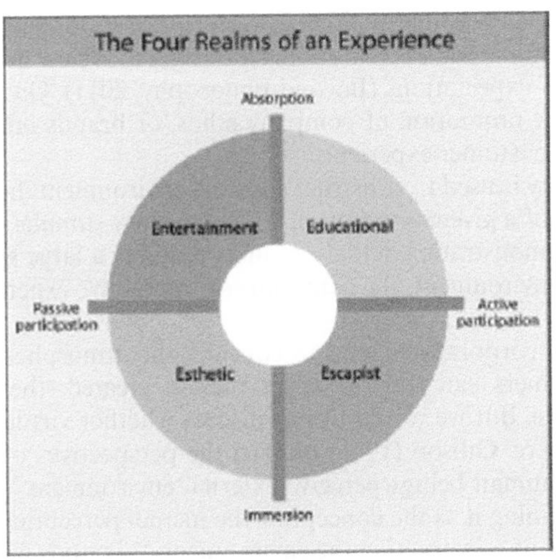


Figure 1. Schematic diagram of the customer experience.

## 2.2 *Presence*

Lombard and Ditton (1997) proposed the concept of "presence" as a crucial factor to evaluate in the virtual environment. Stanney (2002, 2003) and Thalmann (1994a, 1994b) both also thought that if virtual reality can induce presence more, the corresponding evaluation is higher.

Teresa et al. (2008) collected and arranged all the relevant literature about "presence." In the study, they cite the definition of presence from the International Society for Presence Research (ISPR), indicating that presence is a state of mind or subjective feelings. Even when individuals are in a context or experiences created by technology such as computers, they will mistakenly feel they are in a real environment, forgetting the presence of partial or complete media. Currently, "presence" is often evaluated with PQ-Presence Questionnaire.

Witmer and Singer (1998) designed a questionnaire to survey presence in the virtual environment. The questionnaire contains four dimensions, including control factors, sensory factors, distraction factors, and realism factors. Control factors concern the controls and environmental responses during the interaction with the environment. Sensory factors are feelings perceived via senses (vision and hearing) or equipment. Distraction factors aim to observe whether equipment, external objects, and interests would cause distraction, and the fourth dimension is realism factors, considering the reality of the constructed virtual environment and operation in the real environment, also concerning whether a sense of direction and reality will be lost once returning to reality.

Several studies (Botella et al., 2006; Choi et al., 2001; Difede et al., 2002; Walshe et al., 2003; Rothbaum et al., 1999; Price et al., 2007) currently consider "presence" and "highly anxious immersion" to be equal. These studies aim to evaluate the presence created by virtual reality or assigned missions by computers. These authors utilized Subjective Ratings of Distress (SUDs) or SUS Questionnaire to define the increase of the sense of presence.

# 3 METHOD

This study aims to adopt the customer virtual environmental experience patterns developed by Pine et al. (1998). The evaluation dimensions developed by Witmer and Singer (1998) are also adopted to conduct the study of the presence in the control room of a nuclear power plant. Considering the difficulty of visiting the control room in a nuclear power plant and its danger and seriousness, this study constructed a 3D simulation scene, discussing the presence with customers in different experience patterns.

## 3.1 *The construction of a virtual scene*

This study adopted Cinema 4D and Unity 3D techniques to construct the customer experience pattern in the control room in a nuclear power plant. The active and passive experiences proposed by Pine et al. (1998) are utilized to build two experience patterns, in order to understand which environment offers a higher sense of presence to customers. Apart from that, to create zero barriers for customers to participate in a 3D virtual environment, this study constructs an additional pure control room scene for test takers to experience the operation beforehand. This study simultaneously provides voice and conversation window guidance to carry out customer experience patterns.

Active customer experience pattern mainly offers an interactive conversation window in the virtual environment for customers to interact with virtual supervisors, simultaneously informing test takers their virtual identities as operators step by step, so as to rule out the barriers for virtual supervisors. Test takers could repeatedly watch the conversation window in the scene before reporting to virtual supervisors. In addition, test takers can freely adjust the observing angle in the virtual scene, and sound and effects are also played to cooperate with the experience pattern, as shown in Figure 2.

Passive customer experience pattern allows customers to respond to the supervisors in the virtual control room. The difference between the active and passive patterns lies in the

Figure 2. Active customer experiences.

guidance. The passive pattern adopts a guiding approach to proceed to the next step. An "OK" button is the only way to execute the next step. No buttons need to be touched, and no data need to be reported, test takers simply need to confirm their positions based on their subjective recognition to proceed to the next step, as shown in Figure 3.

### 3.2 *Questionnaire design*

This study refers to the dimensions of presence proposed by Witmer and Singer (1998) to design a questionnaire. In the questionnaire, control factors, sensory factors (sight and hearing), distraction factors, and realism factors are the bases. Taking the control room in a nuclear power plant as an example, the questionnaire contains 31 questions in terms of active pattern, and 26 in passive pattern, researching the presence of active and passive patterns.

### 3.3 *Experiment design*

According to the previous research goal and considerations, this study utilized between-subjects design to conduct experimental analysis, and used a random-groups design approach to randomly allocate test takers to different groups, comparing and analyzing the presence in active and passive experiences.

This study invites students in Yuan Ze University to participate. The statistical approach adopted to analyze data includes: descriptive statistics such as mean and standard deviation. T test is also adopted to discuss the evaluation of the presence in the control room of a nuclear power plant in two different experiences. Four dimensions are included in the questionnaire to analyze presence: control factors, sensory factors, distraction factors, and realism factors.

#### 3.3.1 *Subjects*

The subjects of this study were composed mainly of university students, aged 21–25. Gender and major are not specified, but subjects have to be familiar with computer operations, and those who are familiar with the design of the control room in nuclear power plants are not allowed to participate in this study. During the experiment, test takers were divided into groups A and B, with 18 subjects each group. The genders in each group were equal. Group A experienced active pattern, while Group B the opposite. This study aims to discuss the level of presence via active and passive customer experiences.

#### 3.3.2 *Experimental variables*

Before the virtual simulation experience, test takers were given sufficient time to practice and adapt to the operation of the virtual 3D interface. Experimenters would ask several operational questions to check test takers' understanding of the operation. There was no time limit for the experiment; the experiment ends once the test takers rule out all the barriers encountered. After the experiment, the test takers were required to fill in the evaluation sheet and

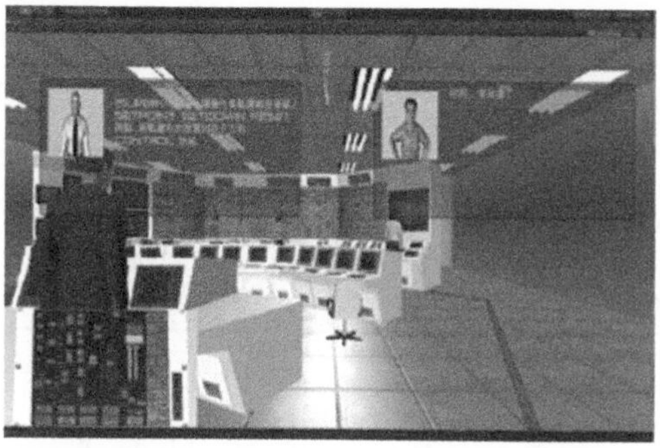

Figure 3. Passive customer experiences.

provide relevant suggestions. Active and passive customer experiences are the independent variables in this experiment, while the dependent variables are the evaluation of presence, including control factors, sensory factors, distraction factors, and realism factors.

3.3.3 *Procedure*

Test takers were informed of the goal, background, procedure, and notifications of this study prior to the experiment, and after confirming that test takers had no further questions, each test taker was required to fill in a consent form. The content of the form includes basic personal information, eyesight, and experience in operating a 3D interface.

Experimenters would set up the experimental equipment and facility beforehand, and divide test takers into two groups randomly. The active customer experience was offered to Group A. Test takers could propose any question during the procedure, and experimenters would give moderate assistance or answers. Group B would undertake the passive customer experience, and similarly, questions are welcomed during the procedure.

There was no time limitation for the experiment. Test takers could repeatedly review the experiment according to the steps provided. Finally, test takers would fill in the evaluation form about the feelings of presence, and the experimenters would answer any questions proposed by the test takers. In addition, experimenters would also record opinions about improvement proposed by test takers.

## 4 RESULTS

This study utilized the evaluation form of presence proposed by Witmer and Singer (1998), discussing the presence of customer experiences based on active and passive perspectives, and four dimensions such as control, sensory, distraction, and realism. Test takers were randomly chosen to undertake active and passive experiences, and the numbers of male and female are equal. The measurement of the evaluation form is based on "ratio," which uses a "line" as the overall sensory score, and test takers would intuitively mark on the line. The mark would then be the feelings for the test.

In this study, the highest score would be at end of the 12 centimeters line, so if test takers mark on the point of 7.65 centimeters, then the ratio score of it would be 7.65/12 = 0.6375. Inferably, the experience is the independent variable in this study, and the evaluation of presence is the dependent variable. The results are detailed below.

### 4.1 *The evaluation results of presence in two experience patterns*

This study aims to discuss the presence in the control room in a nuclear power plant based on experiences. Analyzing the overall evaluation score ratio and mean of two customer experience patterns, the results show that the presence of active customer experience is ($0.5762 \pm 0.118$),

Table 1. Evaluation results of presence based on four dimensions in different experience patterns.

| | Active | Passive | t test |
|---|---|---|---|
| Control factors | 4.706 ± 0.914 | 4.541 ± 0.710 | 0.7576 |
| Sensory factors | 5.927 ± 1.327 | 6.058 ± 1.307 | 0.7804 |
| Distraction factors | 2.785 ± 0.813 | 2.472 ± 0.865 | 0.3004 |
| Realism factors | 2.270 ± 1.036 | 2.301 ± 0.867 | 0.9278 |

and passive customer experience is (0.5378 ± 0.121). On top of that, t test shows that for test takers, there is no significant difference between the two kinds of customer experiences, and the value is 0.3038 ($p > 0.05$).

### 4.2 *The evaluation results of the four dimensions in two experience patterns*

This questionnaire of presence is composed of four dimensions, which are control factors, sensory factors, distraction factors, and realism factors. The results show that the four factors are not significantly different, as shown in Table 1.

An item in control factors stated: "I feel the virtual scene responds quickly," and for this the results of active and passive experiences show significant differences. The t test value shows 0.040 ($p < 0.05$).

## 5 DISCUSSION AND CONCLUSION

Contrary to traditional mediums for offering customer experience, this study utilized a 3D virtual environment to provide customer experiences. Whether virtual interaction can impress customers on sight or hearing and further induce "presence" is what this study aims to discuss. But this study also strives to understand how to provide a better sense of presence by demonstrating interactions.

Thus, this study used the control room in nuclear power plants to construct virtual customer experience patterns, discussing which pattern offers better immersion. Though no overall statistical significance was found, one question in the questionnaire: "Do I feel the virtual scene responds quickly?" shows significant differences. During experience procedures, interactively receiving information transmission in active pattern offers a more significant sense of presence, and customers responded quicker in interaction and received information quicker; hence, it helps to elevate the presence of customers.

In addition, when analyzing the four dimensions, sensory factors (sight and hearing), distraction factors, and realism factors show no significant differences. Sensory factors focuses on sight and hearing. Active and passive groups were all experienced in 3D virtual environments, so there was no significant difference between them. In terms of hearing, active and passive groups all received dubbed guidance, so there was no significance there either.

Sensory factors have no seeming differences, showing that the passiveness and activeness of virtual environments would not cause different levels of presence because of different interaction patterns.

In terms of distraction factors, active and passive environments show similar results. Both active and passive environments are equipped with dubbed instructions, and during the experiment, both of the environments were silent, indicating that the presence of customer experience is not relevant to active and passive patterns, and dubbed instruction will not influence the distraction factors of the presence of customers.

When it comes to realism factors, both active and passive environments were constructed with the 3D virtual environment, and the only difference is their interaction patterns. Thus, there is no significant difference. The data analysis done with t test shows no significance, so interaction patterns will not affect presence.

There is one thing worth noticing. In control factors, "response" has significant differences, indicating the information interaction in active experience has more sense of presence for customers, which means when customer experience requires information feedback, such as exhibition, active experience provides a better sense of presence.

This study utilized a virtual environment to discuss the presence of customer experience. Generally, active and passive patterns have no significant differences, showing that in customer experiences, interaction experiences will not affect presence. This result is retrieved from the analysis of post-test questionnaire, so customers could not feel the differences of presence during procedure. This study hopes to add physical measurement in the future, and use cross-examination to further research on the issue.

## ACKNOWLEDGEMENTS

This study is subsidized and supported by project NS 0991053 and NS 0991108 of the nuclear research center, and we are especially grateful to them. This study is thankful for the participation of students in Yuan Ze University. This study also appreciates the opinions offered by the judge committee of Ergonomics Society of Taiwan.

## REFERENCES

Botella, C., Baños, R.M., Guerrero, B., García-Palacios, A., Quero, S. & Alcañiz, M. (2006). Using a Flexible Virtual Environment for Treating a Storm Phobia. *PsychNology Journal*, Vol. 4, pp. 129–144.

Building Great Customer Experiences, (2011), Beyond philosophy.

Cameirão, M.S., Badia, S.B., Mayank, K.,Guger, C. & Verschure, P.F.M.J. (2007). Physiological Responses during Performance within a Virtual Scenario for the Rehabilitation of Motor Deficits, *Proceedings 10th Annual Intl Workshop on Presence,* pp. 85–88.

Choi, Y.H., Jang, D.P., Ku, J.H., Shin, M.B. & Kim, S.I. (2001). Short-Term Treatment of Acrophobia with Virtual Reality Therapy (VRT): A Case Report. *CyberPsychology and Behavior,* Vol. 4, pp. 349–354.

Difede, J. & Hoffman, H.G. (2002). Virtual Reality Exposure Therapy for World Trade Center Post-traumatic Stress Disorder: A Case Report. *CyberPsychology and Behavior*, Vol. 5, pp. 529–535.

Gibson, J.J. (1966). The Senses Considered as Perceptual Systems. Boston: Houghton Mifflin Co.

Pine II, B.J. & Gilmore, J.H. (1998). Welcome to the Experience Economy, Harvard Business Review, pp. 97–105.

Price, M. & Anderson, P. (2007). The Role of Presence in Virtual Reality Exposure Therapy. *Journal of Anxiety Disorders*, Vol. 21, pp. 742–751.

Rand, D., Kizony, R., Feintuch, U., Katz, N., Josman, N., Rizzo, A. & Weiss, P.L. (2005). Comparison of Two VR Platforms for Rehabilitation: Video Capture versus HMD. *Presence: Teleoperators and Virtual Environments*, Vol. 14, pp. 147–160.

Rothbaum, B.O. & Hodges, L.F. (1999). The use of Virtual Reality Exposure in the Treatment of Anxiety Disorders. *Behavior Modification,* Vol. 23, pp. 507–525.

Simon, D. (2004). Customer Experience. *Greater China CRM*, (http://www.gccrm.com).

Sponselee, A.M., Kort, Y. & Meijnders, de A. (2004). Healing Media: The moderating role of presence in restoring from stress in a mediated environment, *Proceedings 7th Annual Intl Workshop on Presence*, pp. 197–203.

Teresa, T., Anna, S., Chery, C.B. & Bridget, R. (2008). How Real Is It? The State of (Tele) Presence in Therapy with Mediated Environments. *Proceedings of the 11th Annual International Workshop on Presence*, Vol. 3280–284.

Wald, J. & Taylor, S. (2003). Preliminary Research on the Efficacy of Virtual Reality Exposure Therapy to Treat Driving Phobia. *Cyber Psychology and Behavior*, Vol. 6, pp. 459–465.

Walshe, D.G., Lewis, E.J., Kim, S.I., Sullivan, K.O' & Wiederhold, B.K. (2003). Exploring the Use of Computer Games and Virtual Reality in Exposure Therapy for Fear of Driving Following a Motor Vehicle Accident. *CyberPsychology and Behavior*, Vol. 6, pp. 329–334.

Witmer, B.G. & Singer, M.J. (1998). Measuring Presence in Virtual Environments: A Presence Questionnaire, *Presence,* Vol. 7, No. 3, pp. 225–240.

*Ergonomics in Asia – Shih & Liang (eds)*
*© 2012 Taylor & Francis Group, London, ISBN 978-0-415-68414-9*

# The effects of tactile types on the identification of tactile symbols

Yung-Hsiang Tu, Yu-Tzu Liao, Wan-Ning Huang, Pei-Chun Liao, Huan-Lung Lo, Wei-Ling Lu, Tsao-Ting Hung & Yi-Lin Wang
*WangGraduate School of Industrial Design, Tatung University, Taipei, Taiwan*

ABSTRACT: The effects of three different tactile types on the performance of identifying six tactile symbols are reported. Thirty participants were asked to identify eighteen randomly assigned tactile symbols which differed in their tactile modality (Solid-face, Slant-line, Swell-dot), and their tactile shape (Square, Circle, Triangle, Inverted-Triangle, Rhombus, Cross). The accuracy of identification and the mistaken pairs of tactile symbols were recorded. Repeated measure of ANOVA shows that the Solid-face has the higher identification accuracy than the Slant-line and Swell-dot type. Within those six tactile symbols, the Square, Triangle, and Inverted-triangle have a higher accuracy than the other three shapes. From the mistaken-pair table, Circle is easily mistaken for Rhombus; and Rhombus is significantly mistaken for Circle, and the Cross symbol is mistaken for the Circle or Rhombus. The implications of this result are discussed.

*Keywords*: touch, visual impairment, tactile drawing, symbol, tactile map

## 1 INTRODUCTION

Almost half of our learning experience comes from vision. The visually impaired people have to make use of their other senses in order to know the world better. As many researchers have mentioned, touch sensation might be the major information input source for the blind, so a lot of visually graphical information can be transferred into tactile graphics for the blind to sense. Most of the tactile materials are produced on physical media through different methods, including collage, embossed paper, thermoforming, microcapsule paper, high-density Braille printing, and 3D printing, and 3D pin arrays (Lévesque & Hayward, 2008; McCallum & Ungar, 2003; Sjostrom et al., 2003; Yu & Brewster, 2002; Walsh & Gardner, 2001; Edman, 1992); these are used for geographic or orientation maps, mathematical graphs, and diagrams. The tactile perception becomes a major study field for understanding the effectiveness and efficiency of those tactile tools. The perception of tactile objects can be a diversity of fields in which the effects might interrelated with each other. Within those factors, the geometric forms and surface textures have been examined. The sensation of the tactile form was studied from tactile character recognition (Vega-Bermudez et al., 1991; Johnson & Phillips, 1981); Braille word recognition (Heller, 1985, 1986, 1987; Phillips et al., 1983); and tactile pictures identification (Megee & Kennedy, 1980; Kennedy & Fox, 1977). It seems that blind people have some advantages from the large experience of exploration of simple tactile objects in many ways but, in the situation of lacking the help of vision, they have difficulty in complex picture recognition. The judgment of visual texture might be less complex than judging the tactual texture, as claimed by Bjorkman (1967). How the roughness of texture was judged was discussed by researchers following different methods (Hollins & Risner, 2000; Miyaoka et al., 1999; Connor & Johnson, 1992; Jones & O'Neal, 1985; Heller, 1982; Lederman et al., 1982). Heller (1989) said that touch seems rather poor at the identifying 2D patterns or pictures. He also found that touch has a higher accuracy of surface identification than vision in finer texture identification. Other studies pictured the requirement of tactile graphics, such as Frascara et al. (1993) who said that the tactile symbols can be separated into three kinds, which were Point type, Line type and Area type. And he suggested that the tactile height of the symbols should be higher than 0.5 mm at least for Area symbols, 1.0 mm for Line symbols, and 1.5 mm for point ones.

This present study was focused on the tactile texture and tactile form. The following experiment was to investigate the effect of different surfaces on the identification of different tactile symbols.

## 2 METHOD

Thirty subjects were recruited, 15 were sighted persons who were university students of Tatung University, aged from 22 ~ 58 years old and whose mean age was 31, and the other 15 were from Taipei School for the Visually Impaired, aged from 13 ~ 18 years old, whose mean age was 15 years old, including two blind students and eleven with low vision. All of the subjects self report that they had no injury on their hands or fingers, and they were all right-handed.

There were two independent variables in this study; one was the tactile modality which included three levels: Solid-face, Slant-line, and Swell-dot, and the other was the tactile shape. The Solid-face type was a full convex face within the assigned symbol. The Slant-line type was a symbol filled with 45 degree of 1.0 mm lines and 1.0 mm intervals. And the Swell-dot type of tactile modality was a symbol filled with dots which had a diameter of 1.0 mm and the gap between their two centers was 1.0 mm also. The six tactile shapes were: Square, Circle, Triangle, Inverted-Triangle, Rhombus, and Cross. The Square symbol had a size of 8 mm × 8 mm, the diameter of the Circle symbol was 8 mm, the Triangle symbol had an 8 mm side length, the Inverted-Triangle symbol had the size of Triangle symbol but its horizontal side located on the top, the Rhombus symbol had an 8 mm long axis, and the size of the Cross symbol was also 8 mm × 8 mm with 4 mm of side length (Figure 1). We randomly arranged those symbols three times into three stimuli lines for the subject to explore (Figure 2).

We prepared three different arrangements and randomly assigned them to the subjects in order to prevent the possible learning effect. The tactile symbols were produced on swell paper by the Tactile Image Enhancer. The stimuli were three pieces of A4 size swell paper and each of them was stuck on a strawboard to prevent unintentional movement. The subject was blindfolded and asked to sit comfortably in front of a task desk. In the first stage of the experiment, a demo swell sheet which had the six tactile symbols printed on it was provided for the subject to explore; the experimenter verbally told the subject the name of each tactile symbol, then the subject had two minutes to explore all of the symbols as he wished. Later, the subject removed the blindfold and took three minutes break. When the subject returned to his seat and was blindfolded again, the formal experiment began with randomly assigned stimuli which were prepared for the experiment. The stimulus was installed on the desk at a fixed standard position, and the experimenter led the subject's right hand

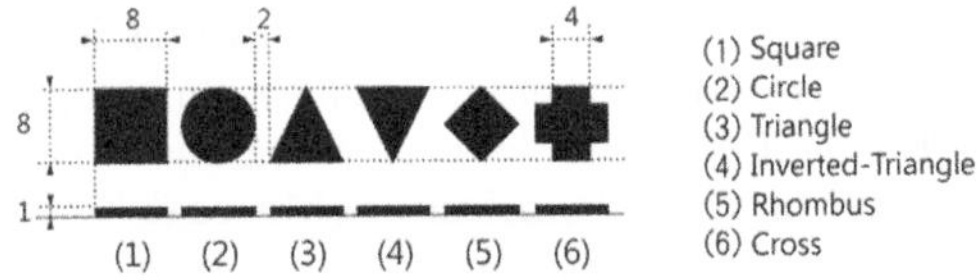


Figure 1. The dimensions of those six symbols in this study.

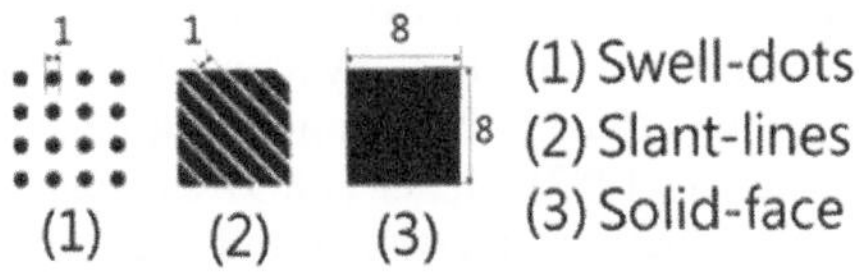


Figure 2. An example of the three stimuli employed.

to touch the standby position which was the left end of the top tactile line. The subject was asked to start the exploring process on the first tactile symbol; he had to tell the experimenter the name of the symbol he identified. The experimenter recorded the answer and told him to proceed to the next exploration. To finish the whole three sheets of tactile stimuli took the subject about 30 minutes. A gift was given to the subject for his contribution. The accuracy of identification was recorded and the mistaken pairs were also included in a table.

## 3 RESULT AND DISCUSSION

With a repeated measure of ANOVA, the accuracy of identification among those three tactile types were significantly different: $F(2, 28) = 22.677$, $p < 0.001$, the Solid-face type of tactile symbols had the highest accuracy compared to the others. The Solid-face had an accuracy of 77.67%, the Slant-line had 66.30%, and the Swell-dot had an accuracy which was lower than half, 44.07%. This result shows that the full filled way (solid) tactile symbol is a better design for the user on the tactile symbol identification. This result verified the finding of many researchers about the solid type of tactile symbol having the better recognition rate than other types. The solid type of tactile symbols had significant borders (or edge) to its shape, and the fingertip might sensitively catch the difference between the border and its concave base. A possible disadvantage of the solid type could be the information loss of its inner shape. In the case of one shape has to be represented in a way with inner and outer parts, e.g. for a castle, cup or donut, the inner shape of the object might be misunderstood by the fingers or hands.

Although studies indicated that blind subjects had the superior performance because of their learning effects on the Braille reading training (Craig, 2002; Grant, Thiagarajah & Sathian, 2000), the Braille-dot like stimuli still had the least accuracy rate compared to the others. This suggests that Braille training can help the Braille reading but it is not helpful for identifying different shapes of object. One possible reason might be that the 6 dots of Braille array is a small region for the fingers, and the Braille character recognition is based on the shapes outlined by these dots to encode the 'word' (Loomis, 1981). The outlined shape had only localized dots and the fingers have to transfer the combination of these dots into a boundary-like shape. Subjects could not transfer those Braille-like dots into a suitable or possible shape in a short time. It might be valuable to formulate the density of dots, the pitch of them and the exploration strategy for the Braille user in utilizing their Braille training in other tasks. Some researchers argued that the advantage of the blind on haptic tasks reflected the results of using cues that are normally ignored by the sighted, or the efficiency of their exploration strategies (D'Angiulli, Kennedy & Heller, 1998; Shimizu, Saida & Shimura, 1993).

Within those six symbols, repeated measure of *ANOVA*, $F(5,25) = 36.590, p < 0.001$, found that the Square (80.74%), Triangle (81.11%), and Inverted-Triangle (79.63%) had higher accuracy than Circle (64.07%) and Rhombus (51.85%), and the Cross (26.67%) had the least accuracy. This result showed that the Cross symbol (also call the Plus symbol) was the last symbol to choose in the selection of tactile symbol for its low recognition rate. It had been found that a symbol with an obvious shape was easy to recognize. In the study of testing a refreshable Braille display by L'evesque & Hayward (2008), the Plus symbol (similar to our Cross symbol) was the least one to be recognized among six designs (Square, Circle, Trigon, Inverted-Triangle, Rhombus, Plus). The size of their symbols had two categories which were 2 cm (small) and 3 cm (large), and the recognition rate for the two categories were 55% and 75%, respectively, and for the all sizes of Plus symbol was 65%. Those rates were all larger than our result (26.67%), in which the Cross symbol had a size within 8 mm. It might be interesting to see the size effect in any future study.

To look at the inside of these identification errors, Table 1 was prepared which reported the total mistaken amount of these six shapes from each other. The last column of this table was arranged to show the ambiguous identification. The unknown might tell us the information about which tactile shape might cause confusion most. The items of rows showed the original symbol and the column items represented the response symbol. The stars at the

Table 1. The mistaken-pair table for this study.

| Original \ Mistake | | ■ | ● | ▲ | ▼ | ◆ | ✚ | Unknow | *p*-value |
|---|---|---|---|---|---|---|---|---|---|
| ■ | | | 1 | 4 | 1 | 4 | 0 | 2 | |
| | | | 1 | 2 | 3 | 2 | 2 | 9 | |
| | | | 6 | 1 | 0 | 0 | 3 | 13 | * |
| ● | | 3 | | 0 | 0 | 0 | 2 | 0 | |
| | | 2 | | 2 | 0 | 4 | 6 | 18 | |
| | | 11 | | 0 | 0 | 20 | 13 | 27 | *** |
| ▲ | | 0 | 0 | | 3 | 1 | 0 | 0 | |
| | | 0 | 0 | | 1 | 3 | 1 | 8 | |
| | | 2 | 2 | | 1 | 2 | 4 | 20 | * |
| ▼ | | 1 | 0 | 6 | | 2 | 0 | 0 | |
| | | 0 | 1 | 8 | | 8 | 1 | 6 | |
| | | 6 | 2 | 1 | | 1 | 3 | 15 | * |
| ◆ | | 1 | 19 | 4 | 1 | | 10 | 3 | * |
| | | 0 | 25 | 3 | 0 | | 8 | 8 | *** |
| | | 8 | 16 | 7 | 2 | | 9 | 16 | |
| ✚ | | 3 | 28 | 2 | 0 | 30 | | 2 | *** |
| | | 5 | 20 | 2 | 0 | 23 | | 19 | *** |
| | | 30 | 11 | 3 | 0 | 13 | | 26 | *** |
| *p*-value | | *** | *** | | | *** | | - | |

right column or lowest row meant the *p*-value in a repeated measure was lower than 0.05 (* < 0.05; ** < 0.01; *** < 0.001) which represented the significant differences among the relative shapes.

Within these rows the effects of texture are reviewed in four points. Firstly, Circle in dots is significantly mistaken for Rhombus, or an unknown item. The Rhombus in dots has the same situation, often being confounded with a Circle. So it is best to prevent showing these together in a close situation. Secondly, the Rhombus in slant is mistaken for Circle too, which hints that the Rhombus is frequently misjudged by the finger. The third, Cross in solid and in slant are all mistaken into Circle or Rhombus, which hints that the sensation of Cross by finger is very near to that of those two shapes. It might be caused by the design of the Cross symbol, in which the nick of the Cross symbol was 2 mm on each side (half of the symbol edge length) in this study. Certainly, it is desirable to obtain further observation on the nick effect for similar symbols. Lastly, the dot type of Cross symbol was significantly mistaken for a Square symbol, which suggests the gap distance (1.0 mm) of the Cross was too tight for the finger to distinguish whether the ends of Cross were linked with a dot.

Gibson (1962) found that straight lines among lots of curved lines would be perceived as curves by vision. Vision illusions of this type have a basic difference from illusions of touch, which is the way of searching the object information. In vision, the information is caught entirely by our eyes, while the touch sensor (with no vision assistant) gathers information one at a time by hands or by fingers. There are some illusions caused by the vision or touch which make a person produce an error estimation. Lots of graphical illusions about shape and size by vision were proposed and discussed. Many researchers found that some of the illusions are also occasioned by touch (e.g., Gentaz & Hatwell, 2004; Millar & Al-Attar, 2000, 2002; Day & Avery, 1970). The illusion effect might be one of the reasons affecting symbol identification.

By long periods of observation, Kleinman (1979) concluded that exploration strategies can be listed into several kinds which include thoroughness of searching; tracing the path; comparing the features in between; comparing the congruent perimeter of objects; and to mirror the image and tracing its edge. In the experiment, we asked the participant to follow an arranged order and orientation to explore those symbols, which might be a factor of affecting the accuracy rate.

The tactile map is designed for the visually impaired people to reach a local environment in order to build a mental image about the space. If the symbol in the tactile map tells the wrong story to the user, the user might make a wrong decision for his movement. To advance study the possible impact dynamic for this field is inflated and apprehensive for the users.

## REFERENCES


Bjorkman, M. (1967). Relations between intra-modal and cross-modal matching. *Scandinavian Journal of Psychology,* Vol. 8, pp. 65–76.

Connor, C.E. & Johnson, K.O. (1992). Neural coding of tactile texture: comparison of spatial and temporal mechanisms for roughness perception. *J Neuroscience*, Vol. 12, pp. 3414–3426.

D'angiulli, A., Kennedy, J.M. & Heller, M.A. (1988). Blind children recognizing tactile picture respond like sighted children given guidance in exploration. *Scandinavian Journal of Psychology*, Vol. 39, pp. 187–190.

Day, R.H. & Avery, G.C. (1970). Absence of the horizontal-vertical illusion in haptic space. *Journal of Experimental Psychology,* Vol. 83, pp. 172–173.

Edman, P.K. (1992). Tactile Graphics. AFB Press, New York.
Foulke, E.M. & Warm, J.S. (1967). Effects of complexity and redundancy on the tactual recognition of metric figures. *Perceptual & Motor Skills,* Vol. 25, pp. 177–187.
Frascara, J. & Sadler-Takach, B. (1993). The Design of Tactile Map Symbols for Visually Impaired People, *Information Design Journal*, Vol. 7, No. 1, pp. 67–75.
Gentaz, E. & Hatwell, Y. (2004). Geometrical haptic illusions: the role of exploration in the Muller-Lyer, vertical-horizontal, and Delboeuf illusions. Psychonomic Bulletin & Review, Vol. 11, pp. 31–40.
Gibson, J.J. (1962). Observation on active haptics. *Psychological Review,* Vol. 69, pp. 477–490.
Grant, A.C., Thiagarajah, M.C. & Sathian, K. (2000). Tactile perception in blind Braille readers: A psychophysical study of acuity and hyper acuity using gratings and dot patterns, Attention, Perception & Psychophysics, Vol. 62, No. 2, pp. 301–312.
Heller, M.A. (1982). Visual and tactual texture perception: intersensory cooperation. *Percept & Psychophysics,* Vol. 31, pp. 339–344.
Heller, M.A. (1985). Tactual perception of embossed Morse code and Braille: The alliance of vision and touch. *Perception,* Vol. 14, pp. 563–570.
Heller, M.A. (1986). Active and passive tactile Braille recognition. *Bulletin of the Psychonomic Society,* Vol. 24, pp. 201–202.
Heller, M.A. (1987). The effect of orientation on visual and tactual Braille recognition. *Perception,* Vol. 16, pp. 291–298.
Heller, M.A. (1989). Picture and pattern perception in the sighted and the blind: the advantage of the late blind. *Perception,* Vol. 18, pp. 379–389.
Hollins, M. & Risner, S.R. (2000). Evidence for the duplex theory of tactile texture perception. *Percept & Psychophysics,* Vol. 62, pp. 695–705.
Johnson, K.O. & Phillips, J.R. (1981). Tactile spatial resolution. (I) Two-point discrimination, gap detection, grating resolution, and letter recognition. J. Neurophysiol, Vol. 46, pp. 1177–1192.
Jones, B. & O'Neal, S. (1985). Combining vision and touch in texture perception. *Perception & Psychophysics,* Vol. 37, pp. 66–72.
Kennedy, J.M. & Fox, N. (1977). Pictures to see and pictures to touch. In D. Perkins & B. Leondar (Eds.), The arts and cognition (pp. 118–135). Baltimore: J. Hopkins Press.
Kleinman, J.M. (1979). Developmental changes in haptic exploration and matching accuracy. *Developmental Psychology,* Vol. 15, No. 4, pp. 480–481.
Lederman, S.J., Loomis, J.M. & Williams, D.A. (1982). The role of vibration in the tactual perception of roughness. *Perception & Psychophysics,* Vol. 12, pp. 109–116.
Lévesque, V. & Hayward, V. (2008). Tactile graphics rendering using three laterotactile drawing primitives, *Proc. 16th Symposium on Haptic Interfaces For Virtual Environment And Tele-operator Systems,* pp. 429–436, Reno, Nevada, March 13–14.
Loomis, J.M. (1981). Tactile pattern perception. *Perception,* Vol. 10, pp. 5–27.
Mcallum, D. & Ungar, S. (2003). An introduction to the use of inkjet for tactile diagram production. *British Journal of Visual Impairment,* Vol. 21, No. 2, pp. 73–77.
Megee, L.E. & Kennedy, J.M. (1980). Exploring pictures tactually. *Nature,* pp. 283, 287–288.
Millar, S. & Al-Attar, Z. (2000). Vertical and bisection bias in active touch. *Perception,* Vol. 29, pp. 481–500.
Millar, S. & Al-Attar, Z. (2002). The Muller-Lyer illusion in touch and vision: implication for multisensory processes. *Perception & Psychophysics,* Vol. 64, pp. 353–365.
Miyaoka, T., Mano, T. & Ohka, M. (1999). Mechanisms of fine-surface-texture discrimination in human tactile sensation. J Acoust Soc Am, pp. 2485–2492.
Phillips, J.R., Johnson, K.O. & Browne, H.M. (1983). A comparison of visual and two modes of letter recognition. *Perception & Psychophysics,* Vol. 34, pp. 243–249.
Shimizu, Y., Saida, S. & Shimura, H. (1993). Tactile pattern recognition by graphic display: Important of 3-D information for haptic perception of familiar objects. *Perception & Psychophysics,* Vol. 53, pp. 43–48.
Sjostrom, C., Danielsson, C., Magnusson, C. & Rassmus-Grhn, K. (2003). Phantom-based haptic line graphics for blind persons. *Visual Impairment Research,* Vol. 5, No. 1, pp. 13–32.
Vega-Bermudez, F., Johnson, K.O. & Hsiao, S.S. (1991). Human tactile pattern recognition: active versus passive touch, velocity effects, and patterns of confusion. J Neurophysiol, Vol. 65, pp. 531–546.
Walsh, P. & Gardner, J.A. (2001). *TIGER, a new age of tactile text and graphics,* In Proc, CSUN 2001.
Yu, W. & Brewster, S. (2002). *Comparing two haptic interfaces for multimodal graph rendering*, In Proc, HAPTICS 2002, pp. 3–9.

*Ergonomics in Asia – Shih & Liang (eds)*
*© 2012 Taylor & Francis Group, London, ISBN 978-0-415-68414-9*

# Activities implementation in house kitchen interior

I. Nyoman Artayasa
*Interior Design, The Faculty of Fine Arts and Design, Indonesian Institute of the Arts Denpasar, Denpasar, Indonesia*

ABSTRACT: The purpose of interior design is to solve human problems related to carrying out activities in spaces, in order to achieve comfort, safety, effectiveness and the development of productivity in accordance with human character and culture. People are regarded as the starting point of an interior design and also become the main character so they should get special attention, especially in the discussion of all things related to the design plan, since the design will be used by them to conduct their activities. Human activities are varied, depending on the time and place, which are restricted by the norms of human behaviour and sense as well as the position and role of the person carrying out the activity. One activity that is conducted in a house with special character devoted to subsistence activities is cooking, which is carried out in the kitchen. Cooking activities can be classified as half of all interior work and also as a tiring job. Cooking consists of three main activities namely preparation, washing and cooking. There are several things we have to prepare for cooking activities which can be accommodated well in the kitchen by undertaking the activities with special care and attention to the inhabitant's social and cultural rights. The study was carried out by using descriptive qualitative research to describe the traditional activity in the area of a traditional Balinese kitchen. The purpose of this research is to find out what activities are really worthy to be accommodated in an appropriate facility to create a safe and comfortable kitchen interior and to support an increase of productivity in cooking activities. The data or the populations used in this study are 40 traditional houses in Labaksuren village, but there are only 10 houses which are eligible to be used as the subject of the study. The result of the study showed that the activities carried out in the house kitchen interior in the Semaja Bengkelsari village can be described as follows. Things should be prepared during the preparation activities: cabinets for food may be hung (*langki*) or not, and the washing of tubs, buckets, table and tools where cutting activities are done (*talenan*/cutting board), and other equipment such as *dingklik*/stools for seating and also the water store. In processing activities, things that should be prepared are; furnace; pans, place for processing the food/*bale-bale*, and things needed during the food presentation activities, which have to be provided: the bench; *bale-bale*; cabinet food store which may be hung or not.

## 1 INTRODUCTION

An interior design is applied to solve human problems related to the implementation of activities in a space to achieve comfort of safety, effectiveness and to increase productivity based on the character and culture belonging to human life. Besides being the main subject, people also become the main characters of the interior design and have the important role in confirming all things related to the planning of room design, since the room will be used by them to perform all of their activities. The range of human activities are diverse and are influenced by time and place, and also restricted by norms of behaviour and the status and role of the persons carrying out the activities. Activities which are conducted in houses also vary depending on the person who is doing the activity, timing, place and how often and how important are the activities to be done. One special activity which is done in a house kitchen is cooking. As one of the daily activities, cooking is carried out from the morning up until night-time, which starts with the preparation of breakfast, lunch and also dinner.

Time estimated for doing such activities is about 8 hours a day. Suptandar (1985) stated that in general women work from 7 a.m to 7 p.m, and 378 minutes for carrying out the activities in the kitchen. Cooking can be categorized as semi-hard work (Asri, 1991) and also a tiring job. Cooking consist of three main activities namely; preparation, washing and cooking. These three activities can also divide into; mix, wash, flavor mix, the food ready to be served and eaten, wash and store away. Usually these activities are associated with facilities such as the food processing table, kitchen sink, store table, and table to serve the food. However, not all of the activities can be associated with the facilities form. There are several things that can be done in order to accommodate the activities which are done in the kitchen, by conducting the activities carefully and giving attention to the social and cultural norms belonging to the persons who are living in the house. This study is a descriptive and qualitative research, in which, in the first year of the research, is aiming at the observation of traditional activity in the area of a traditional house kitchen. Meanwhile, in the second year, the research is focused on the activity analysis with conclusions about the activities which needs to be accommodated in the traditional Balinese kitchen. The aim of the study is to find out the most important activity to be incorporated into an appropriate facility to produce a safe and comfortable house kitchen interior in order to increase productivity.

Based on the explanation given above, the problem of the study can be formulated as the following; what kind of activity can be conducted in the kitchen and what can be incorporated in the form of the kitchen facility?

## 2 MATERIALS AND METHOD

Materials: The population of the study consists of 40 traditional Balinese house kitchens located in Br. Semaja, Bengkel Sari Village, Tabanan Bali, but only 10 house kitchens are eligible to be used as the sample of the study

Methods: The study is descriptive qualitative, which is aimed at the observation of activities which are conducted in traditional Balinese house kitchens.

## 3 RESULTS AND DISCUSSION

### 3.1 *Activities in the kitchen*

Generally, activities in traditional house kitchens in Br. Semaja, Bengkel Sari village, Selbar district, Tabanan Bali, are not much different from traditional house kitchen in other villages. The activities can be distinguished accordance with the frequency with which the activity is done. Routine activity which is performed during the day consists of:

1. Morning activity: food preparation for the whole family, which is done by the mother. The activity consists of: preparation, processing, and serving. Breakfast preparation may be done inside or outside the kitchen.
2. Afternoon activity: lunch for the family members without any preparation.
3. Evening activity: preparation for dinner/supper, starts with preparation to serve the food which is cooked in the morning.

Ritual activities based on *wewaran* are the *saiban* offering, offering based on *tri wara a*nd *panca wara* once in a month, offering based on *tri wara, panca wara* and *sapta wara* on Saturday, and offering based on *purnama*/full moon and *tilem*/dark (new) moon. Those ritual/religious activities are prepared in the kitchen after the cooking activity. The activity pattern of the ritual activities has the same pattern as the cooking activity, such as preparation, processing and serving (ready to be offered as ritual/religious offering). Besides routine and religious activities on certain days, there is also incidental religious activity: *Panca Yadnya; Dewa Yadnya, Rsi Yadnya, Pitra Yadnya, Manusa Yadnya* and *Butha Yadnya* which are also done with the support of activities which are done in the kitchen. *Panca Yadnya* is a ritual activity which needs the support, not only from the family members, but also from other people outside the family members, so there will be increased activities that require additional facilities.

From the description above, we can see that the basic pattern of the activities in the traditional kitchens of Semaja Village are similar to the common kitchens in other places, such as preparation, processing and serving (Neufert, 1980). Meanwhile for a greater level of activity, for instance Panca Yadnya, additional facilities and spaces will be needed.

| | Routine/ Everyday | Tri wara and Panca Wara | Tri wara Panca wara and Sapta Wara | Purnama and Tilem | Panca Yadnya |
|---|---|---|---|---|---|
| Morning | Cooking breakfast | Offering preparation | Offering preparation | Offering preparation | Preparation and ritual implementation |
| Afternoon | | | Lunch | | |
| Evening | Cooking dinner | Offering activity | Offering activity | Offering activity | |

For each activity pattern or steps, there are sub-activities such as:

1. Preparation: consists of storage, washing, materials cutting. To carry out this activity, food storage cabinets are needed, a washing basket, a small board for cutting the food material (talenan) and other equipment such as dingklik used for seating.
2. Processing: food processing or cooking consists of frying, boiling and roasting, and the facilities needed to support this processing activity are furnace, pans, pots, etc.
3. Serving: in serving the food, a space to serve the food is needed on the dining table or serving table. To support this activity, the facilities needed are a place to keep the plates, bowls, etc. In the matter of space allocation, the serving area should be near to the dining room or the dining table. The activity following the serving is the cleaning activity. The facilities to be used in this activity are those required on the 1st activity.

From the 3 activities mentioned, there are activities which can be done in the same area which is usually at bale-bale, an average-sized single bed located in front of the furnace. There, several activities can be conducted on the bale-bale, such as the preparation of the food material, pouring of spices, or take a rest while waiting for the food to be ready to be served. Beside dingklik on a high dining table, in the village kitchen there is also small dingklik functioned as a place for sitting while keeping the flame of the stove, firewood splitting, and breaking of coconut shells.

A bale-bale shall be owned by each household, which can be used in greater level of ritual/religious activity. Therefore the bale-bale can also be loaned out to other households that are conducting a ritual/religious ceremony as a multifunctional additional facility.

## 4 FACILITIES IN HOUSE KITCHEN INTERIOR

### 4.1 *Preparation*

The facilities which are needed on this stage are: food storage cabinet which may be hung or not (*langki*), tub/washing basket, a small board for cutting food material, and other equipment such as stools or seating.

4.2 *Processing*

Processing: to process or cook the food, the processes are frying, boiling and roasting, and the facilities are furnace pans, pots, etc.

4.3 *Serving*

Serving: to serve the food, plates are needed on the dining or serving table, and the facilities needed are plates, bowls etc. In the matter of space allocation, the serving area should be located near to the dining room or dining table.

4.4 *Additional facilities*

The spirit of cooperativeness in the village, which is very high, can be seen from the ritual and custom activity which needs lots of human power and should be done together. The facilities used in performing the spirit of cooperativeness is manifested in the form of *bale-bale* that is prepared by each household to be used by each community member who conducts a ritual/religious ceremony. *Bale-bale* is a very important item used in the kitchen by its owner because on *bale-bale* the owner can do many things such as preparing the food, spices poured, or take a rest while waiting the food ready to be served.

## 5 CONCLUSION

The above description can be summarised as follows:

The activities carried out within the facilities of interior house kitchens in Br. Semaja, Bengkel Sari village are:

1. Preparation: food storage cabinet that may be hung or not (langki), tub/washing basket, a small board for cutting food material (talenan), and other equipment such as dingklik to be used as place for sitting, a water storage.
2. Processing facilities needed; stove, pans, pots, bale-bale as the place for processing the food.
3. Serving: a dining table, bale-bale, food storage cabinet that may be hung or not.

## SUGGESTION

A *Bale-bale* that is multi-functional needs to be developed in order to make the facilities of the kitchen more efficient.

## REFERENCES


Anoraga, P. (1998). Psikologi Kerja. Jakarta: Rineka Cipta.

Artini, K. (2000). Dimensi Estetika pada Karya Arsitektur dan Dizain. Jakarta: FSRD. Univ. Trisakti.

Ching, F.D.K. (1985). Arsitektur: Bentuk Ruang dan Susunannya. Jakarta: Erlangga.

Depdikbud. (1992). Kamus Besar Bahasa Indonesia. Jakarta: Balai Pustaka.

Grandjean, E. (1973). Ergonomics of home. London: Taylor & Francis Ltd.

Grandjean, E. (1988). Fitting The Task to The Man: A Textbook of Occupational Ergonomics. 4th. Edition. London: Taylor & Francis Ltd.

Helander, M. (1995). A Guide to The Ergonomics of Manufactuting. London: Taylor & Francis Ltd.

Mangunwijya. (1980). Pasal-Pasal Pengantar Fisika Bangunan. Jakarta: PT: Gramedia.

Manuaba, A. (1977). Pengetrapan Ergonomi dalam rangka Peningkatan Kegiatan usaha pendidikan dan Pembangunan Masyarakat Desa. Ceramah Keliling Pendidikan Masyarakat. Tanggal 24–29 Maret 1977. di Bali.

Manuaba, A. (1988). Gizi Kerja dan Produktivitas. Denpasar: Bagian Faal, Fakultas Kedokteran Universitas Udayana.

Manuaba, A. (1992). Upaya memberdayakan Ergonomi di PTP XXI-XXII. Surabaya: Seminar membudayakan Ergonomi di Pabrik Gula PTP XXI-XXII, 30 November 1992.

Manuaba, A. (1998). Bunga Rampai Ergonomi: Vol I. Program Pascasarjana Ergonomi-Fisiologi Kerja Universitas Udayana, Denpasar.

Manuaba, A. (2006). Total Approach in Evaluating Comfort Work Place. Preseted at UOEH International Symposium on Confort at the Workplace. Kitakyushu, Japan, 23–25 Oct 2005.

Parwata, I. & Wayan. (2008). Intervensi Ergonomi Meningkatkan kenyamanan dan Menghemat Energi Listrik Rumah Type 36/120 di Perumahan Nuansa Kori Sading Mengwi Badung. (disertasi). Denpasar: program Pascasarjana Universitas Udayana.

Prabu, W. (2005). Prinsip Desain Interior. Bandung: ITB.

Santosa, A. (2005). Pendekatan Konseptual dalam Perancangan Interior. Dimensi Interior. Vol. 3 No. 2. Desember 2005: 111–125.

Suprayogo, I & Tobroni. (2001). Metodologi Penelitian Sosial—Agama. Bandung: PT. Remaja Kosdakarya.

Suptandar, P. (1985). Perancangan tata Ruang Dalam. Jakarta: FSRD Univ. Trisakti.

Suwardi, E. (2003). Metodologi Penelitian Kebudayaan. Yogyakarta: Gajah Mada University Press.

*Ergonomics in Asia – Shih & Liang (eds)*
*© 2012 Taylor & Francis Group, London, ISBN 978-0-415-68414-9*

# Redesigning tractors for reduced soil cultivation and increased productivity in the agricultural sector in Bali Indonesia

I. Ketut Widana
*Engineering Department, Bali State Polytechnic, Bali, Indonesia*

ABSTRACT: This research is about the implementation of ergonomics in the agricultural field. It is the result of a seminar that involved various stakeholders, such as agricultural field officer, kelian subak abian, kelian desa, farmers (owners and cultivators of the fields), and some prominent figures of Batunya village of Baturiti Tabanan. It reveals some problems as follows: 1) high cost of farming; 2) low productivity; 3) farmers feel an increase in work load; 3) farmers frequently feel ache on some parts of body after working; 5) tiredness after working; and 6) lack of work motivation. Based on the various problems listed above, the work equipment has been redesigned, namely by making the plough blade wider than its normal width, by modifying the steering wheel, shock breaker and sound, also by improving of the working conditions of farmers by giving them a short rest, and extra nutrition as well as work outfits. To measure the effects of the improvements for the raising of health, productivity and reduction of production cost of vegetables, measuring will be done either pre-test or post-test through an experimental research by design *Treatment by Subject Design*. The subjects are farmers, male, aged between 28–53 years old, with equivalent work experience. The variables that will be measured are the cost of production, productivity, workload, musculoskeletal disorders, tiredness and work motivation. The data analysis will be done with the SPSS program with the meaning level $\alpha = 0.05$. The data from the research results show that the implementations of ergonomics results in a decrease in workload, musculoskeletal disorders and tiredness. The implementation of ergonomics is also able to increase work motivation and productivity as well as lowering the production cost of vegetables. Production costs which from the beginning were Rp. 621.381 later became Rp. 434.967 or equal to a 30% reduction. Work productivity increased by 24% and work load reduced by 8%. A performance indicator which also improved is musculoskeletal disorders (MSD), that is 46.62%, later becoming 38.81% or a reduction of 17%; similarly fatigue was reduced by 19%. The final variable, work motivation, also increased by 14%, to reach a level of 67.76.

*Keywords*: improvement of work condition, production cost, work health

## 1 PREFACE

The agricultural fields that become the heart of the economic growth of the Bali region especially and Indonesia in general, now have begun to be less attractive to the local population. The young generation has been reluctant to engage in farming and prefer to have careers in other fields of work. Some people consider farming is equivalent to dirtiness, tiredness and sickness, therefore it is becoming less interesting as various restrictions result from engaging in farming. The decline of the people's interest in farming has resulted in a high level of changes in land use.

From the data of Bali region agricultural conditions, there are 1000 ha of land that have a function change every year. The farmer's cultivating lands have decreased from year to year. In the beginning of the 1980s Bali still had 120,000 ha of farming lands, and now it is predicted only less than 80,000 ha are left. In 2004 there were 82,095 ha of farming land. The following year, namely in 2005, it had decreased to 81,210 ha or a reduction of 885 ha. A decrease also occurred in 2006 namely by 213 ha, consequently only 80,997 ha remain.

Farming as it should be still has a great attention from all sides. The amount of farm land, as much as can be maintained, should be increased, and also the enthusiasm of the community to be more deeply involved in the farming sector has to be accelerated. The gap between reality and the actual condition is worth being an object of study in order to find solutions to ensure that the community's interest in engaging in farming is constantly maintained, so that the rapid rate of land function change can be stopped.

Various kinds of solutions deserve to be considered, to make the agricultural fields interesting again. The improvement in work organization, such as supplying nutrition, applying active rest, the using of work outfits, as well as work tools that are appropriate to the needs and the anthropometric of farmers is one of alternatives that can be applied to overcome the problem.

The preliminary study shows that the average frequency of work pulse is 104.76 ± 8.73 beats/minute, therefore the work load according to Christensen (1991) is considered medium. The score of the musculoskeletal complaints is 46.62 ± 6.74 or increasing by 64% and the score of tiredness is 62.43 or increasing by an amount of 90%. The results above illustrate that the musculoskeletal complaints and tiredness are still increasing above 20%, therefore these need to be reduced. Also the medium work load still needs to be reduced to the 'light' category. From the results of the seminar that involved various parties, such as agricultural field officer, kelian subak abian, kelian desa, farmers (owners and cultivators of the fields), and some prominent figures of Batunya village of Baturiti Tabanan, some problems are revealed as follows: 1) high cost of farming; 2) low productivity; 3) farmers feel an increase in work load; 4) farmers frequently feel ache on some parts of body after working; 5) tiredness after working; and 6) lack of work motivation. The preliminary study (2010) indicated that the dominant hazard sources were the excessive conscription of muscle power and forced work behavior since the means and facilities of work are not ergonomic. The farmers who work rely on muscle power and work tools, namely standard tractors, which are not appropriate to the anthropometric of farmers. When steering the tractors, farmers have to continuously stand, which is less natural work behavior as the steering wheel of tractor is too high and too wide for the average range of the farmers. Also, the blade of plough is not suitable for the needs of the dry land. The amount of shaking that is transmitted to the arm of the steerer and the high-level of noise immediately make the work condition of farmers worse. Based on the measurements conducted in the preliminary research, the relative humidity condition of Batunya Village,Baturiti, was 76%, the average speed of wind was 18 m/dt and dry ball temperature was 27°C and wet ball was 24°C, meanwhile the sound intensity was 95 dB (A), exceeding the threshold limit level, namely < 85 dB (A). The time between work and rest is not balanced and also the non-availability of drinking water is a problem. Based on the various problems above, some effort have been made, namely redesigning work tools by making a plough blade which has a wider work width, by doing some modifications on the steering wheel, shock breaker and sound attenuation as well as by upgrading the working conditions of farmers by giving them a short rest, extra nutrition and work outfits.

## 2 REVIEW

Agriculture is the mainstay of the national economy, thus, government and all sides must be concerned with the agriculture sector. However, the reality is that only some people care, even in most regions of Indonesia, especially Bali, and the reduction in the amount of the farming land indicated something surprising. All people seem to care, however in fact it is only a policy, there is a lack of implementation.

Since the era of 1970s, the development of Indonesia has been shifting from agriculture to industry sectors. Various industries have been developed and a variety of high technologies have been adopted from advanced countries, however, it was not followed by a process of proportional technology changing and there was a lack of consideration of the limited potential including the readiness of human resources, such as in the process of mechanization in the agriculture sector. To fulfill the demand of markets, both local and international, a massive mechanization has been carried out. Almost all aspects of agriculture have applied high tech machines imported from advanced countries. One important thing that seemed to be forgotten is that those machines are designed based on the body size (anthropometric) of the workers, work culture and the environmental conditions of the exporter countries, which are different from the community and work environment in Indonesia. Besides, the industrialization process is less balanced in Indonesia with the availability of human resources. The Indonesian community that previously lived as a traditional agriculture community, have been forced to adapt to the conditions above. As a result, some problems have emerged such as the rise of work complaint levels due to farmers not understanding the operational standard of the tools and the right working conditions. The emergence of various illness due to the physical environment conditions, such as the heat levels, high wind and noise that exceed threshold levels abilities and limits of workers, can disturb the physical functions of body that can result in problems of breathing function, hearing, muscles, skin and also early tiredness which can lower the body's health level, work productivity and finally decrease the level of wealth and quality of life of the farmers.

Other things that are also important that become the cause of subjective complaints from workers and the cause of illness due to work are the character and social-culture condition of the community of Indonesia. The awareness level of various sources of accidents and illness, which is low and the social condition of workers in the agriculture sector that are in general grouped in the lowest economic group, has lowered the level of discipline and alertness in working activity. Generally, they only work with one consideration namely fulfilling the needs of family therefore unconsciously they have ignored their work health. The characteristics of "gugon tuwon" (just the way it was) of Indonesian people, has caused the farmers community to be reluctant to strive for their rights to have a work environment that is more secure, comfortable and healthy with the appropriate level of wealth, in accordance with what is written in the law no 1,1970 about workers protection. On the other side, up until now, most of our community still considers that implementation of ergonomics in certain activities means an increase in the cost of production and can reduce the farmers' income. However actually the case is that implementation of ergonomics is an investment which can double their income as well as increase wealth and can help the farmers to reach a better quality of life. A lot of ergonomics researches have proved that the improvement of working conditions through ergonomics implementation is an investment that is cheap but can result in high productivity.

Stepping from the range of problems above, hence, it is necessary to have intensive efforts to develop a work culture that is safe, comfortable, healthy, efficient and productive. Those efforts are not enough with only a socialization through counseling, slogans or symbols, however, it needs a proof that is sizeable and easy to understand by the farming community. One of the ways that can be done is the improvement of work conditions through ergonomics implementation with the participatory approach, as well as by measuring their effects, therefore it is able to give a comprehension and real benefit that can affect directly towards the improvement of security, comfort, efficiency, and work productivity, increase farmers income as well as motivate farmers to develop a work culture that is secure and healthy, that finally increases the level of health, productivity and farmers wealth, and therefore a better quality of life is reachable.

Furthermore, for socialization and continuity of the program, based on the measuring of various problems and the positive effects of the implementation of ergonomics, a report can be arranged into a standard reference book about conditions and ergonomic work methods for the agriculture sector, that can be a reference for the government, agriculture counselor in general or educational institutions such as agriculture polytechnic, in order to develop a work culture that is secure, comfortable, healthy, efficient, and productive, which is able to take farmers and owners of the farming land to reach a better quality of life.

## 3 RESEARCH METHODOLOGY

This research is an experimental research by equal subjects design (Treatment by Subjects Design). Based on the design, the measurement is conducted twice, namely before treatments and after conducting improvements of work and redesigning processes. The target population is all farmers in Batunya Village Baturiti Tabanan Bali, meanwhile the reachable population is farmers in Banjar Batunya and Banjar Taman Tanda. The sample are aged 28–53 years old consisting of 16 peoples.

## 4 ANALYSIS AND RESULT

The data collected, such as work load (pulse), musculoskeletal disorders, tiredness, and work motivation are analyzed with a normality test, namely the Shapiro-Wilk test on a level of significance 5% ($\alpha = 0.05$). All data are analyzed with the application of SPSS 15.0 program for Windows, descriptively. As all data are distributed normally, a difference test is used namely Two Pair Sample T-test on the significance level of 5% ($\alpha = 0.05$), towards the previous conditions before and after the implementation of ergonomics.

Based on the measuring conducted with simple tools, such as stop watch, the data collected indicate that the work load is decreasing following the implementation of ergonomics in farming. From the Nordic Body Map questionnaire, it indicates that musculoskeletal complaints decrease after the farmers have carried out their activities by implementing ergonomics, and also the tiredness experienced by the farmers was lessened. From the observing conducted, the farmers' motivations tend to increase as the result of ergonomics touching on the activities done. It is strengthened by the subjective data on farmers collected using a motivation questionnaire.

By using the analytical formula of benefit cost ratio, there can be seen a decline that is quite significant in the cost of spending by the farmers for each hectare of land cultivated. This also at once indicates an increasing of productivity for the farmers. The cost indicator is also verified or strengthened by the productivity data using the physiological approach, namely work pulse. Production cost which from the beginning was Rp. 621.381 declined to become Rp. 434.967 or equal to a reduction of 30%. Work productivity increased 24% and workload reduced 8%. A performance indicator which also increased is musculoskeletal disorders (MSD), that is from number 46.62 reduced to become 38.81 or a reduction of 17%, and there was a similar reduction in fatigue equal to 19%. The last variable, job motivation, also increased to 14%, that is, reaching the number 67.76.

## 5 CONCLUSION AND SUGGESTION

Based on the discussion above, it is concluded that the production cost of vegetables declines and the productivity improves after the implementation of ergonomics. Also, other variables such as workload, musculoskeletal problems and tiredness are significantly decreasing after the implementation of ergonomics in farming. The work motivation of farmers is also improving. The improvement of motivation is due to workers in farming using the ways suggested in ergonomics.

As a closing statement, let the researcher give some suggestions for the farmers, government, and other researchers in the future, namely: 1) the implementation of ergonomics is very helpful for the farmers in maintaining conditions of health which are secure, comfortable, effective and efficient as well as productive, therefore do not hesitate to adopt ergonomics and implement it in the agricultural sector; 2) the cost of investment to begin working ergonomically will soon be gained back from the improvement in work productivity, because of that point, the government is hoped to become a pioneer for the implementation of ergonomics in farming activity; 3) for the researchers who will study about the same problem, pay a great attention to the non-physiological aspects of the subject therefore they can complete the results of this research. The final variable, job motivation, also increase by an amount equal to 14%, that is reach the level of 67.76%.

## REFERENCES


Adiputra, N. 2002. Denyut Nadi dan Kegunaannya dalam Ergonomi. Jurnal Ergonomi Indonesia (The Indonesian Journal of Ergonomics), Vol: 3, No. 1. Juni 2002. pp. 22–26.

Adiputra, N., Sutjana, D.P., Suyasning, HIS. & dan Tirtayasa, K. 2001. Gangguan Muskuloskeletal Karyawan Beberapa Perusahaan Kecil di Bali. Jurnal Ergonomi Indonesia (The Indonesian Journal of Ergonomics), Vol: 3, No. 2. Desember pp. 22–26.

Christensen, E.H. 1991. Physiology of Work. Dalam: Parmeggiani, L. Editor. Encyclopaedia of Occupational Health and Safety, 3rd (revised) Ed. Genewa: ILO. pp. 1698–1700.

Colton, T. 1985. Statistika Kedokteran. (Rossi Sanusi, Pentj). Yogyakarta: Gadjah Mada University Press.

Cook; Campbell. 1979. Quasi Experimentation. Design & Analysis Issues for Field Settings. London: Houghton Mifflin Company.

Daniel, W.W. 1999. Biostatistics, A Foundation for Analysis in the Health Sciences. New York: John Wiley & Sons, INC.

Grandjean, E. 2000. Fitting the Task To The Man. A Textbook of Occupational of Ergonomics. 4th Ed. London: Taylor & Francis.

Machfoedz, M. 2006. Kewirausahaan, metode, manajemen dan implementasi. Yogyakarta: FE UGM.

Manuaba, A. 1999. Ergonomi Meningkatkan Kinerja Tenaga Kerja dan Perusahaan. (Makalah). Bandung: Symposium dan Pameran Ergonomi Indonesia 2000, 18–19 Nopember.

Pheasant, S. 1991. Ergonomics, Work and Health. London: Macmillan Academic Profesional Ltd.

Pulat, B.M. 1992. Fundamentals of Industrial Ergonomics. New Jersey: Prentice. Hall, Englewood Cliffs.

Suma'mur, PK. 1992. Higiene Perusahaan Dan Kesehatan Kerja. Jakarta: CV. Agung.

Susan, J. Hall. 1995. Basic Biomechanics. Toronto: Mosby-Year Book, Inc.

Sutjana, D.P. 2000. Penerapan Ergonomi Meningkatkan Produktivitas dan Kesejahteraan Masyarakat. Orasi Ilmiah Pengukuhan Guru Besar Fisiologi Universitas Udayana. Denpasar, 11 November.

Sutjana, D.P. dan Sutajaya, I.M. 2000. Penuntun Tugas Lapangan M.K. Ergonomi Fisiologi Kerja. Denpasar: Program Pasca Sarjana Ergonomi Fisiologi Kerja Universitas Udayana.

Tirtayasa, K. 1992. Pengaruh Sandaran Pinggang pada Kursi Kerja Terhadap Keluhan Subyektif pada Tukang Bordir di Perusahaan Garmen. Denpasar: Majalah Kedokteran Udayana No. 75, Januari.

Walhi, 2008. Moratorium konversi lahan, demi keberlanjutan hidup di Bali, [cited 2011 January 12]. Available from: URL: http://www.beritabumi.or.id/degradasi+jumlah+petani+Bali

Wignyosoebroto, S. 2003. Ergonomi, Study Gerak dan Waktu. Teknik Analisis untuk Peningkatan Produktivitas Kerja. Edisi Pertama. Cetakan Kedua. Surabaya: Guna Widya.

*Ergonomics in Asia – Shih & Liang (eds)*
 *ISBN 978-0-415-68414-9*

# How does the central Badung market play its role for the 24-hour society of Denpasar?

Made Sri Putri Purnamawati
*The Hindu Dharma State Institute, Denpasar Bali, Indonesia*

Nyoman Adiputra
*Faculty of Medicine, Udayana University Denpasar, Bali, Indonesia*

ABSTRACT: The Central Badung Market is located in the central part of Denpasar city. Traditionally, this market was designed to function in the day time only. But, due to the development of Denpasar Municipality, unpredictably the market has grown into a 24-hour market which supplies the multi-sector demands of people. The vendors, consumers, and workers who are mostly the Balinese, are affected. The workers are segmented into morning-, evening-, or night-time workers. The night-time workers were originally from Denpasar only, but nowdays they are young people from outside of Denpasar, such as from Gianyar, and Badung subprovinces. Some of them still at formal schooling. The 24-hour functioning of this market is affecting the normal activity of the Balinese. Is it well designed or has it just accidently occured? One thing to remember is that this 24-hour market seems to have been prepared without planning, but instead has it grown automatically. The problems encountered are: a) formal schooling for the young Balinese in Denpasar is a disturbed, due to the need for night-time workers as carriers in the market till late evening; b) disturbance of the normal biological clock of people nearby the market, due to it being busy and noisy for 24 hours. The positive impacts are: a) a 24-hour market opening up job opportunities for people nearby; b) strengthening the accessibility to food stuffs in all parts of the small market at the grass roots levels in Denpasar Municipality; c) the stakeholders involved in the market automatically synchronize themselves into 3 shift works, morning-, evening- and night-shift work. The conclusions drawn are: the 24-hour Badung market is built based on filling the need and bridging the gap; this market creates job opportunities for local people nearby; and strengthening supplies of good and fresh food stuffs to the lower level of the market which is more accessible to people in general. It is suggested that the government should be actively involved in monitoring the negative impacts as well as the positive benefits of this market to the Balinese.

*Keywords*: 24-hour society; night-shift work; traditional market, central Badung market

## 1 INTRODUCTION

The central Badung market is located in the central part of Denpasar town. Before, it belonged to Badung sub-provincial region, and Denpasar was a city of Badung regency. But after Denpasar became a municipality, the central market belonged to Denpasar Municipality Government. Therefore, this market had never been planned to be a 24-hour market. Because the development of Denpasar city is so advanced, with associated urbanisation, people need to be supplied in their demands in a more rapid way. In fact the Denpasar Municipality consists of 4 district areas namely Eastern Denpasar, Western Denpasar, Northern Denpasar and Southern Denpasar. In every district there is at least one market, accordingly. The Badung market, therefore became the central one. All of the commodities are centralized in this market, then distributed into 4 other lower level markets.

The commodity inputs come from Java, or from producer centres such as vegetable farms in Bedugul and Kintamani. Therefore, there is a need to manage the central market as a 24-hour market.

The activities in this market consist of the following: morning shift till evening (05.00 am to 18.00 pm) for supplying the normal daily needs of Denpasar inhabitants. From 19.00 pm till 05.00 am the market activities are concentrated as a central market to distribute all goods to the adjacent smaller markets. The vendors for both market activities come from different areas and the buyers also. For the central market the vendors are bigger companies and the buyers are varied, from big companies to smaller companies. In the normal market activity, the vendors are the local people or owners of market stands and the buyers are Denpasar inhabitants. The Denpasar inhabitants buy for themselves, not like buyers in the night market, most of whom are big buyers. By doing that, the Badung market really becomes a 24-hour market. Does the market make everyone happy? How do the local people react to it? Are the losses and the benefits equally weighted from this change? Those are some questions which arise in regard to the process of change at Badung market. A change process is absolutely time-bound, as observed by Costa (2002a; b) and Adiputra (2002; 2003; 2007).

## 2 MATERIALS AND METHOD

Subject of this study is the central Badung market. Data and information on of this market was gathered from the respondents, who consisted of vendors, workers, business men or women, the management, and customers, in total about 36 persons.

A walk-through survey was applied, followed by an interview with the respondents. Observation technique was also conducted on the commodities sold. The data obtained were analyzed descriptively.

## 3 RESULTS

Table 1 represents the respondents, who consisted of workers, vendors, managers and buyers. Table 2 depicts the shift work, time allocated and people involved in central

Table 1. The respondents interviewed in Central Badung market.

| No. | Shift work | Respondent | Total number |
|---|---|---|---|
| 1. | Morning shift | 3 carriers, 4 vendors,<br>4 buyers, 1mng. | 12 persons |
| 2. | Evening shift | 3 carriers, 4 vendors,<br>4 buyers, 1mng. | 12 persons |
| 3. | Night shift | 3 carriers, 4 vendors<br>4 buyers, 1mng. | 12 persons |

mng = manager

Table 2. Central Badung market: its activity and workers involved.

| No. | Shift-work | Allocated time | People involved |
|---|---|---|---|
| 1. | Morning shift | 06.00 am–14.00 pm | I,FC,V, Mng |
| 2. | Evening shift | 14.00 pm–19.00 pm | I,FC,V, Mng |
| 3. | Night shift | 19.00 pm–06.00 am | V,B,FC, Mng |

Nb.I = Inhabitants; FC = Female Carriers; V = Vendors, Mng = Management; B = Buyers

Table 3. The commodities sold in central Badung market.

| No. | Shift-work | Commodities sold |
|---|---|---|
| 1. | Morning shift | all goods for daily needs; tansaction in small portion. |
| 2. | Evening shift | all goods for daily needs; transaction in small portion. |
| 3. | Night shift | all goods for lower level markets; big buyers and transactions in bigger portion, between vendors and buyers |

Badung market. Table 3 depicts the commodities sold and the level of transaction in every shift work.

## 4 DISCUSSION

In every shift, the market activities are similar; mainly consist of selling and buying as an interaction between the vendors and the buyers. But the difference is the level of transaction on both sides. In the morning shift there is transaction between Denpasar inhabitants and the seller (owners of local market stands) who are selling all goods for the daily needs such as food stuffs (rice, vegetables, meats, eggs, cococnut oil, all kinds of tubers, fruits, or spices) and any kind of goods for religious ceremony. The main characteristics of the business are the small transactions, purchasing for one day's to three days' needs (Suyadnya, 1997).

Every evening shift is similar to the morning shift, but rather more people come and also more vendors are involved. The buyers are Denpasar inhabitants, who are looking for more fresh food stuffs (vegetables, meats/fish and fruits). Also they would like to buy the flowers used for daily offerings. The people involved are almost similar in number to the morning shift, but less than in the night market. The workers, in terms of the carriers, (Suyadnya, 1997) are involved in, just servicing the people who buy a small number of goods not more than 10 kg of weight.

In every night shift the activities of the people involved are quite different. The buyers are mostly the owners of local stands in this market, owners of market stands from smaller market levels. The vendors come from Java or from producer centres such as for vegetables, fruits, spices and other producer centres, from other sub-provincial regions in Bali. The transactions are mostly on a bigger scale. The most market activity is between the big vendors and big buyers. More people are involved in these transactions for carrying out or transporting the commodities from the vendors to the buyers. The workers in this case are responsible for carrying out commodities ranging from one small truck to big trucks of goods. Therefore, more people are needed as carriers. That is supported by the other researchers (Suyadnya, 1997; Hutagalung, 2009).

Looking at the people involved at these three shift-works at central Badung market, it is interesting to say that in the night shift more people involved. They come from nearby, or from Badung areas, or from outside of Badung regency such as from Gianyar regency (Suyadnya, 1997; Hutagalung, 2009). They come to Badung market at night and go back home in the early morning at 5.30 am. Some of them are young girls; they earn money for supporting their formal schooling at the junior or senior high school. That is also true, for female workers from Denpasar or Badung regency near the market. For mature female workers it does not make any problem. But for the young girls, it has to be questioned. How they will manage the time for formal schooling, while they are working at night till early morning? Do they have enough time for their schooling?

The other problems encountered are the 24-hour business at this market. It disturbs people who live nearby who are not well adaped yet. More air pollution will result due to very large numbers of vehicles which go in and out. Of course, the activities of central Badung market which had been changed into a 24-hour market, also give a benefit for local inhabitants. It provides more opportunity to people to be more actively involved as local vendors, as carriers or workers, or as sellers by opening a market stand for business. All of those are considered as pro-active responses of local people nearby to the market.

But, it is fair to say that the growth of this market into a 24-hour acitivity was not well planned. It grew occasionally, as a response to the people's need especially for Denpasar Municipality as a consequence of urbanisation (Suyadnya, 1997; Costa, 2002a). Therefore, obviously some weaknesses are still found in regards to central Badung market operating as a 24-hour market. Environmental sanitation and hygiene need to be improved, that is a must (Hutagalung, 2009). Safety and security for everyone who visits this market should be taken into consideration. The local vendors also should be little bit more curious and be more patient in servicing the customers. Among the costumers it is already known that vendors of this market are very unfamiliar in offering their services to customers.

Therefore, customers know how to overcome the bad habits of sellers in this market, that is, buying something in Badung central market is something one does not do in a hurry. Everything should be bargained, as low a price as possible. In doing that the vendors' behavior is often not so familiar, and makes one irritated and makes the customer to feel angry. This was reported also by Suyadnya (1997) and Hutagalung (2009). Those problems are still happening now.

The conclusions drawn are: 1) Badung central market has automatically grown into a 24-hour market without any proper planning; 2) it grows just in response to the demand; 3) in any case, the central market gives positive benefits as well as negative effects for the local people nearby; 4) there are still some aspects which need to be improved for a better performance of this market, environmental sanitation and hygiene in particular.

In the future, in overcoming the existing weaknesses for a better quality of life, a continuous improvement needs to be endorsed. In doing that, a SHIP (systemic, holistic, interdisciplinary and participatory) approach is a must as is suggested by some researchers (Kogi, 2004; Manuaba, 2007).

## REFERENCES


Adiputra, N. 2002. Pengaruh Globalisasi terhadap Kualitas Kehidupan (The effect of globalization on Quality of Life). Editorial. Udayana Medical Journal. October 2002. 33(118): 205–206.

Adiputra, N. 2003. Akankah Globalisasi Melahirkan Masyarakat 24 jam? (Does globalization create the 24 hrs society?). Editorial. Udayana Medical Journal. October 2003. 34 (122): 211–212.

Adiputra, N. 2007. The quality of working life (Editorial). MEDICINA. Mei 2007. 38 (2): 73–74.

Costa, G. 2002a. The 24-H Society Between Myth and Reality. Proceedings. XI National Congress and The XIII Seminar of The Indonesian Physiological Society and International Seminar on Ergonomics and Sports Physiology, Denpasar, October 14–17, 2002: 55–60.

Costa, G. 2002b. Ergonomics and Shift Work: Problems and Solutions. Proceedings. XI National Congress and The XIII Seminar of The Indonesian Physiological Society and International Seminar on Ergonomics and Sports Physiology, Denpasar, October 14–17, 2002: 41–47.

Hutagalung, R. 2009. Perbaikan Kulaitas Kerja dengan Menerapkan Pendekatan Ergonomi Meningkatkan Kinerja Buruh Angkat-Angkut Tradisional di Pasar Badung, Denpasar. Dissertation. Postgraduate School, Udayana University Denpasar, Bali, Indonesia.

Kogi, K. 2004. Participatory methods effective for ergonomic workplace improve—ment. Applied Ergonomics 37(4) July: 547–554.

Manuaba, A. 2007. A total approach in ergonomics is a must to attain humane, competitive and sustainable work system and products. J Hum Ergol 36(2): 23–30.

Suyadnya, IGM Oka. 1997. Work Posture of Female Load Carrier in Denpasar Market. Proceedings. Asean Ergonomics-97. 5th SEAES Conference. Kuala Lumpur, Malaysia. 6–8 November, 1997: 491–495.

*Ergonomics in Asia – Shih & Liang (eds)*
*© 2012 Taylor & Francis Group, London, ISBN 978-0-415-68414-9*

# Modern hotel management versus traditional village values: The case of Bali

D.P. Sutjana & N. Adiputra
*Faculty of Medicine, Udayana University Denpasar, Bali, Indonesia*

ABSTRACT: Bali is considered to be one of the top tourist destinations and the centre of growth of tourism in Indonesia. It is associated by a development process of hotels and other tourist facilities. At present, there are more than 25,000 rooms of star hotel accommodation available in Bali. Looking at the interface between tourists and hotel employees, there is an interesting thing to note about the Balinese, especially those who become hotel workers or employees. They normally work full day with strict rules and regulations, either a morning-, evening-, or night-shift. At the same time they have to participate in social activities at their home village. They are still considered to be members of local society. In doing their task as employees in the hotel, they often have to make a choice between two conflicting priorities. They either work in the hotel as normal, without do anything in their society; or they do their social tasks in the village, and have to be absent from hotel duty. These conditions frequently occur, and make life unhappy for the Balinese. This is a matter of conflict management; without any solution it is becoming a stressor for the Balinese worker. It is especially true if the hotel management does not understand the local culture. This article reports how the Balinese hotel employees are solving their problems, on the basis of a win-win solution. Based on past experiences, the solutions practiced are as follows: a) there is an exchange of shift-work between the Balinese and non-Balinese employees, based on individual agreement; b) if all the employees are Balinese, they manage in a such way that a Balinese who does not have any business in his village society is willing to replace/exchange for his colleague who has a societal task; c) the Balinese is asking his right to have a day-off on the day for his village societal festival; but this must be requested in advance. By doing that, work satisfaction is achieved; the business goals and the societal need can be maintained sustainably in a parallel way. That is as an example of how a pro-contra problem is solved wisely, in a win-win solution.

*Keywords*: modern management; traditional value; hotel conflict management, win-win solution

## 1 INTRODUCTION

Bali province, is considered to be one of the top tourist destinations area in Indonesia. It is hoped that the growth of tourist visits to Bali will have a significant distribution effect on the adjacent areas such as Lombok island and East Java province. It was true for several years before there was an increase in the numbers of tourists coming to Bali from year to year. But, unfortunately, after the first and second Bali bombings, tourist visits dropped till the tourism collapsed. Therefore all the tourist facilities were unfortunately unoccupied. From past experience it is found to be true that tourism is a fragile industry. It is easily affected by so many things such as security problems, war, epidemics, etc.

On the other hand, for most of the hotels in Bali, where most of the workers are Balinese; there is also another problem encountered. The Balinese even if they are hotel workers, are at the same time are also considered to be members of their original community village.

Regarding that, it very often occurs that at the same time the Balinese are faced with two difficult choices to make. They should be present in their hotel working place, but they should also be present for their community tasks. If they are attending the community event, it means they would be absent from their hotel duty. If the hotel management is not aware of this problem, this will become a risk factor. At the end, it produces a work-related stress for the worker (Houtman & Jettinghoff, 2007).

Therefore, it is important for the hotel manager to know the Balinese culture, then it can be applied in hotel management for the goals of business and work satisfaction. This article reports how modern management is practiced in hotels by taking into account the local customs in an appropriate manner for the benefit of the hotel and for the working life of the hotel employees as well (Houtman & Jettinghoff, 2007).

## 2 MATERIALS AND METHOD

Subject of the study was a four star hotel in the Sanur area. This hotel is not the first hotel built in Bali, but it can be used as an example on the matter. The existing management style is subjected to study. A triangulation method (Axelsson, 2000) was applied in gathering the data, in terms of interviews, observation, and collection of secondary data from the relevant agency. The data obtained was descriptively analyzed.

## 3 RESULTS

The hotel studied is presented in Table 1, which includes also the number of workers and the number of worker according to their religions. Table 2 depicts how the management is tackling the issue in regards to off-days for the Balinese hotel workers.

## 4 DISCUSSION

The subject of this study is the Hotel X in the Sanur area. This is a four star hotel with 200 emplyees. Most of the employees are Balinese, at about 80.0%, and a small number of workers (20.0%) are non-Balinese, as presented in Table 1. In its daily operation from the grand opening up to now, this hotel has always been fully-booked with guests. Therefore, this hotel management is able to practice occupational safety and health at a better level compared with other similar class hotels. The hotel's Polyclinic was opened and headed by a part-time

Table 1. The hotel studied and numbers of hotel workers by religion.

| No. | Hotel | Number of workers | Worker by religion |
|---|---|---|---|
| 1. | Hotel X, Sanur | 200 | Balinese Hindu 80.0%;<br>Non-Hindu Balinese 20.0% |

Table 2. How the hotel management handles the issue for the Balinese workers off-days.

| No. | Hotel | Hotel solution in managing the issue |
|---|---|---|
| 1. | Hotel X | 1. Non-Balinese workers are willing to replace the Balinese<br>2. The Balinese are replaced by other Balinese who are not involved in a community ceremony<br>3. In advance, the Balinese puts his/her off days exactly on the days of community festivals |

medical doctor and staffed by a nurse, as permanent personnel. By this, it can be reasonably argued that the program of occupational safety and health is well implemented in this hotel (Adiputra and Sutjana, 2000).

Even so, there is still a problem encountered. The Balinese worker is faced with a difficulty in selecting the day-off during the hotel's operation. It arises from the fact, that all Balinese workers, are at the same time still considered to be members of their original community village. As a worker at the hotel the Balinese should obey the rules and regulations of hotel. Otherwise, good services cannot be performed. The hotel needs also a clean governance for both aspects of management and the workers themselves. In doing this, both sides, worker and manager, should be able to manage the existing conflict wisely (Anonim, 2011; McNamara, 2011). The hotel's operations are at the exact time from morning to next morning. It is a 24-hr operation, which is organised into morning-, evening-, and night-shifts. Everybody should be ready in their offices, according to their shift-work. The conditions are sometimes very difficult for the Balinese. Why? It is a fact that the Balinese in the same time have to do double functions at the same time. As a member of local community there are certain duties and responsibilities to do. That is a must formal and informally. They should be present in community meetings, which are conducted every month. They should present also in every religious ceremony at community level or at clan level. They should also be present in every ceremony conducted by members of the community on an invitation basis. That is a part of best practice in Balinese culture. As a Balinese, if they are attending the community event, it means they have to be absent from their hotel work and vice versa. Everyone is worried about being dismissed from the workplace or from their community. Therefore the Balinese try to solve this problem based on a win-win solution. The management should take the matter into consideration for the well-being of everybody in the business (Wisner, 1997; Manuaba, 1997a; Brown, 2000), and follow with management commitment and involvement. It is true that, the management have an important role in the shaping of an organisation's culture. In case of hotel employees, the Balinese try to solve the problem in such a way without resulting in any bad consequences.

In the hotel where the attendance of manpower is planned every month, the Balinese workers submit the proposal for their attendances in the month. In doing that, of course, the Balinese has looked at the calender of the community festivals in their home village. By doing that, the Balinese can work normally in the hotel, without being absent from community duty. Their rights for off-days in the hotel are taken during the days when the community festivals will be carried out. In other hotels, where worker's attendance is done based on weekly planning, the Balinese solve the problem by substituting their shift-work schedules with non-Balinese workers, in the same department; this is commonly arranged among themselves (Manuaba, 1997a; Widana et al., 1997; Sutjana et al., 2002). In the case where there is no non-Balinese worker, the current problem is solved by substitution also, but among the Balinese who do not come from the same village. It is well understood, due to the fact that festivals in one community are conducted on different days from other villages. By doing that, happiness for every party (hotel worker or management) could be achieved and produce work satisfaction at the end. The business of the hotel is run well, and community tasks can also be conducted, and there is no need for conflict to become a big issue (Sutjana et al., 2002), with the result that every party tries to solve the conflict, peacefully, at the worksite (Anonim, 2011; McNamara, 2011).

That is evendence that this culture can be applied in modern management without losing the identity of Balinese culture. But the implementation of the intended method is much affected by the manager's views (Manuaba, 1997a,b; Wisner, 1997), particularly those on Balinese culture. If it is possible to follow a systemic, holistic, interdisciplinary, and participatory (SHIP) approach wisely (Manuaba, 2007), this will fulfil the notion that "good economics is good economics" as frequently suggested by Hendrick (2002).

From the discussion, it can be concluded that: 1) the Balinese as subject and object of a hotel work schedule, can be managed carefully with the end result being a win-win solution; 2) Balinese culture in this case can be adopted into modern management techniques which are practiced in hotels.

For further study, it is recommended that other methods that are practiced in other hotels need to be studied for a better solution, in regards to the existing problem for the Balinese as an hotel worker.

## REFERENCES


Adiputra, N. & Sutjana, D.P. 2000. Canteen versus Food Allowances for Ensuring the occupational Health of Hotel Employee. Proceedings of the Joint Conference of APCHI 2000 ASEAN Ergonomics 2000. Singapore. November 27-December 1, 2000: 461.

Anonymous. Conflict Management. Available at: etu.org.za/toolbox/docs/.../ conflict. html. Accessed at 03-03-2011.

Brown, Jr. O. 2000. Editorial: XIV Triennial Congress of the International Ergonomics Association and 44th Annual Meeting of the Human Factors and ergonomics Society: "Ergonomics for the New Millenium" Ergonomics 43(7): 829–832.

Hendrick, H.W. 2002. Good Ergonomics Is Good Economics. Proceedings. The International Seminar on ergonomics and Sports Physiology. Denpasar, October 14–17, 2002: 16–40.

Houtman, I. & Jeetinghoff, K. 2007. Raising Awareness of Stress at Work in Developing Countries. A modern hazard in a traditional working environment. Advice to employers and worker representatives. Protecting Workers' Health Series No. 6. TNO. WHO.

Manuaba, A. 1997a. Improving the Effectiveness of a Hotel's Management Team in Bali. Proceedings. Asean Ergonomics-97. 5th SEAES Conference. Kuala Lumpur, Malaysia. 6–8 November, 1997: 485–490.

Manuaba, A. 1997b. Ergonomics challenges in Southeast Asian region. A vision to the 21th Decade. Proceedings. Asean Ergonomics-97. 5th SEAES Conference. Kuala Lumpur, Malaysia. 6–8 November, 1997: 681–686.

Manuaba, A. 2007. A Total Approach in Ergonomics is a Must to Attain Humane, Competitive and Sustainable Work System and Products. J Hum Ergol 36(2): 23–30.

McNamara, C. Management. Available from: managementhelp.org/intrpsnl/basics. htm. Accessed at 03-03-2011.

Sutjana, D.P., Suardana, E. & Swamardika, I.B.A. 2002. Kiat Manajemen untuk Mengatasi Krisis dalam Pengelolaan Hotel: Studi Kasus pada Hotel P di Sanur. Proceedings. The International Seminar on ergonomics and Sports Physiology. Denpasar, October 14–17, 2002: 594–598.

Widana, K., Manuaba, A. & vanwonterghem. 1997. The physical workload of laundry man at GBB Hotel. Proceedings. Asean Ergonomics-97. 5th SEAES Conference. Kuala Lumpur, Malaysia. 6–8 November, 1997: 530–535.

Wisner, A. 1997. What ergonomics is needed for the development of ASEAN Countries in the 21th Century? Proceedings. Asean Ergonomics-97. 5th SEAES Conference. Kuala Lumpur, Malaysia. 6–8 November, 1997: 677–680.

*Ergonomics in Asia – Shih & Liang (eds)*
 *ISBN 978-0-415-68414-9*

# A method for clarifying cognitive structure by integrating sentence completion test and formal concept analysis

Hideyuki Matsuo & Toshiki Yamaoka
*Faculty of Systems Engineering, Wakayama University, Japan*

ABSTRACT: The purpose of this paper is to propose a method that integrates Sentence Completion Test (SCT) and Formal Concept Analysis (FCA). Furthermore we would like to introduce the software we developed which helps the analyst to practice the method. A feature of the method is to collect participants' cognitive data structurally by using SCT. FCA enables us to interpret general structures of these individual data from SCT. Hence the method is applicable to analyze participants' cognitive structures hierarchically and deeply.

*Keywords*: FCA, SCT, cognitive analysis, questionnaire, interview

## 1 INTRODUCTION

### 1.1 *Background*

FCA is useful to reveal concept structures by analyzing data which consists of objects and those attributes. Therefore FCA is applicable to analyze the data of questionnaires which contain elements of objects and evaluations of them.

SCT enables us to acquire data of participants' opinions structurally and systematically, so we started to develop a method that integrates these two methods.

### 1.2 *What is FCA?*

Formal Concept Analysis was proposed to restructure the lattice theory as a data analysis by Rudolf Wille in 1981.

The concept of FCA is called the Formal Concept. Formal Concept consists of objects and their attributes. These elements each work as extension and intention. A schematic diagram of Formal Concept is shown below in Figure 1.

Although FCA treats data which consists of objects and their attributes, it does not matter in FCA whether the attributes are objective facts or subjective personal valuations. We would like to use FCA to analyze the personal cognition of participants.

A Formal Context is represented as cross table describing relations between objects and attributes. An example of Formal Context is shown in Figure 2.

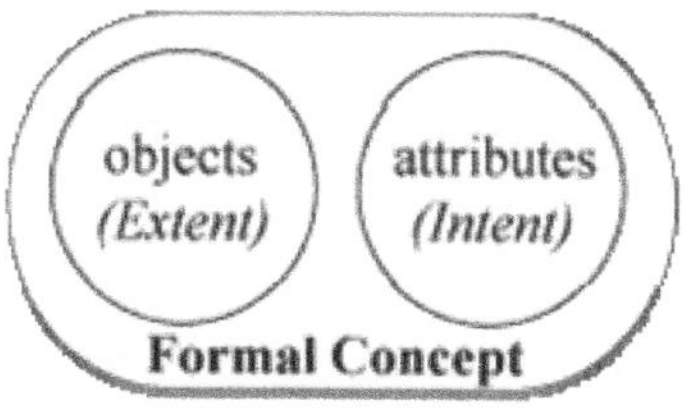


Figure 1. A schematic diagram of Formal Concept.

| attributes / objects | viviparous | oviparous | terrestrial | aquatic |
|---|---|---|---|---|
| Human | × | | × | |
| Dolphins | × | | | × |
| Frog | | × | × | × |
| Bird | | × | × | |
| Fish | | × | | × |

Figure 2. A example of Formal Context. It is a cross table which represents relations between objects and attributes.

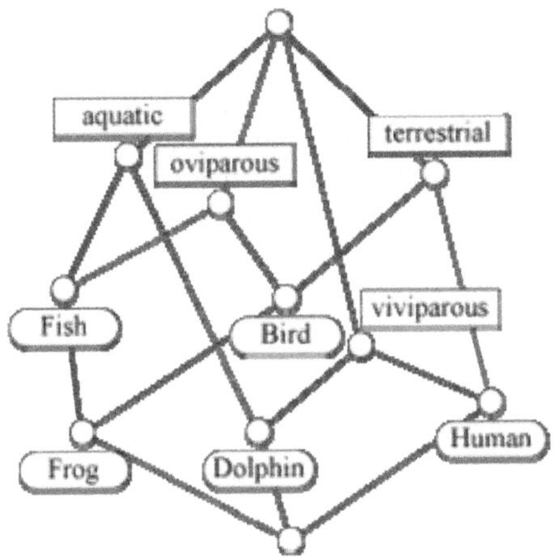


Figure 3. A example of a Concept Lattice. The Hasse diagram consists of the nodes of Formal Concept.

The purpose of FCA is to reveal structures of Formal Concepts. The structures of FCA shows are inclusive relations of Formal Concepts.

The result of FCA is the Hasse diagram called a Concept Lattice. A example of a Concept Lattice is shown in Figure 3. The Nodes of a Concept Lattice are Formal Concepts. The Hasse diagram is used here to represent inclusive relations between Formal Concepts.

On the Concept Lattice, the upper nodes have attributes that many objects contain, and lower nodes have special attributes. In contrast, lower objects have many attributes, upper objects have special attributes. It explains which concept is more general or special.

Interpretations of FCA are conducted from these relations between Formal Concept and Concept Lattice.

A summarized process of FCA is as follows:

1. Define Formal Contexts.
2. Make Formal Concepts.
3. Structurize Formal Concepts.
4. Draw Concept Lattice.
5. Interpret Concept Lattice.

There is some software that conducts the procedure from (2) to (4) automatically.

### 1.3 *What is SCT?*

Sentence Completion Test (SCT) is a projective test. SCT reveals the participant's recognition by having them answer incomplete sentences. Although there are a lot of questionnaire forms in SCT, simple examples of questions are as follows:

I like ( ).
I don't like ( ).
I want to ( ).
etc.

Participants are requested to complete these blanks freely. SCT is effective at extracting the personalities of participants because a result of SCT shows their beliefs, attitudes, motivations, or other mental states. SCT is also used to reveal a participant's cognitive structures by using conjunctions which control sentence structures. To take an example:

I like ( ) because ( ).

The methods of SCT like this allow us to get both the participant's recognitions and their reasons. It can be said that this method is a Laddering Method. On the proposed method here, we use this form of SCT.

## 2 METHODS

### 2.1 *Define formal contexts from data of SCT*

On the method we propose here, Formal Contexts of FCA are defined from data obtained by using SCT. Specifically, after having participants answer the structurized form (shown in Figure 4), Formal Contexts are created from relations between elements filled in by participants on the questionnaire. A example of an answered form of the SCT questionnaire is shown in Figure 5, which evaluates three cars called A, B and C. The remarkable feature of this process is that we can obtain three or more Formal Contexts from one questionnaire as you can see from Figures 6–8.

The elements of the form we call *O*, *A1* and *A2* are structured by the words "is" and "so". The word "is " is used to get attribute elements (*A1* and *A2*) of *O*. The word "so" is used to get cause element (*A1*) of *A2*. Here, we can define *A2* as an attribute of *A1* because generally the causal elements of something can be regarded as attribute elements of it.

Therefore we can define the Formal Contexts as shown in Figures 6–8. Figure 6 represents the Formal Context in relations between *O* and those attributes that we give.

*Level 1 Formal Context*. Figure 7 represents the Formal Context in relations between *A1* as objects and *A2* as those attributes that we give.

Figure 4. The form of the SCT questionnaire. Structurized with the words "is" and "so".

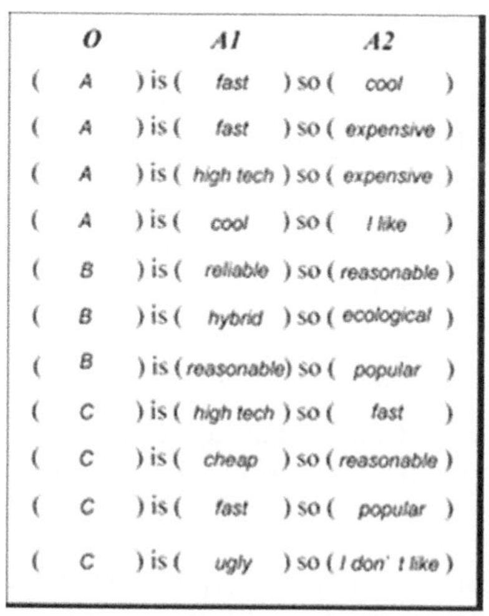

O A1 A2
( A ) is ( fast ) so ( cool )
( A ) is ( fast ) so ( expensive )
( A ) is ( high tech ) so ( expensive )
( A ) is ( cool ) so ( I like )
( B ) is ( reliable ) so ( reasonable )
( B ) is ( hybrid ) so ( ecological )
( B ) is ( reasonable) so ( popular )
( C ) is ( high tech ) so ( fast )
( C ) is ( cheap ) so ( reasonable )
( C ) is ( fast ) so ( popular )
( C ) is ( ugly ) so ( I don' t like )


Figure 5. A example of result of the SCT questionnaire. It is an evaluation of three cars. The elements of *A1* are cause elements of *A2*.

attributes

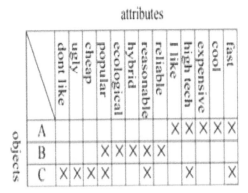

| objects | dont like | ugly | cheap | popular | ecological | hybrid | reasonable | reliable | I like | high tech | expensive | cool | fast |
|---|---|---|---|---|---|---|---|---|---|---|---|---|---|
| A | | | | | | | | | X | X | X | X | X |
| B | | | | X | X | X | X | X | | | | | |
| C | X | X | X | X | | | X | | | X | | | X |

Figure 6. The *Level 1 Formal Context* from the result of the SCT questionnaire shown in Figure 5.

attributes(A2)

| objects(A1) | cool | expensive | I like | reasonable | ecological | popular | fast | dont like |
|---|---|---|---|---|---|---|---|---|
| fast | X | X | | | | X | X | |
| high tech | | X | | | | | X | |
| cool | X | | X | | | | | |
| reliable | | | | X | | | | |
| hybrid | | | | | X | | | |
| reasonable | | | | X | | | | |
| cheap | | | | X | | | | |
| ugly | | | | | | | | X |

Figure 7. The *Level 2 Formal Context* from the result of the SCT questionnaire shown in Figure 5.

attributes(A2)

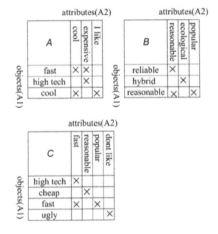

| A — objects(A1) | cool | expensive | I like |
|---|---|---|---|
| fast | X | X | |
| high tech | | X | |
| cool | X | | X |

attributes(A2)

| B — objects(A1) | reasonable | ecological | popular |
|---|---|---|---|
| reliable | X | | |
| hybrid | | X | |
| reasonable | X | | X |

attributes(A2)

| C — objects(A1) | fast | reasonable | popular | dont like |
|---|---|---|---|---|
| high tech | X | | | |
| cheap | | X | | |
| fast | X | | X | |
| ugly | | | | X |

Figure 8. The *Level 3 Formal Contexts* from the result of the SCT questionnaire shown in Figure 5.

*Level 2 Formal Context*. The attributes of *Level 2 Formal Context* consists of all of elements of *A2* that appeared in result of SCT. On the other hand, Figure 8 represents Formal Contexts in relations between *A1* and *A2* that appeared for a specific *O* that we give.

*Level 3 Formal Context*. Therefore, every object has its own Level 3 Formal Context.

These Formal Contexts that are created in many levels provide us with many results of FCA, so it helps us to read the result of SCT hierarchically and deeply.

## 2.2 *Interpret concept lattices*

Examples of a Concept Lattice created from Formal Contexts defined from data of SCT, are shown in Figures 9–11. The Concept Lattice created from *Level 1 Formal Context* is called *Level 1 Concept Lattice*. Similarly, the *Level 2 Concept Lattice* and the *Level 3 Concept Lattice* are defined.

An example of a brief interpretation of each Concept Lattices is as follows.

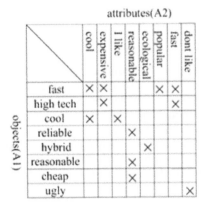

According to the *Level 1 Concept Lattice* shown in Figure 9, the paths to the three nodes which contain objects (A, B or C) are divided from the two nodes. It means that each two nodes of concept represents superordinate concepts of each concept which contains an object. Therefore the objects are categorized into two clusters by the two concepts. Additionally, the

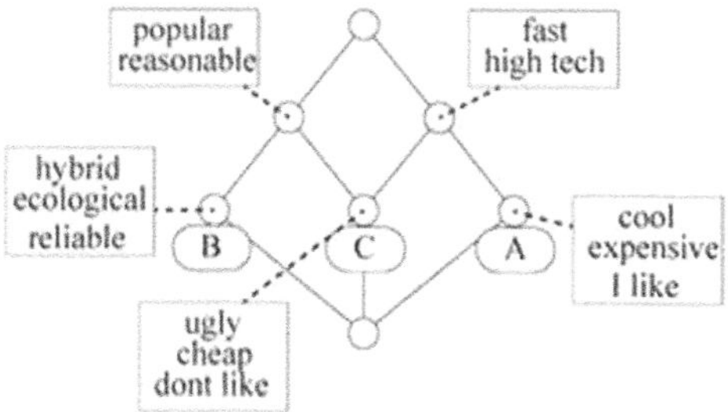


Figure 9. The *Level 1 Concept Lattice* created from the *Level 1 Formal Context* shown in Figure 6.

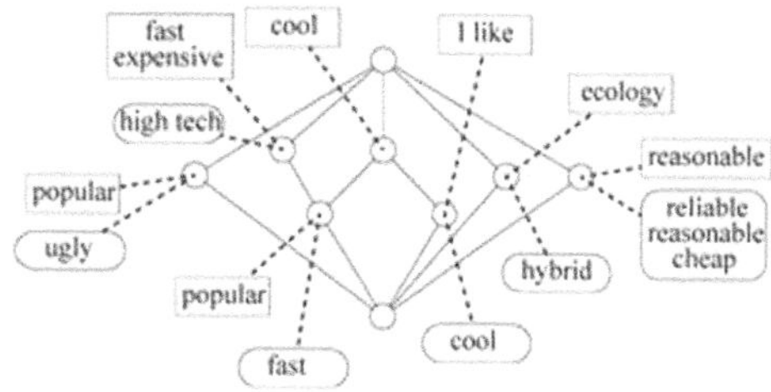


Figure 10. The *Level 2 Concept Lattice* created from the *Level 2 Formal Context* shown in Figure 7.

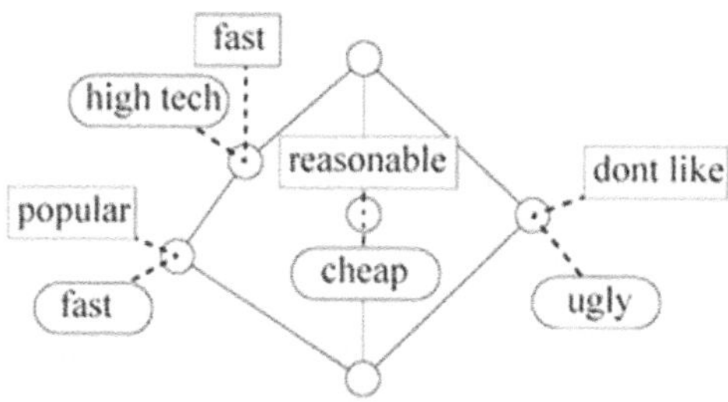


Figure 11. The *Level 3 Concept Lattice* created from the *Level 3 Formal Context* of C shown in Figure 8.

attributes integrate with the object in each node are regarded as very characteristic attributes of these objects.

According to the *Level 2 Concept Lattice* shown in Figure 10, if you take notice of the node which contains the object "fast," you can find out that "popular" is the most characteristic attribute of "fast." Additionally, the concept which consists of "fast" and "popular" is divided into two concepts because of the "high tech." It means that the "high tech" is the element which categorizes into "cool" or "expensive" cars among the "popular" cars.

In Figure 11, we show the *Level 3 Concept Lattice* created from a *Level 3 Formal Context* in the Figure 8. It is about car C. Although you can create every *Level 3 Concept Lattice* of each object, here we show only one of these for simplicity. The *Level 3 Concept Lattice* represents structures of attributes of specific object. If you take notice of the left path which connects "high tech" and "popular", you can find out that "high tech" is closely related with "popular" in the evaluation of C. The relation between "high tech" and "popular" is unclear in the result of SCT or the *Formal Context*.

Thus, each level of *Concept Lattice* helps us to interpret participant's cognitive conceptual structures hierarchically and deeply.

## 3 THE SOFTWARE

### 3.1 *Introduction*

Practicing the method proposed in this paper is hard work. Analysts have to make a list of objects and attributes, and make a lot of cross tables. Therefore, we developed software which conducts the series of operations in the method automatically. The function of the software is to make Formal Contexts in each level from input sentences and draw various Concept Lattices from these Formal Contexts. The software was programmed with Action Script 3.0. Therefore we can provide the copy of the software as a FLASH, iOS or Android applications.

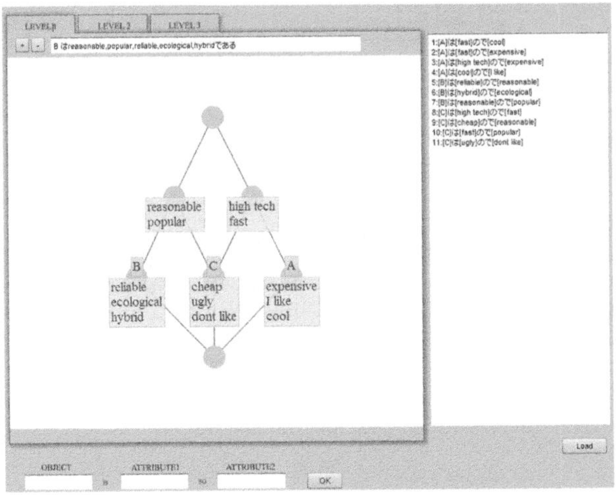


Figure 12. A captured image of the software of the method.

## 3.2 *How to use*

A captured image of the software is shown in Figure 12. As you can see in the image, the software has input boxes in the bottom for the sentences of SCT. Analysts only have to input the words of results from SCT. If you input enough sentences to draw Concept Lattices, the diagrams will be drawn automatically. You can switch the windows of The *Level 1–3 Concept Lattices* by clicking tabs, then, interpret these Concept Lattices.

# 4 APPLICATION FIELD

## 4.1 *Product development*

For the project managers or ergonomic engineers who work in product development, this method is useful in the sections of defining requirements and evaluations. For example, it is possible to apply this method to UI design. When conducting user tests to some prototypes, the form of SCT can be used. Then the method software can be used and the structures of the problems of the prototypes can be interpreted. Therefore, it helps to define the specifications of your products.

## 4.2 *Cognitive analysis*

This method is useful for revealing human personalities. Comparing differences in cognitive structures of various participants helps us to grasp someone's characteristics. For example, comparing a participant's cognitive structures of knowledge about computers helps us to understand what they know, or don't know about computers.

# REFERENCES

Bernhard Ganter (1984): Two Basic Algorithms in Concept Analysis.

Ganter, B. & Wille, R. (1999): Formal Concept Analysis, Springer.

Rudlof Wille (1981): Restructuring Lattice Theory: An Approach Base On Hierarchies Of Concept.

Rudlof Wille (2005): Formal Concept Analysis as Mathematical Theory of Concepts and Concept Hierarchies.

Sacks, Joseph M & Sidney Levy (1950): The Sentence Completion Test. Projective psychology: Clinical approaches to total personality.

Symonds, Percival M. (1947): The sentence completion test as a projective technique.

*Ergonomics in Asia – Shih & Liang (eds)*
*© 2012 Taylor & Francis Group, London, ISBN 978-0-415-68414-9*

# A guide to constructing an interview

Yusuke Morita
*Graduate School of Systems Engineering, Wakayama University, Japan*

Toshiki Yamaoka
*Faculty of Systems Engineering, Wakayama University, Japan*

## ABSTRACT

*Purpose*: The purpose of this study is to produce a guide to constructing an interview.
*Background*: Interviews are used in product development to survey user's *kansei* and images today. Interviewers think how they construct the interview and how they can elicit wanted answers from interviewees during the session. How well they draw good answers is up to the words they throw to the interviewees, in other words, the vocabulary of interviewers (which comes from their experience in interviewing) could change the outcome of the interviews. It takes years for interviewers to gain experience in order to be better interviewers. It would be a great help if there were any tools or hints that could help inexperienced interviewers.
*Method*: Finding common flows and skills used in 11 interview methods in 7 fields.
*Results*: The flows of those 11 methods could be divided into 3 groups, and a total of 23 different skills found could fall into 8 then further into 3 by concepts. We researched each of the flows and skills to find the relationships among them and which skill(s) is used when (in the flows).
*Conclusion*: Those interview flows and skills that were found and confirmed in this study can be used as supplements to an insufficient vocabulary for interview construction. We made a guide that covers those flows and skills and hope it helps interviewers to manage interviews regardless of their experience, and we believe this guide also could be a text for them.

*Keywords*: interview, guide, interview flow, interview skills, Formal Concept Analysis (FCA)

## 1 INTRODUCTION

Interviews are used in product development to survey a user's *kansei* and images today. Interviewers think how they construct the interview and how they can elicit wanted answers from interviewees during the session. How they draw good answers is up to the words they throw to the interviewees, in other words, the vocabulary of interviewers (which comes from their experience in interviewing) could change the outcome of the interviews.

The purpose of this study is to make a guide to constructing an interview.

## 2 METHOD

Finding common flows and used skills in 11 interview methods in 7 fields.

### 2.1 *11 Interview methods in 7 fields*

The 11 methods of interview and 7 fields where those interview methods are used are as follows.

(Fields: Methods)

1. Police: 1. Cognitive Interview (C.I).
2. Justice: 2. Examination.
3. Medical: 3. Medical Interview, 4. Nursing Interview.
4. Counseling: 5. Psychological counseling.
5. Journalism: 6. Collecting data.
6. Social Study: 7. Life Story Interview, 8. Ethnography.
7. Marketing Research: 9. Depth Interview 10. Laddering Research, 11. Focus Group Interview.

## 3 EXTRACTING FLOWS AND SKILLS

The flows of those 11 methods could be divided into 3 groups. (Figure 1)

### 3.1 *Interview flow groups*

1. Fact Finding Group
Interview methods in this group focus on eliciting truth from interviewees. Interviewers pay full attention to any inventions or false reports.

In the interviews of this group, several techniques based on cognitive psychology are used for some purposes including encouraging interviewees to restore photographic memories and leading them to blurt out the facts. However, the psychological burden incurred to the interviewee is not considered. CI and Examination which could potentially change the rest of their lives for interviewees and other involved people belong to this group. Therefore, interviewers are required to have high concentration and full attention during interviews.

2. Problem Solving Group
The methods in this group are to find interviewees' physical and psychological problems, to make accurate diagnosis and to offer suitable treatments. Medical Interview, Nursing Interview and Counseling are included in this group. The feature of this group is that the interviewers try to express the message, "I understand you" to the interviewees. This helps interviewees to feel satisfied to be understood as well as to be treated. Due to above reasons, interviewers are required to grab information as much as and as correct as possible.

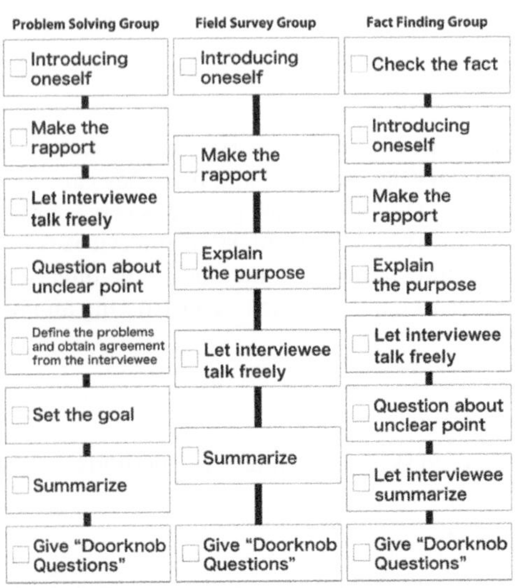


Figure 1. 3 Groups of interview flow.

3. Field Survey Group

The main purpose of this group is researching the events that interviewees experienced by asking questions of them. To collect on-target answers in a short time, interviewers are often required to conduct a preliminary survey to prepare a framework for the interviews. This group includes Journalism, Social Study and Marketing Research. Rapport should be built to make interviewees to disclose as much information as they have.

1. Introduce oneself.
   This is to build a foundation of rapport.
2. Make the rapport.
   This is the step to make rapport actually by talking about non-related things to relax and to know each other well.
3. Explain the purpose.
   This is to let the interviewees to understand the outlines and "rules" of the interviews. Interviewers explain about the purpose and give precautions of the interviews to interviewees in this step.
4. Let interviewee talk freely.
   Let interviewee talk freely about the topics. Interviewers can stop it if required.
5. Define the problems and obtain agreement from the interviewee.
   The purpose of this step is to define the problem by referring to past cases, and to obtain agreement from the interviewee to confirm if interviewer understands correctly.
6. Question about unclear points.
   To clarify unclear points.
7. Set the goal.
   This is to set the goal where they need to go to solve the problem based on interviewee's feelings and thoughts.
8. Summarize or let interviewee summarize.
   Summarize the interview and confirm if understandings are correct.
   Or, let interviewees summarize and check the statements by themselves. New information can be added in this step.
9. Give "Doorknob Questions".
   Finish interview after making sure if interviewees don't have anything more to say.

### 3.2 *Interview skills*

23 skills were used in the studied interviews and they were separated into 8 types depending on the effects as below.

1. Concession.
   To reduce the burden of interviewees by showing that interviewers can compromise.
2. Pressure.
   To put pressures on interviewees to collect information.
3. Summarize.
   To give the feedback—"I understand the points you give"—to the interviewees.
4. Drawing out.
   To obtain all information they have even if it looks unrelated (which could be something related) to see the whole picture.
5. Stress.
   To raise the stress level and keep the interviewees concentration on the interview.
6. Thaw.
   To lower the stress level and let them feel relaxed.
7. Sympathy.
   To be closer psychologically.
8. Dissonance.
   To get interviewees into *"cognitive dissonance"* by taking unexpected actions
   Furthermore these "Effects" can be sorted into 3 concepts.

1. Collaboration.
   Skills to use to lower level of stress and make interviewees feel freer to speak, this helps eliciting real voices and truths.
2. Indispensability.
   Must-have skills for interviewers that are required in the interview.
3. Compression.
   Skills to use to raise the stress level for interviewees to see their reactions and answers.

## 4 RELATIONSHIPS AMONG FLOWS AND SKILLS

Formal Concept Analysis (FCA) was conducted by using the context chart to study the relationships among flows and skills found. The result is a concept lattice which is shown below as Figure 2.

The result of FCA clearly shows that the Problem Solving Group contains skills including "Concession", "Sympathy", and "Summarize". Furthermore, all groups use skills of "Drawing out" and "Thaw".

The following figure (Figure 3) show the detailed results.

Table 1. Interview skills.

| Effect | Skills to use |
|---|---|
| Concession | Giving options, Taking a break, Accepting "don't know", |
| Pressure | Getting closer, Silencing, Questioning without pauses, |
| Summarize | Rewording, Summarizing, Illustrating, |
| Drawing out | Reproducing the situation, Change the time axis, Changing perspectives, Doorknob Question, Illustrating, |
| Stress | Questioning without pauses, Getting closer, Advancing to false reporting and fiction, |
| Thaw | Letting interviewees copy interviewers' attitude (so that they can relax, "Mirroring"), Reflecting the emotions, Talking about non-related things, |
| Sympathy | Setting the common purpose, Reflecting the emotion, Working together, Implying efforts, |
| Dissonance | Silencing, Smiling for nothing, Showing restlessness. |

Table 2. Interview skills.

| Concept | Effects |
|---|---|
| Collaboration | Concession, Summarize, Sympathy |
| Indispensability | Thaw, Drawing out |
| Compression | Pressure, Stress, Dissonance |

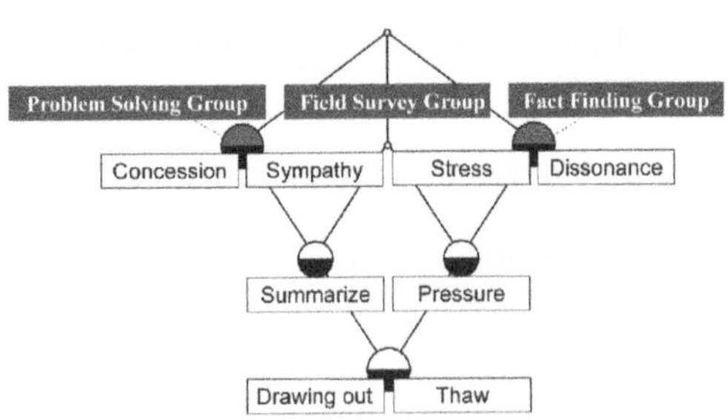


Figure 2. Result from FCA (Concept lattice).

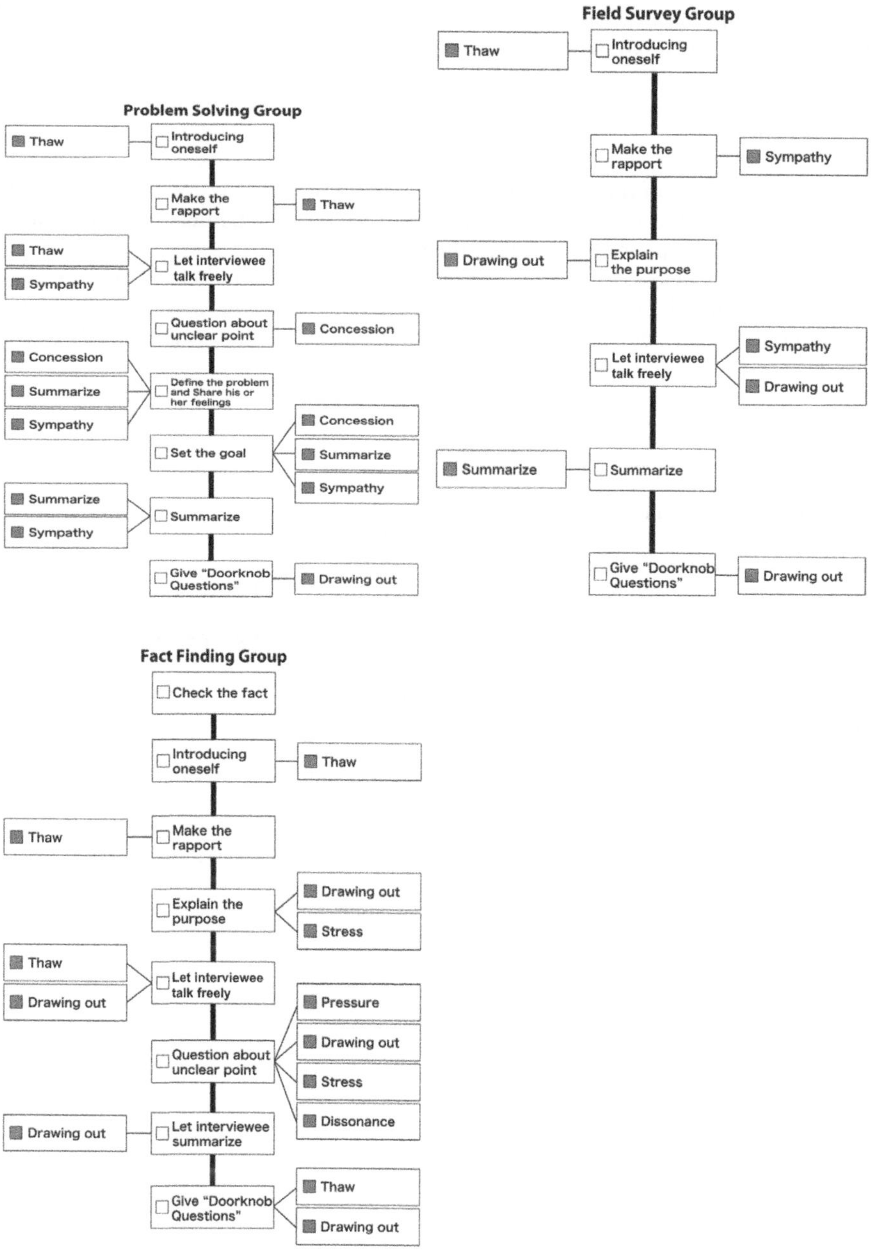


Figure 3. Skills adapted for each step of the flow.

## 5 CONSIDERATION

Problem Solving Group includes "Collaboration" skills a lot to get along with interviewees to elicit as much as possible from them.

Fact Finding Group includes many "Compression" skills. This would be for the purpose of increasing pressure on interviewees on purpose so that they speak in a careful manner and not let them to tell lies.

Interviews in Field Survey Group aim to lead to one result from the answers. Possibilities of faked stories or lies are out of consideration. The reason why there are fewer skills for each steps comparing to other interviews in different groups would be because flexible correspondences are expected in this group.

## 6 CONCLUSION

A guide to constructing an interview which covers the flow patterns and the skills of interviews is made as a result of this study. This guide is expected to help interviewers to construct fruitful interviews by him/herself regardless their experience.

However the effectiveness of this guide hasn't been verified yet and needs to be examined. The next step that we, the researchers, intend to take is comparing the result of interviews done by experienced interviewees with their own vocabulary and non-experienced interviewees with the guide.

## REFERENCES


Alfred Benjamin, Helping Interview, Houghton Mifflin, 1981.

Banister, P., Burman, E., Parker, I., Taylor, M. & Tindall, C. Qualitative Methods in Psychology, A Research Guide, Open University Press, 1994.

Ernestine Wiedenbach & Caroline E. Falls, Communication Key to Effective Nursing, Tiresias Press, 1978.

Gladys, B., Lipkin, R.N., M.S., Roberta, G. & Cohen, R.N., M.S., Effective Approaches to Patients'Behavior, Springer Pub., 1998.

Joy Ann Duxbury, Difficult Patients, Elsevier Health Sciences, 2000.

Knight, C. & Aldrich, M.D. The Medical Interview: Gateway to the Doctor-Patient Relationship, Parthenon Pub. Group, 1999.

Marion Nesbitt Blondis & Barbara, E. Jackson, Nonverbal Communication with Patients, Back to the Human Touch, Wiley, 1977.

Owen Harige, The Handbook of Communication Skills, Taylor & Francis, 2006, pp. 531–546.

Rebecca Miline & Ray H.C. Bull, Investigative Interviewing: Psychology and Practice, Wiley, 1999.

Uwe Flick, Qualitative Forschung, Rowohlt, 1998.

Willig, C. Introducing Qualitative Research in Psychology, McGraw-Hill, 2008.

# *Part II: Aging and work ability*

*Ergonomics in Asia – Shih & Liang (eds)*
 *ISBN 978-0-415-68414-9*

# Aging and work ability: Their effect on task performance of industrial workers

C. Theppitak, Y. Higuchi, D.V.G. Kumudini, V. Lai, M. Movahed, H. Izumi & M. Kumashiro

*Department of Ergonomics, Institute of Industrial Ecological Sciences, University of Occupational and Environmental Health, Kitakyushu, Japan*

ABSTRACT: To examine the effect of aging and work ability on the task performance of industrial workers, twenty four subjects including 12 younger and 12 older male workers were selected. The average age of the young and old groups was 28.75 years and 61.25 years, respectively. Work ability was measured using the Work Ability Index (WAI) and subjects were divided into 'moderate' and 'excellent' according to WAI. Task performances were examined by using the function of maintaining concentration (TAF). Subjects did a TAF test, 2 trials per day (before and after work) on Monday, Wednesday and Friday. The result showed an aging effect on task performances on Monday and Wednesday. The excellent work ability group has higher task performance than the moderate group in both young and older groups. However, we could not find significant work ability effect on task performance. Aging effect seems to have a higher impact on task performances than work ability effect.

*Keywords*: aging, WAI, TAF, concentration

## 1 INTRODUCTION

The average age of the working population is getting higher, with increasing numbers of middle-aged and older workers employed in many different jobs. It is known that age-related declines can occur in important mental and physical abilities. Fatigue is a state resulting from work imposed upon the biological body. It can be construed as a failure in the coordination of organically related biological functions, both physical and mental. In actual form, it appears as a decrease in enthusiasm for work and as a lowering of efficiency. Theoretically, it is defined as the combined output of mental and physiological function (Takakuwa, 1982). Fatigue is caused from insufficient sleep, prolonged mental or physical work, or extended periods of stress or anxiety. Acute fatigue results from short-term sleep loss or short periods of heavy physical or mental work. Bryson & Forth (2007) indicated that the day of the week can have an effect on productivity: it caused productivity decline over the course of the week as a result of increasing fatigue. High levels of fatigue cause reduced performance and productivity, and increase the risk of accidents and injuries. The work ability is characterized by the balance between a worker's individual resources and demands of the work. The Work Ability Index (WAI) is a prevalent method to assess perceived work ability (Ilmarinen et al., 1997, Tuomi et al., 1998). The WAI is affected by several work-related and individual factors (Ilmarinen et al., 1997). If the work ability of worker is lower or higher than the required performance, this situation tends to induce operator boredom or stress respectively. In order to improve the design and organization of work to achieve a balance for the worker's work ability, it is important to know whether job performance is higher or lower for older workers in comparison with younger workers. Therefore, the aging and work ability effect on task performance was examined in this study.

## 2 METHOD

### 2.1 *Subject*

Twelve young and twelve old male healthy workers were selected as subjects. The average of age for young and old group was 28.75 ± 2.83 years and 61.25 ± 1.05 years, respectively.

### 2.2 *Procedure*

To investigate any aging effect on task performance, subjects were categorized by age into younger (20–35 years old) and older (> 60 years old) groups. Work ability of subjects was measured with the Work Ability Index (WAI). This is an assessment of the ability of a worker to perform his/her job. The index consists of a questionnaire on seven dimensions which are rated and the summative index ranges from 7 to 49, which is classified into poor (7–27), moderate (28–36), good (37–43), and excellent (44–49) work ability (Berg, 2009). The result from the WAI measurement classified subjects into 'excellent' and 'moderate' work ability groups. The members in each group consisted of 6 young and 6 older workers. Task performance was examined by using the function of maintaining concentration (TAF). TAF is the equipment that is used to evaluate objective fatigue on the concept of concentration maintaining ability (Takakuwa,1982). Fatigue is a feeling resulting from a decrease in work enthusiasm and slow efficiency, according to the concept of TAF. It is the combined output of mental activity and physiological functions. The concentration ability was assumed to represent the mental activity, and maintaining that concentration can be illustrated as representing physiological functions.

The subjects were informed and trained about using TAF equipment before starting the experiment. After practice, subjects have to do the TAF test in 2 trials per day before and after work on Monday, Wednesday and Friday. In each trial, subjects have to do the TAF test 3 times. In each test, the subject has to match the cross of a target and the cross of a telescope as accurately as he can and keep or maintain this situation in one minute then rest for 10 seconds. A standardized curve of TAF tests was obtained to show the task performance. The TAF-L indicated the level of concentration while the TAF-D showed the degree of fluctuation in maintaining that concentration.

We employed a paired sample t-test for analyzing the difference of concentration and maintain attention ability between the younger and older groups. To examine the aging and work ability effect on task performance we used two-way ANOVA for statistical analysis.

## 3 RESULTS

### 3.1 *Comparison of concentration ability between older and younger groups*

TAF-L indicated the level of concentration ability, and showed that the trend of concentration ability of the younger group was to have higher concentration ability than older group on all 3 days (Monday, Wednesday and Friday). However, we could find a significant different of concentration ability between younger and older groups only on Monday before work (A.M.). We could observe that the younger group showed their highest concentration ability on Wednesday before work (A.M.) and lowest concentration ability on Friday after work (P.M.). The concentration ability of the younger group was significantly decreased after work on Wednesday. For the older group, they have their highest concentration ability on Wednesday before work and lowest on Monday before work. However, we could not find a significant difference of concentration ability in the older group between Monday Wednesday and Friday.

### 3.2 *Comparison of the ability to maintain concentration between older and younger groups*

TAF-D indicated the degree of fluctuation in maintaining concentration ability and shows the trend of maintaining concentration ability of the younger group was better than the older group on all 3 days (Monday, Wednesday and Friday). However, we could find no significant difference in the ability to maintain concentration between the younger and older groups

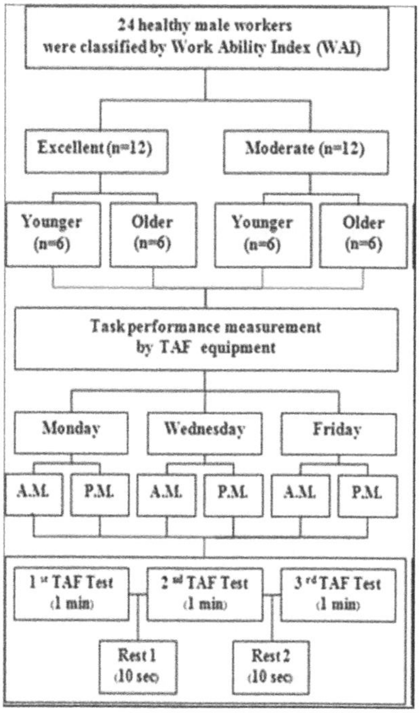


Figure 1. 24 male workers were classified by Work Ability Index (WAI) and divided into excellent and moderate group. Each group consists of 6 younger and 6 older workers. Task performance of each group was measured by using TAF on Monday, Wednesday and Friday at before work (A.M.) and after work (P.M.). Each trial, subject has to do the TAF test 3 times.

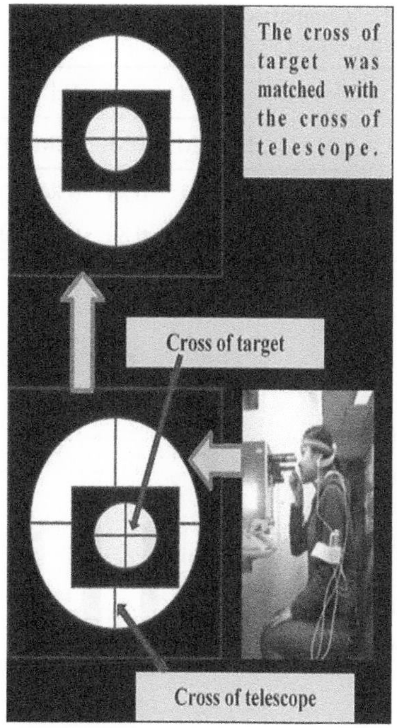


Figure 2. Task performance was measured by TAF equipment. Subjects look through the telescope to find the target try to match the cross of the target and the cross of telescope by moving the handle of TAF equipment.

on Monday after work and Wednesday both before and after work. We could observe that the younger group showed the best ability to maintain concentration on Wednesday before work and lowest ability to maintain attention on Friday before work. For the older group, they have the best ability to maintain concentration on Friday after work and the lowest on Monday after work. We could not find any significant difference in the ability to maintain concentration before work and after work on Monday, Wednesday and Friday in both older and younger groups.

### 3.3 *Comparison of concentration ability between the excellent and moderate groups*

The ‘excellent’ work ability group showed significantly higher concentration ability than the ‘moderate’ group on Monday before work. The excellent group showed a significant decrease in their concentration ability on Wednesday. Moreover, we could observe the highest concentration ability on Wednesday before work and lowest concentration ability on Wednesday after work in the excellent group. We could not find a significant difference in the concentration ability between before work and after work in the moderate group on any of the 3 days. For the moderate group, we could observe that they have the highest concentration ability on Wednesday before work and the lowest on Friday after work.

### 3.4 *Comparison of the ability to maintain concentration between the excellent and moderate groups*

We could not find any significant difference in the ability to maintain concentration of the excellent and moderate groups in all 3 days. The excellent group showed the best

Table 1. Comparison of the concentration ability between younger and older groups.

| | Mon | | Wed | | Fri | |
|---|---|---|---|---|---|---|
| TAF-L | A.M. | P.M. | A.M. | P.M. | A.M. | P.M. |
| Young | 0.88 | 0.87 | 0.89 | 0.83 | 0.85 | 0.82 |
| Old | 0.80 | 0.81 | 0.83 | 0.80 | 0.82 | 0.81 |

Table 2. Comparison of the ability to maintain concentration between younger and older groups.

| | Mon | | Wed | | Fri | |
|---|---|---|---|---|---|---|
| TAF-D | A.M. | P.M. | A.M. | P.M. | A.M. | P.M. |
| Young | 0.06 | 0.06 | 0.05 | 0.05 | 0.07 | 0.06 |
| Old | 0.09 | 0.10 | 0.09 | 0.09 | 0.08 | 0.08 |

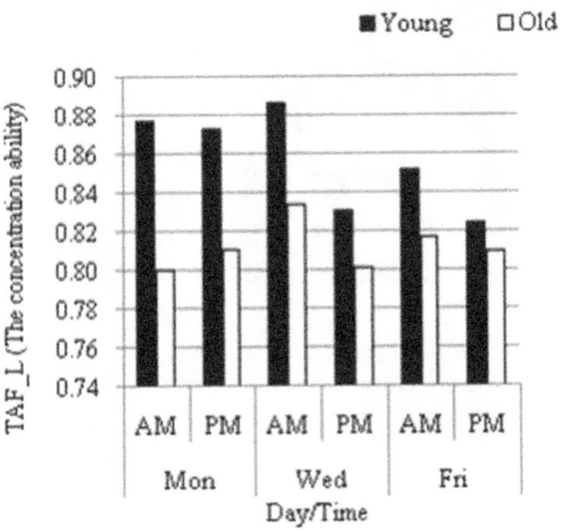


Figure 3. The concentration ability (TAF-L) of younger group was compared with older group on Monday, Wednesday and Friday before work (A.M.) and after work (P.M.).
*p-value < 0.05 (Paired sample T-Test)

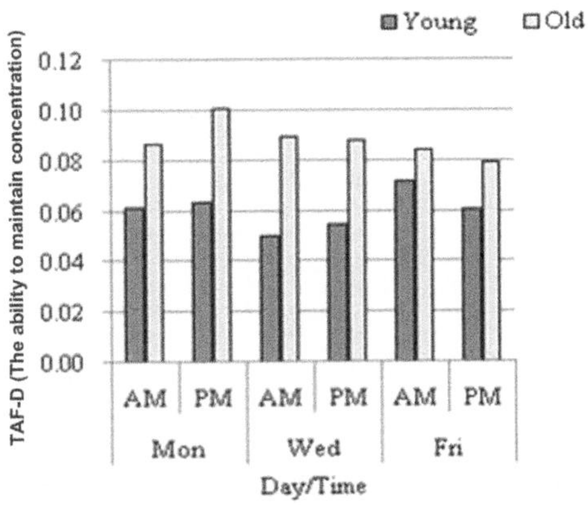


Figure 4. The ability to maintain concentration (TAF-D) of young group was compared with older group on Monday, Wednesday and Friday both before work (A.M.) and after work (P.M.).
*p-value < 0.05 (Paired sample T-Test)

ability to maintain concentration on Friday (after work) and the lowest on Monday after work. This trend was also found in the moderate group. The ability to maintain concentration of the excellent group was significantly decreased after work on Monday. We could not find any significant difference in the ability to maintain concentration between before work and after work on Monday, Wednesday and Friday in the moderate group.

### 3.5 *The effect of aging and work ability effect on concentration ability and the ability to maintain concentration*

The effect of aging and work ability on task performance was investigated by two-way ANOVA. The result showed an aging effect on the concentration ability of subjects on Monday before work. However, we could not find any work ability effect on concentration ability. When we consider the interaction between aging and work ability effect, the result indicated that the difference in the concentration ability of subjects was not caused by any aging and work ability effect.

The result from Table 6 showed an aging effect on the ability to maintain attention of subjects on Monday after work and Wednesday both before and after work. However, we could not find any work ability effect on the ability to maintain attention. When we consider the interaction between aging and the work ability effect, the result indicated that the difference in the ability of subjects to maintain concentration was not caused by any aging and work ability effect.

Table 3. Comparison of concentration ability between excellent and moderate work ability groups.

| | Mon | | Wed | | Fri | |
|---|---|---|---|---|---|---|
| WAI | A.M. | P.M. | A.M. | P.M. | A.M. | P.M. |
| Excellent | 0.86 | 0.86 | 0.87 | 0.81 | 0.84 | 0.83 |
| Moderate | 0.81 | 0.83 | 0.88 | 0.82 | 0.83 | 0.81 |

Table 4. Comparison of the ability to maintain concentration between excellent and moderate work ability groups.

| | Mon | | Wed | | Fri | |
|---|---|---|---|---|---|---|
| TAF-D | A.M. | P.M. | A.M. | P.M. | A.M. | P.M. |
| Excellent | 0.07 | 0.10 | 0.07 | 0.07 | 0.07 | 0.06 |
| Moderate | 0.08 | 0.09 | 0.07 | 0.07 | 0.08 | 0.07 |

Table 5. Analysis of variance related to concentration ability (TAF-L) by age and work ability index (WAI).

| | Mon | | Wed | | Fri | |
|---|---|---|---|---|---|---|
| Factors | F | Sig | F | Sig | F | Sig |
| TAF-L | | | | | | |
| AM | | | | | | |
| Age | 6.58 | 0.018 * | 3.76 | 0.067 | 1.77 | 0.198 |
| WAI | 2.77 | 0.112 | 0.49 | 0.49 | 0.01 | 0.926 |
| Age/WAI | 0.20 | 0.657 | 0.12 | 0.731 | 0.54 | 0.470 |
| TAF-L | | | | | | |
| PM | | | | | | |
| Age | 4.54 | 0.460 | 1.01 | 0.326 | 0.25 | 0.620 |
| WAI | 1.26 | 0.276 | 0.01 | 0.918 | 0.62 | 0.441 |
| Age/WAI | 0.29 | 0.604 | 0.06 | 0.806 | 0.01 | 0.918 |

* p-value < 0.05

Table 6. Analysis of variance related to ability to maintain concentration (TAF-D) by age and work ability index (WAI).

| | Mon | | Wed | | Fri | |
|---|---|---|---|---|---|---|
| Factors | F | Sig | F | Sig | F | Sig |
| TAF-D | | | | | | |
| AM | | | | | | |
| Age | 6.31 | 0.089 * | 6.90 | 0.016 | 0.97 | 0.337 |
| WAI | 2.01 | 0.172 | 0.91 | 0.351 | 01.17 | 0.292 |
| Age/WAI | 0.44 | 0.755 | 0.37 | 0.849 | 02.76 | 0.112 |
| TAF-D | | | | | | |
| PM | | | | | | |
| Age | 6.31 | 0.021 * | 7.36 | 0.013 * | 20.7 | 0.165 |
| WAI | 0.10 | 0.755 | 1.06 | 0.316 | 1.90 | 0.184 |
| Age/WAI | 0.03 | 0.864 | 1.43 | 0.245 | 0.16 | 0.918 |

* p-value < 0.05

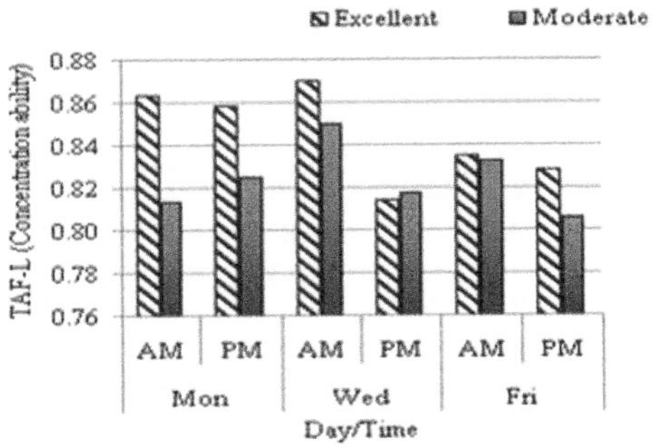


Figure 5. The concentration ability of excellent group compared with moderate group on Monday, Wednesday and Friday both before work (AM) and after work (PM).
* p-value < 0.05 (Paired sample T-Test)

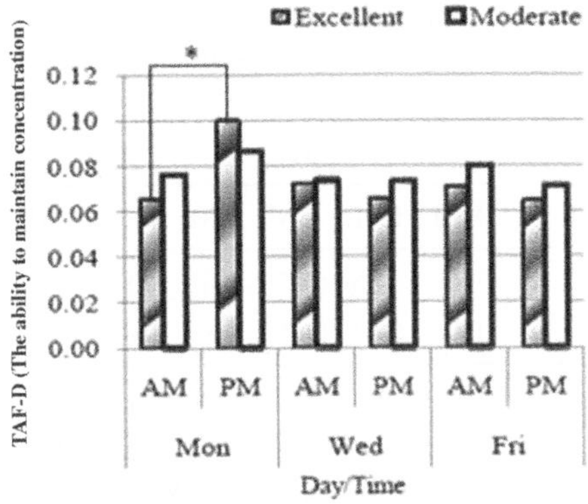


Figure 6. The ability to maintain concentration of excellent work group compared with moderate group on Monday, Wednesday and Friday both before work (A.M.) and after work (P.M.).
* p-value < 0.05 (Paired sample T-Test)

## 4 DISCUSSION

Several previous studies have showed that as age increases, changes in health and various aspects of work capacity also change (de Zwart et al., 1995, Broersen et al., 1996, Reilly et al., 1997). An age-related imbalance between workload and work capacity was suggested to result in a chronic over workload, increasing the risk to long-term health. (de Zwart et al., 1995). The short-term effects are signs of fatigue, which are mostly experienced during or immediately after a day of working (Kiss et al., 2007). In this study we examined the effect of aging and work ability effect on task performance on different days of the week. The results indicated an aging effect on task performance on Monday and Wednesday. Especially, on Monday the concentration ability of the older group was lower than the younger group on Monday morning (before work). Moreover, the ability to maintain concentration of older adults was lower than younger adults on Monday after work and Wednesday both before and after work. It seems that Monday is an important day that we should pay more attention to than other days in the week. Bryson and Forth (2007) indicated that worker productivity is lower on Mondays. It has been suggested that Monday performance is worse than performance on other days of the week for psychological reasons. Pettengill, 1993 cites behavioral science suggesting mood swings during the week, with Monday associated with negative moods. Interestingly, Facebook's Global Happiness Index shows Monday to be the least happy day of the week based on an analysis of emotional words from the users of the site. Long-term studies have found Monday to be associated with a disproportionately high percentage of stroke incidences, suggesting that low productivity may also be due to physiological reasons associated with the Monday effect (Pettengill, 1993). Bryson and Forth (2007) indicated that productivity may be lower on Mondays because workers need to be reoriented to the work environment after the week-end or may lack motivation to work given the length of time until the next weekend. Fatigue after daily work is thought not to be a problem if enough recovery time is offered between two periods of work. If there is not enough time to recover from this fatigue, the cumulated effects of this fatigue will lead to long-term adverse health effects (Sluiter et al., 1999). Previous studies found that the mean recovery score of older workers was significantly higher than younger workers. These findings which we mentioned above could support our result that the older group have the lowest concentration ability on Monday morning (before work). The work ability effect on task performance was not found in this study. We could see a significant difference of concentration ability in the excellent work ability group when compared with the moderate group on Monday before work.

This may be caused from the limitation of the small numbers in the excellent and moderate groups that we used to examine the work ability effect. Previous studies indicated that factors associated with poor work ability, as defined by WAI, were lack of leisure-time, vigorous physical activity, poor musculoskeletal capacity, older age, obesity, high mental work demands, lack of autonomy, poor physical work environment, and high physical work load (Berg et al., 2009). During the experiment, we could observe that the younger group in this study seem to have a higher workload than older group. Although they might have high work ability, their higher workload also could be a risk factor for maintaining good work ability.

## 5 CONCLUSIONS

The younger group have higher concentration and ability to maintain concentration than the older group. Especially on Monday, younger adults have a higher concentration ability than the older group on Monday morning (before work). We could find no significant difference in the ability to maintain concentration between younger and older groups on Monday evening (after work) and Wednesday both before and after work. We could not find any clear-cut work ability effect on task performance between the excellent and moderate work ability groups due to the limitation of the small numbers of participants.

It seem that the aging effect has more impact on task performance than work ability effect. The most interesting result is that Monday is the most important day for workers that employers should consider in designing or managing the job or workload to achieve a balance for their work ability.

## REFERENCES


Berg, T.I.J. van den, Elders, L.A.M., de Zwart, B.C.H. & Burdorf, A. (2009). The effects of work-related and individual factors on the Work Ability Index: a systematic review. *Occup Environ Med.* Vol. 66, pp. 211–220.

Broersen, J.P., de Zwart, B.C., van Dijk, F.J., Meijman, T.F. & van Veldhoven, M. (1996). Health complaints and working conditions experienced in relation to work and age. *Occup Environ Med.* Vol. 53, pp. 51–57.

Bryson, A. & Forth, J. (2007). *Are There Day of the Week Productivity Effects*? Manpower Human Resources Lab Discussion Paper No. 4, London School of Economics.

de Zwart, B.C.H., Frings-Dresen, M.H.W. & van Dijk, F.J.H. (1995). Physical workload and the ageing worker: a review of the literature. *IntArch Occup Environ Health*, Vol. 68, pp. 1–12.

Ilmarinen, J., Tuomi, K. & Klockars, M. (1997). Changes in the work ability of active employees over an 11-year period. Scand, J. *Work, Environ. Health*, Vol. 23 Suppl 1, pp. 49–57.

Kiss, P., Meester, M.D. & Breaceckman, L. (2007). Differences between younger and older workers in the need for recovery after work. Int Arch Occup Environ Health.

Pettengill, G.N. (1993), "An experimental study of the 'Blue Monday' hypothesis," Journal of Socio-Economics, Vol. 22, pp. 241–257.

Reilly, T., Waterhouse, J. & Atkinson, G. (1997). Aging, rhythms of physical performance, and adjustment to changes in the sleep-activity cycle. *Occup Environ Med.* Vol. 54, pp. 812–816.

Sluiter, J.K., Van der Beek, A.J. & Frings-Dresen, M.H.W. (1999). The influence of work characteristics on the need for recovery and experienced health: a study on coach drivers. *Ergonomics* Vol. 42, pp. 573–583.

Takakuwa, E. (1982). Evaluation of fatigue and the function of maintaining concentration (TAF). Hokkaido University school of Medicine, Sapporo.

Tuomi, K., Ilmarinen, J., Jahkola, A., Katajarinne, L. & Tulkki, A. (1998). *Work Ability Index*, Finnish Institute of Occupational Health, Helsinki.

*Ergonomics in Asia – Shih & Liang (eds)*
 *ISBN 978-0-415-68414-9*

# Age-related changes of posture for pulling exercises in lower position

Masato Sakurai, Kazuyuki Orito & Naoki Tsukahara
*Tokyo University of Science, Japan*

Yuka Yamazaki
*Mitsubishi Electric Corporation, Japan*

Masahiko Sakata & Sakae Yamamoto
*Mitsubishi Electric Group, Central Melco Corporation, Japan*
*Tokyo University of Science, Japan*

ABSTRACT: The angles in forward bending and the area of movement in the center of pressure were measured and a subjective evaluation performed to examine the age-related changes of posture when pulling exercises in the lower position. The task was to pull at a drawer in the lower position while changing the height of the drawer. Three age groups (20, 60, and 70 year olds) participated in this experiment. From the results, the degree of ease in the operation of the task increases with the height of drawers in all the groups. It was found that the posture of the 70 age group while forward bending in performing the task for the lowest drawer is different from those of the 60 and 20 age groups when comparing the angles of bending in the trunk and knee. It suggests that the 70 age group lower their center of the gravity by bending their knees so as to ease the high physical load at their waist for the lower drawers.

## 1 INTRODUCTION

It is well-known that the world population over the age of 65 will increase year by year. In Japan it will be about 33 million, 26.3% of the total Japanese population in 2015 (United Nations Population Division, 2008). Also, elderly people often use and operate many consumer products in daily life. It is important to investigate the posture and its physical load when elderly people operate equipment such as consumer products from the applicative point of view.

A previous study measured the physical load, working posture and center of pressure when pulling at a lever at different heights for both young and elderly subjects (Nishioka et al., 2002). It is suggested that elderly subjects have difficulty in balancing the working posture in pulling exercises compared with that of young subjects because of the decline in the flexibility of muscle and posture. Hence, it is necessary to evaluate quantitatively the posture in pulling exercises with age-related changes and by examining the effect of age-related changes on posture when pulling at a drawer in the lower position we shed light on the decrease of physical ability with aging.

## 2 PURPOSE

The purpose of this paper is to examine the changes of posture for elderly and young subjects when pulling at a drawer in the lower position from measurements of the angle of forward bending, the movement of the center of pressure and a subjective evaluation while pulling a drawer located at different heights.

## 3 METHODOLOGY

### 3.1 *Task*

The task was to pull a drawer from a lower position at different drawer heights. The subjects were instructed to stand up at a designated position on a force sensor plate, to pull out a certain drawer with their dominant hand, and to put it back with both hands.

### 3.2 *Apparatus*

Figure 1 shows the apparatus in this experiment. There were four drawers from 200 to 800 mm in height in 200 mm increments. The distance of pulling the drawer was 200 mm. The task was recorded by the camcorder placed at the side of subjects to measure the angle of the posture in forward bending and the movement of the center of pressure on the force sensor plate during the task.

### 3.3 *Procedure*

Prior to the experiment, all the subjects practiced the task. Each subject performed the task three times for each of four drawers. The subjects were asked to subjectively evaluate how easy it was to perform the task on a scale of 1 (not easy to operate) to 5 (easy to operate). Also, how easy it was to reach and the effect of the weight of the drawer on the pulling exercises were evaluated on a 5 point scale as well.

### 3.4 *Subjects*

Three groups of subjects with ages in the region of 70, 60, and 20 participated in the experiment. Each group consisted of ten subjects for a total of thirty subjects. The average age in each group was 72.8 years (S.D. 1.6), 61.7 years (S.D. 1.7) and 22.6 years (S.D. 1.2), respectively.

### 3.5 *Analysis*

The values of trunk and knee angles while bending forward in performing the task were measured from the images recorded by the camcorder for each subject. Figure 2 shows the

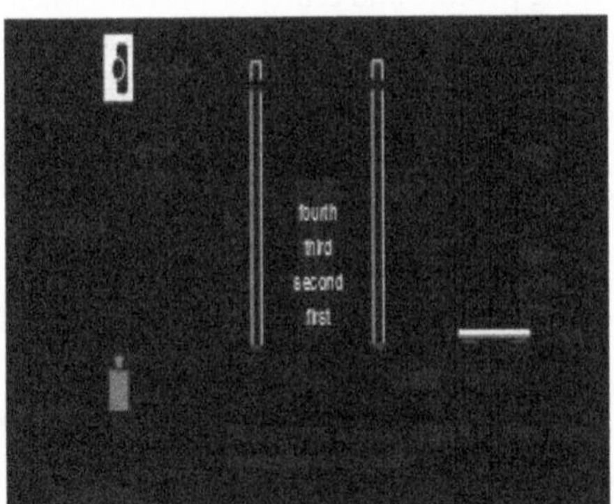


Figure 1. Apparatus. (a) top view, (b) side view.

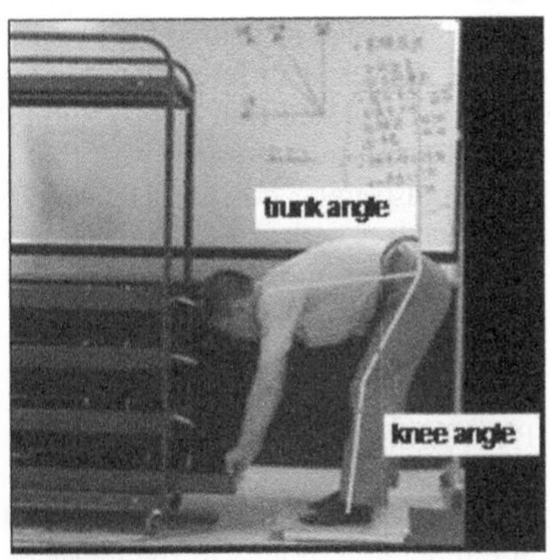


Figure 2. Trunk and knee angles.

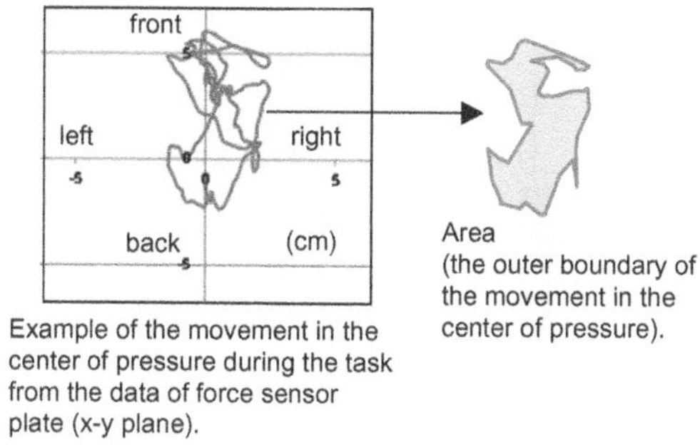
front
left
right
-5
5
back
(cm)
-5
Example of the movement in the center of pressure during the task from the data of force sensor plate (x-y plane).
Area
(the outer boundary of the movement in the center of pressure).

Figure 3. How to calculate the movement in the center of pressure during the task.

trunk and knee angles while bending forward in this experiment. The trunk angle was defined as the angle between the vertical axis and the line from the right shoulder to the right waist of the subject. The knee angle was the angle between the left thigh and leg with the left knee in the vertex. A high physical load at the waist with increasing trunk angle of the subject is indicated. In this paper, the trunk and knee angles can quantitatively represent the posture for pulling a drawer in the lower position and indicate the index of physical load at the waist of the subject.

To evaluate balance when pulling a drawer at different heights, the area of the movement in the center of pressure during the task was calculated from the data measured by the force sensor plate. Figure 3 shows how to calculate this area during the task from the force sensor plate data. The area enclosed by the outer boundary of the movement in the center of pressure was calculated from the data on the x-y plane as measured by the force sensor plate. The result was considered as an index of balance in the posture during the task.

## 4 RESULTS AND DISCUSSIONS

### 4.1 *Subjective evaluation*

Figure 4 shows the results of the subjective evaluation regarding the ease of pulling the drawers averaged over all the subject responses. The horizontal and vertical axes indicate the drawer number and subjective evaluation value, respectively. Each bar represents the average of each group (ages 20, 60, and 70) of the subjects including the standard deviation within each group.

In this figure, the subjective evaluation value increases with the height of drawers in all the groups. In other words, the subjective evaluation of the ease in pulling a drawer increased with drawer height. Drawer number 4 seemed easy to operate for all age groups. Furthermore, there is a tendency that the value for the 70 age group is relatively high compared with those of 20 age and 60 age groups.

Note the style of headings in this instruction sheet. They have been typed and placed in the following manner.

### 4.2 *Trunk and knee angles*

Figure 5(a) and (b) shows the average results of the trunk and knee angles when pulling the drawers for each group. In both graphs, the horizontal axis represents the drawer number and the vertical axis indicates the angles when pulling the drawer. Each bar represents the average values of each group as in Figure 4. The values are the average of the trunk and knee angles based on three measurements for each subject, and the gaps represent the standard deviation for each group.

As shown in Figure 5(a), the values of the trunk angle decrease with the height of drawers in all the groups. In between the second and third drawers, the values remarkably decrease

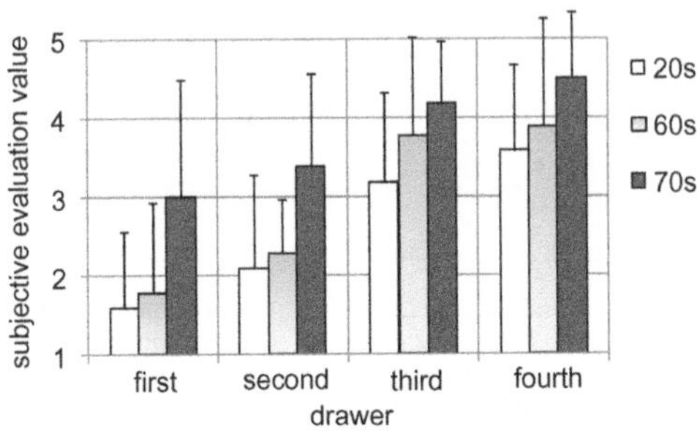


Figure 4. Results of the subjective evaluation regarding the ease of pulling the drawers averaged over all the subject responses.

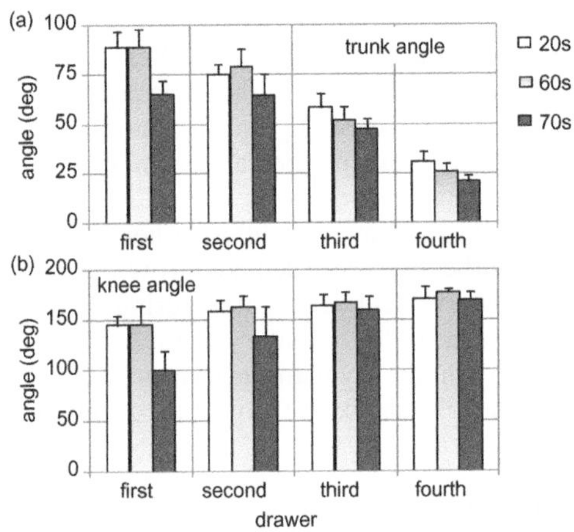


Figure 5. Average results of the trunk and knee angles when pulling the drawers for each group. (a) trunk angle, (b) knee angle.

meaning that the physical load at the waist decreases with the height of the drawers. Therefore, it seems that the posture in the task changes so as to ease the high physical load at the waist. This result corresponds to the results of subjective evaluation in the degree of ease in operating more than the third drawer.

In Figure 5(b), the values of the knee angle increase with the height of the drawers in all the groups. In addition, the value for the 70 age group is relatively small in the lower drawers when compared with the others. Hence, the subjects in the 70 age group bend their knees less when pulling a drawer in the lower position compared with the 20 and 60 age groups, and the subjects in the 70 age group do not bend forward as much when pulling a drawer in the lower position from the results for the trunk angle in Figure 5(a).

## 4.3 *Area of the movement in the center of pressure during the task*

Figure 6 shows the average results of the area of the movement in the center of pressure during the task for each group. The horizontal axis represents the drawer number and the vertical axis indicates the area of the movement in the center of pressure while pulling a drawer. Each bar represents the average value of each group as in Figure 4. The values are the average of the areas based on three measurements for each subject, and the gaps represent the standard deviation for each group.

In this figure, the areas decrease with the height of drawer for all the groups. For the fourth drawer, the values are relatively small and it seems that balance is maintained in all the groups.

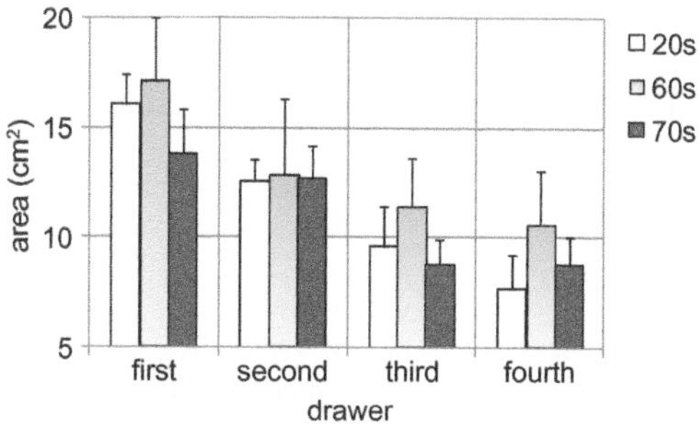


Figure 6. Average results of the area of the movement in the center of pressure during the task for each group.

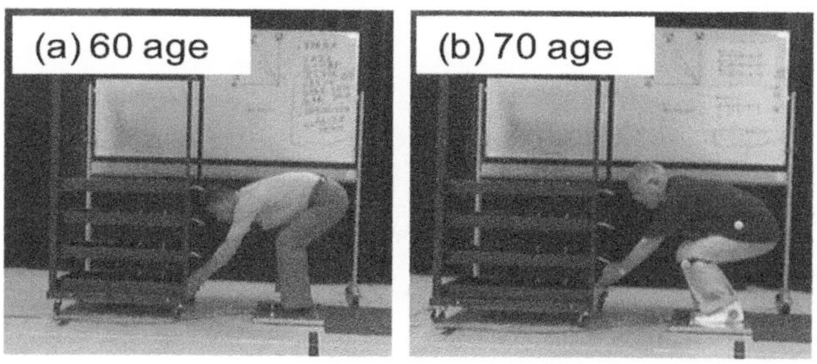


Figure 7. Examples for 60 and 70 age groups while pulling a drawer in the lower position. (a) 60 age group, (b) 70 age group.

For the lower drawers, the area of 70 age group is relatively small compared to the others. It is considered that the subjects of 70 age group lower their center of the gravity with the bending of their knees so as to ease the physical load at the waist for the lower height drawers.

### 4.4 *Comparison of the postures for elderly groups*

As mentioned in the above, the posture while pulling a drawer in the lower position changes with increasing age, and Figure 7(a) and (b) show corresponding examples for 60 and 70 age groups. In those figures, it is clear the 70 age group subject bends at the knees more than the 60 age group. Therefore, it is suggested that the subjects in the 70 age group lower their center of gravity by bending their knees so as to ease the physical load at their waist when pulling a drawer in the lower position in agreement with the finding that muscle mass of all parts of the body decreases past middle age according to a previous study (Tanimoto et al., 2010).

## 5 CONCLUSIONS

The angles of forward bending and the area of movement of the center of pressure were measured and a subjective evaluation performed to examine age-related changes in posture during pulling exercises in the lower position. The task was to pull a drawer in the lower position while changing the height of the drawer. It was found that the posture of the 70 age group while forward bending in performing the task for the lowest drawer is different from those of the 60 and 20 age groups when comparing the angles of bending in the trunk and knee. It suggests that the 70 age group lower their center of gravity by bending their knees so as to ease the high physical load at their waist for the lower drawers. Therefore, it is necessary to take into account age-related changes in posture for the elderly during pulling exercises in the lower position.

## ACKNOWLEDGEMENT

We would like to thank Dr. Miwa Nakanishi from Keio University who provided helpful comments and discussions.

## REFERENCES

Nishioka, M., Okada, A., Miyano, M. & Yamashita, K. (2002). A study on evaluation of physical stress for pulling exercises. *Japan Society of Physical Anthropology,* Vol. 7, No. 1, pp. 49–52. (in Japanese).

Tanimoto, Y., Watanabe, M., Kono, R., Takasaki, K. & Kono, K. (2010). Aging changes in muscle mass of Japanese. *Japanese Journal of Geriatrics,* Vol. 47, No. 1, pp. 52–57. (in Japanese).

United Nations Population Division 2008. World population prospects: the 2008 revision population database, http://esa.un.org/unpp/

*Ergonomics in Asia – Shih & Liang (eds)*
*© 2012 Taylor & Francis Group, London, ISBN 978-0-415-68414-9*

# Effects of mental capacity on work ability in middle-aged factory workers: A field study

D.V.G. Kumudini, Y. Higuchi, C. Theppitak, V. Lai, M. Movahed, H. Izumi & M. Kumashiro
*University of Occupational and Environmental Health, Department of Ergonomics, Kitakyushu, Fukouka, Japan*

ABSTRACT: The aim of the study was to examine the effect of mental capacity on work ability in middle-aged factory workers at a Japanese manufacturing company. The work ability and mental concentration were measured using Work Ability Index (WAI) and Trail Making Test (TMT) respectively. WAI questioner was distributed among the factory workers. 24 moderate and 30 excellent subjects with a mean age of 47.24 years were selected from the sample. The findings substantiated the validity of WAI, as excellent subject's performances were higher than moderate subjects. Also, the excellent subjects in the over 50 group showed better performances than moderate subjects in the 40–49 age group. On the other hand, moderate subjects were affected more by the aging effect than excellent subjects. In summary, there is an interaction effect of aging and WAI within the two age groups. Furthermore, work ability was affected by both mental capacity and aging effect.

*Keywords*: aging, work ability, trail making test, mental capacity, middle-aged

## 1 INTRODUCTION

### 1.1 *Background of the study*

The speed of aging of Japan's population is much faster than in advanced western European countries or the U.S.A. In 2009, the population of elderly citizens (65 years and over) in Japan was 22.7% of the total population, and is the highest in the world (Statistical bureau MIC, 2011).

As elderly populations rise inexorably, it has a considerable impact on the workforce of a country. Consequently, there is a tendency to raise the retirement age which means that workers have to work until an older age. However, from the biological perspective, aging means a progressive deterioration in various physiological systems, which is accompanied by changes in physical and mental capacities of workers (Ilmarinen, 1997). Further, the inevitable policy of keeping workers longer at work can only be successful if the older workers remain in good health (Kiss, et al., 2008). Therefore, evaluating elderly workers, physical as well as mental health is important in industrial settings today.

Although the impact of physical capacity on work ability has been studied by many researchers, there is a lack of information regarding the impact of mental capacity on work ability of workers.

Mental functional capacity is often defined as the ability to perform different tasks that require intellectual and other kinds of mental effort. Cognitive functions, such as perception, memory, learning, thinking, and the use of language have been the primary targets of researches (Ilmarinen, 2001). Moreover, mental capacity needs support, primarily because of the fast and constant change in content, tools at work as a result of the development of information technology, globalization and networks (Tuomi et al., 2001). Among white collar workers in commercial industries, it was found that a worker's physical and mental capacity have an equally important effect on work ability, but only mental health and work ability shared the same determinants (Berg et al., 2008). Another 2-year follow up study about work ability among college educators suggested an, improvement in psychosocial factors which in turn positively influenced the work ability (Marqueze et al., 2008).

Thus, an investigation of workers, mental capacity as well as their work ability is increasingly important today. Hence, the current study attempts to assess mainly the effect of a middle-aged factory worker's mental capacity on work ability. Furthermore, it investigated whether there is an impact of aging and mental concentration ability on their work ability.

Therefore, the main objective of the study was to examine the effect of mental capacity on work ability in middle-aged factory workers in a manufacturing company in Japan.

## 2 METHOD

### 2.1 *Subject*

The subjects comprised fifty four (54) factory workers in a truck manufacturing company in Japan. All subjects were male and had a mean age of 47.24 years, and ranged from 40–58 years. The workers, average employment period at this factory was 25.69 years.

Subjects were divided mainly into two age categories, i.e., 40–49 age group and over 50 groups. According to the civil status of study subjects, they were separated into four categories such as married, unmarried, remarried and bereaved. Furthermore, the Work Ability Index (WAI) was distributed among the workers and according to the results of the WAI, twenty four (24) moderate and thirty (30) excellent subjects were selected for the study.

The detailed characteristics of study subjects are summarized in Table 1 below.

### 2.2 *Work ability*

Work ability of subjects was measured using Work Ability Index. WAI is used to differentiate between workers with good and bad prognosis regarding work ability in the future

Table 1. Characteristics of the subjects (n = 54).

| | n | % |
|---|---|---|
| Age (*years*) | | |
| 40–49 | 37 | 68.50 |
| Over 50 | 17 | 31.50 |
| *Civil status* | | |
| Married | 32 | 59.30 |
| Unmarried | 18 | 33.30 |
| Remarried | 01 | 1.90 |
| Bereaved | 03 | 5.50 |
| *WAI category* | | |
| Moderate | 24 | 44.40 |
| 40–49 | 15 | 62.50 |
| Over 50 | 09 | 37.50 |
| Excellent | 30 | 55.60 |
| 40–49 | 22 | 73.33 |
| Over 50 | 08 | 26.67 |
| *Health condition* | | |
| No problem | 52 | 96.30 |
| Some problems | 02 | 3.700 |
| *Last day alcohol* | | |
| Yes | 28 | 51.90 |
| No | 26 | 48.10 |

(Ilmarien, 2007). This involves a questionnaire, which serves to conduct a self-assessment. The focus is on the employees and their work ability assessed by themselves. Proceeding from the assessments of the employees, an examination is made as to whether restrictions on their work ability are imminent in the future and what need for action there is in order to promote the health of those surveyed over their working lives.

Although standard WAI has four categories i.e., poor, moderate, good and excellent, (Ilmarien, 2007), only two categories i.e., moderate and excellent were used in this study (Table 1), because, after analyzing the data on WAI, it was found that workers with poor work ability were very limited.

### 2.3 *Trail Making Test (TMT)*

To measure the mental concentration ability of the subjects, a Trail Making Test which is one of the most popular neuropsychological tests in measuring mental concentration, visual attention and task switching, was used. Also, TMT provides mainly information on scanning, and speed of processing, mental flexibility and execution of functions. (Tombaugh, 2004).

In general TMT consisted of two parts i.e., TMT-A and TMT-B. However, in the current study, a new version of computer based TMT-A was used to test the factory workers, mental concentration ability, mainly due to the time constraint.

TMT-A consists of encircled numbers from 1 to 25 randomly distributed in space. The object of the test is for the subject to connect the numbers in order, beginning with 1 and ending with 25, in as little time as possible. A commonly reported performance index on the TMT is time to completion.

As TMT is simple and easy to understand, no additional training was given to the subjects. Further, each subject had done five successive trails of TMT. After doing one trail a small break was given and then the next trail started. It was recognized that age, education and intelligence affected TMT performances, (Spreen, 1998).

Thus, TMT performances of middle-aged factory workers were analyzed according to the two age categories (40–49 and over 50) and two work ability (moderate and excellent) categories.

## 3 RESULTS

### 3.1 *Estimated total marginal mean time of TMT*

The Figure 1 illustrates the estimated total time for TMT during five trials. The repeated ANOVA method was used to analyze the marginal mean time for the TMT during five trials (Fig. 1). As depicted in Figure 1 the estimated marginal means for five trials are, 37187.63, 36069.13, 34644.80, 35993.17 and 34,562 (ms) respectively. It can be seen that there was a gradual decrease in mean TMT completion time during the five trials, except in trial four.

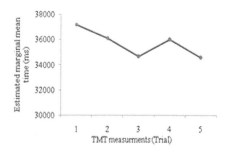

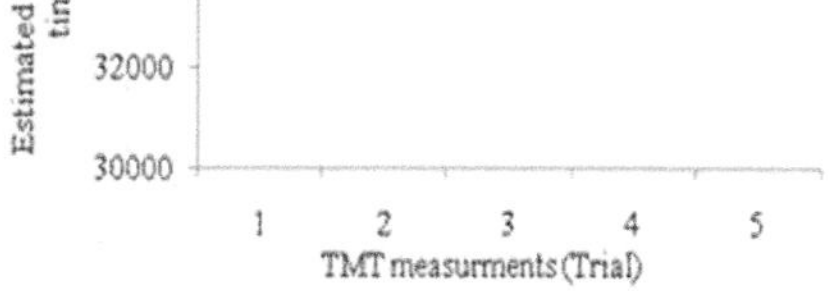

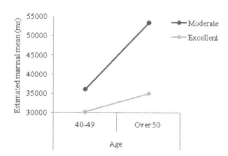


Figure 1. Estimated marginal mean time for five trials of TMT.

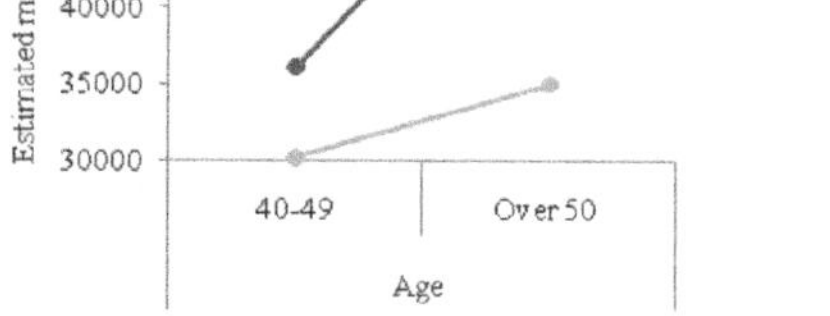


Figure 2. Estimated marginal mean time of TMT for two age categories.

However, the increase in TMT time for trial four was not statistically significant. In summary, it can be seen as a decreasing trend in mean time for the five successive TMT tests.

### 3.2 *Interaction effect of work ability and age*

Interaction effect of aging and work ability is shown in Figure 2. In the moderate and excellent subjects, who belong to the 40–49 group, TMT times were 36080 (ms) and 30187 (ms) respectively. Furthermore, the moderate and excellent subjects who belong to the over 50 age group were 53404 (ms) and 34980 (ms) respectively.

In addition, in both age groups, the moderate subjects' estimated marginal mean time was higher than the excellent subjects. Also, when age increases, the moderate subjects increasing pattern was different from the excellent subjects' pattern. Therefore, the "ordinal" interaction effect of work ability and age on these two age categories was clearly seen.

### 3.3 *Effect of work ability*

Effect of work ability on mean time for TMT is illustrated by Figure 3. For moderate and excellent subjects, estimated marginal means of TMT are 44,742 (ms) and 32,583 (ms) respectively. These results distinguish the TMT performances of moderate and excellent subjects very clearly.

### 3.4 *Effect of aging*

The TMT performances of the two age categories i.e., 40–49 and over 50 years, are shown in Figure 4. The estimated marginal mean time of TMT for the 40–49 age group was 33,133 (ms) and 44,192 (ms) for the over 50 years age group. According to the graph, the

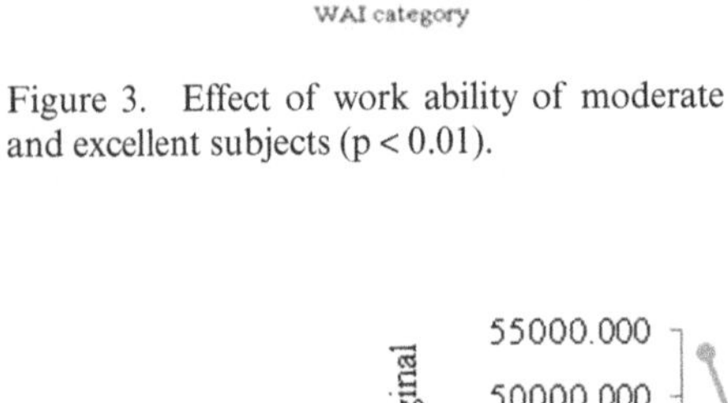

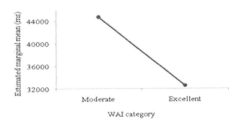


Figure 3. Effect of work ability of moderate and excellent subjects ($p < 0.01$).

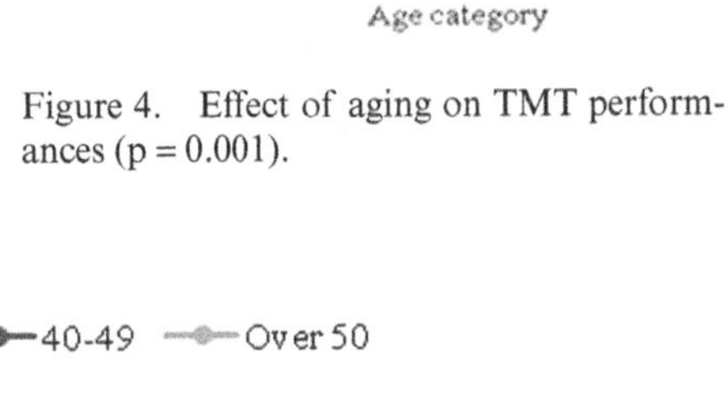

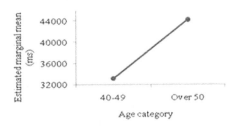


Figure 4. Effect of aging on TMT performances ($p = 0.001$).

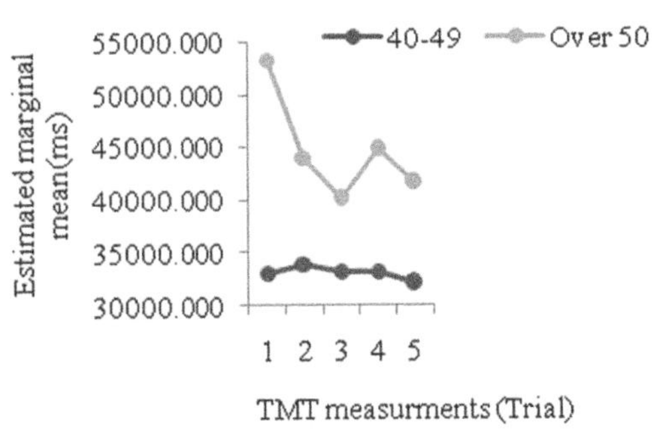


Figure 5. Interaction effect of two age categories.

over 50 years age group demonstrated considerably higher TMT time than the 40–49 years age group.

### 3.5 *Interaction effect of age*

The interaction effect of the two age categories is demonstrated in Figure 5. In the over 50 years age group the mean time was 53,240 (ms) for the 1st trial, which decreased gradually till the 5th trial, except the 4th trial which was not statistically significant (Figure 5). This pattern of over 50 years age group is almost similar with the overall estimated marginal time of subjects (see Figure 1). However, in the 40–49 age group, mean times for five trials of TMT were, 33,033, 33,849, 33,159, 33,186 and 32,043 (ms). Although the over 50 age group showed a gradual decrease in mean time for TMT, the 40–49 age group showed a comparatively stable pattern compared to the over 50 year group. Therefore, it can be seen that there is an interaction effect of two age categories of age.

## 4 DISCUSSION

This study examined the effect of mental capacity on work ability in middle-aged factory workers. Some limitations should be taken into account in this research study. Participants comprised a small sample size of fifty four male subjects. It was rather difficult to find large number of workers in the same work place at the same period of time because of the time consuming nature of the experiment and since it disturbs the free flow of production of the company. Because of these practical difficulties in this field study, subjects were limited to fifty four. Past researches of TMT contained a relatively small number of subjects within a restricted age and educational range (Tombaugh, 2004).

Also, only male subjects were employed for the current study. However, past researchers found that in mental capacity tests, there were no statistically significant differences between men and women (Nygard, et al., 1991). Nevertheless, the results of this field research study are important as they give first hand information on factory workers in their working environment and it could help to get more realistic information.

Results of repeated ANOVA of five trials of TMT demonstrated (Figure 1) that the total time is considerably higher in the 1st trial than the rest of the successive trials. Further, it can be shown that there is a decreasing trend of TMT time from the 2nd to 5th trials, except the 4th trial which was not statistically significant. As the TMT test is simple and easy to understand no additional training was given to workers before the experiment. Also, the other experimental conditions remained unchanged. So, this finding could be explained as the learning effect of the workers.

These results can be further explained by comparing them with the interaction effect of aging (Figure 5). The same pattern can be seen only in the over 50 years age group but a comparatively more stable value of TMT time can be seen in the 40–49 age group. In addition to the learning effect, these results may be influenced by the motivation and skills of the subjects. Spreen and Strauss confirmed that age, education, and intelligence affected TMT performances (Spreen & Strauss, 1998). Further, in most work tasks, speed and precision can be substituted by the high motivation of aging workers and the experience and wisdom they have assembled throughout their working life. Even though the speed of learning may slow with age, the actual learning process is not dependent on a person's age. Strong motivation to learn can also compensate for the slower learning speed (Ilmarinen, 2001). However, in this study, only the elderly group showed the learning effect. Furthermore, it can be argued that the aging effect was influenced by the mental concentration ability of the elderly subjects.

The results of the interaction effect of WAI and age (Figure 2) showed that moderate subjects in both the over 50 group and the 40–49 age group, spent a longer time for TMT than the excellent subjects. Furthermore, for the excellent subjects in the over 50 years group performances were higher than the moderate subjects in the 40–49 year group. In summary, it can be argued that irrespective of age, excellent subjects showed higher performances than moderate subjects. So, these results can be used to justify the validity of the WAI (Figure 2). As described in the previous discussion, this is also affected by the high skills of the excellent subjects than moderate subjects.

According to Figures 4 and 5, it can be clearly seen that the estimated marginal mean of TMT is higher in over 50 year group than the 40–49 group. Horswill et al., 2009, also confirmed that in older drivers (75–84 years) TMT performances were significantly higher than the middle aged group of drivers (35–55 and 65–74) performances. Therefore, the results of this study confirmed the effect of aging in factory workers and are in agreement with the TMT results of Tombaugh (Tombaugh, 2004). Moreover, this result was confirmed by the findings of another recent research which states an association between processing speed and age (a major component of successful TMT performance) (Ashendorf, 2008; Salthouse & Fristoe, 1995). Hamdan & Hamdan (2009), also confirmed that the effects of age and education level had a significant impact on TMT performance.

Generally speaking mental capacity correlated less with work ability than physical capacity did (Nygard et al., 1991). Also the work ability of physical workers, both among men and women was significantly poorer than that of mental workers (Tuomi, et al., 2001). Interestingly, according to the results of the current study, the effect of mental capacity in the older age group of factory workers can be seen clearly (Figure 4). However, in this study subjects did not do mentally demanding work, instead we evaluated mental concentration ability of factory workers whose job is physically demanding.

## 5 CONCLUSION

This study focused on the effect of mental capacity on work ability of middle-aged factory workers. It presented data for TMT time to completion, and compared TMT performance among moderate and excellent workers in two age categories.

The study substantiated the validity of WAI, as excellent subjects' performances were better than the moderate subjects' performances.

The influence of aging among the elderly factory workers can be clearly seen. Further, aging affected both work ability and mental capacity as well as mental capacity affecting work ability. In conclusion, there is a close relation between mental capacity, aging effect and work ability of factory workers.

Therefore, careful attention should be given when assigning tasks which need high mental concentration, ability especially for elderly workers.

## REFERENCES


Berg, T.I.J. van den, S.M.A. Avinia, F.J. Bredt, D. Lindeboom, L.A.M. Elders & Burdrof, A. (2008). The influence of psychosocial factors at work and life style on health and work ability among professional workers. *Int Arch Occup Environ Health,*. Vol. 81, pp. 1029–1036.

Elaine, C. Marqueze, Gustavo, P. Voltz, Flávio, N.S. Borges & Claudia, R.C. Moreno. (2008). A 2-year follow-up study of work ability among college educators. *Applied Ergonomics.* Vol. 39, No. 5, pp. 640–645.

Hamdan, Amer, C. & Eli, Mar, Hamdan, L.R. (2009). Effects of age and education level on the Trail Making Test in A healthy Brazilian sample. *Psychology & Neuroscience.* Vol. 2, No. 2, pp. 199–203.

Horswill, M.S., Pachana, N.A., Wood, J., Marrington, S.H., McWilliam, J. & McCullough, C.M., (2009). A comparison of the hazard perception ability of matched groups of healthy drivers aged 35 to 55, 65 to 74, and 75 to 84 years. Journal of the International Neuropsychological Society. Vol. 15, pp. 799–802.

Ilmarien, J. (1997). Aging and work- coping with strengths and weaknesses. *Scand J Work Environ Health.* Vol. 23, pp. 3–5.

Ilmarien, J. (2001). Aging workers. *Occup Environ Med*: Vol. 58, p. 546.

Ilmarien, J. The work ability index (WAI). (2007). Occupational Medicine.Vol. 57, p. 160.

Lee, Ashendorf, Angela, L. Jefferson Maureen, K.O. Connor, Christine, Chaisson, Robert, C. Green & Robert, A. Stern. (2008). Trail Making Test errors in normal aging, mild cognitive impairment, and dementia. Archives of Clinical Neuropsychology. Vol. 23, pp. 129–137.

Ministry of internal affairs and communications, statistical bureau, statistical re-search and training institute (Statistical bureau, MIC, 2011). (http://www.stat.go.jp/english/data/handbook/c02cont.htm)

Nygard, C.H. Eskelien, L. Suvanto, S. Tuomi, K. & Ilmarinen, L. (1991). Associations between functional capacity and work ability among elderly municipal employees. *Scand J Work Environ Health.* Vol. 17, pp. 122–127.

Philippe, Kiss, Marc, De Meester, Lutgart, Braeckman. (2008). Differences between younger and older workers in the need for recovery after work. *Int Arch Occup Environ Health.* Vol. 81, pp. 311–320.

Salthouse, T.A. & Fristoe, N.M. (1995). Process analysis of adult age effects on a computer-administered Trail Making Test. *Neuropsychology,* Vol. 9, pp. 518–528.

Spreen, O. & Strauss, E. (1998). A compendium of neuropsychological test: Administration, norms and commentary (2nd ed.) New York: Oxford University press.

Tuomi, K. Huuhtanen, P. Nykyri, E. & Ilmarinen, J. (2001). Promotion of work ability, the quality of work and retirement. *Occup,* Med. Vol. 51, No. 5, pp. 318–324.

Tuomi, K. Ilmarinen, J. Jahkola, A. Katajarinne, L. & Tulkki, A. (1998). *Work Ability Index.* 2nd revised edn. Helsinki: Finnish Institute of Occupational Health.

Tombaugh, Tom, N. (2004). Trail making test A and B: Normative data stratified by age and education. *Archives of clinical neuropsychology.* Vol. 19, pp. 203–214.

*Ergonomics in Asia – Shih & Liang (eds)*
*© 2012 Taylor & Francis Group, London, ISBN 978-0-415-68414-9*

# Finger-tapping test for aging workers

Tadao Makizuka
*Kinki University, Iizuka, Japan*

Hiroyuki Izumi, Yoshiyuki Higuchi & Masaharu Kumashiro
*University of Occupational and Environmental Health, Japan*

ABSTRACT: The finger tapping test is one of finger skill measurement, which counts the stroke of finger motion in tapping for unit time. 108 aging workers (50–70 years old), who were all right-handed males, carried out the tapping test. The means of tapping scores were decreased with aging, although with individual differences. Tapping scores were not related to the finger skill reported in worker's self-evaluation in their questionnaires.

*Keywords*: tapping test (speed of finger motion), aging

## 1 INTRODUCTION

Aging workers might be needed in the near future in Japan, because there are severe problems of declining birth rates and the longevity of life. Aging workers would be required with sufficient work abilities.

The work abilities of aging workers comprise skill, motion control and adjustment ability, agility and equilibrium function etc. The skill tests are composed of the burpee test (using the whole body), stepping test (using the legs) and tapping test (using the fingers). The finger tapping test is one of the skill measurement tests. This test is to count the number of finger strokes in tapping for unit time (0–10, 10–20 and 20–30 sec.).

As possible indicators of the work abilities of aging workers, a fundamental study of finger skills by finger tapping test was carried out quantitatively.

## 2 METHODS

The subjects were 108 workers at a particular factory, who were all right-handed males. Subjects were tested for their tapping abilities at the middle (third) finger of the right hand. A tapping tester (Takei, T.K.K. 1347) was used in this study.

Subject data were obtained as first, second and third numerical values each 10 seconds. After a test, all subjects were interviewed about their the finger skills from childhood to the present time.

## 3 RESULTS

### 3.1 *Tapping test*

Correlations between age and tapping test are shown in Figures 1–3. Tapping scores were obviously decreasing with advancing age. The other two scores showed a similar tendency. The mean of the first 10 seconds score was 48.4 (± 8.60 s.d.), the second was 41.4 (± 9.05) and the third was 37.4 (± 9.77) strokes. Reduction of tapping score in time is shown logarithmically in Figure 4. Standard deviations of tapping scores were also obviously increasing with passage of time. Tapping score in aging is shown in Figure 5. Data of under-40s were extrapolated (Kimotsuki, 1967). It seemed that tapping scores from under-40s to 60s were decreasing with aging.

### 3.2 *Self-evaluation of finger skills*

Subjects reported self-evaluation of finger skills at the present day. 13.8% of subjects estimated their skills as 'dexterity', 55.0% of subjects as 'standard level' and another (31.2%)

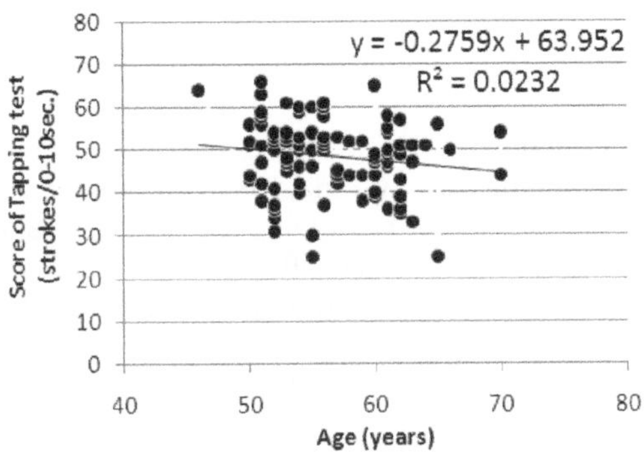


Figure 1. Correlation between age and tapping test (0–10 sec.).

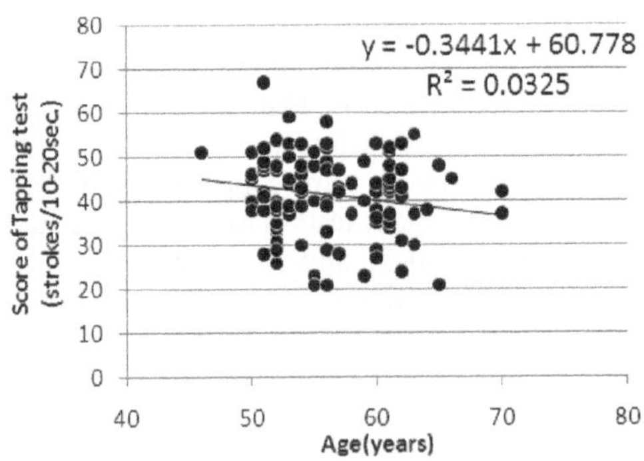


Figure 2. Correlation between age and tapping test (10–20 sec.).

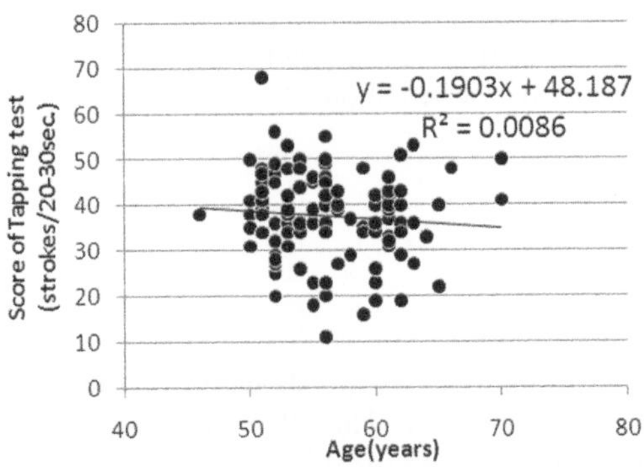


Figure 3. Correlation between age and tapping test (20–30 sec.).

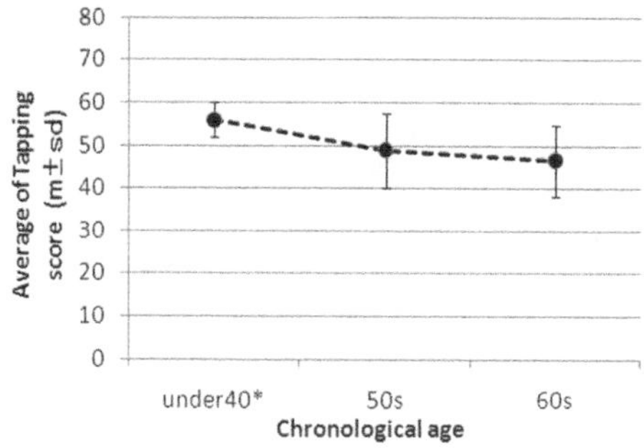


Figure 4. Reduction of tapping score with age.

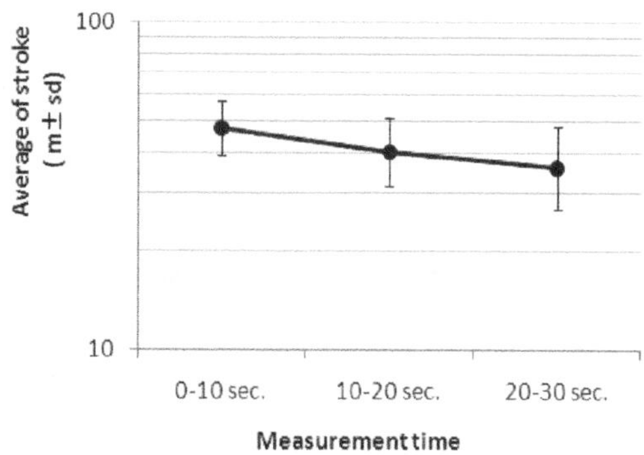


Figure 5. Tapping score in time *data of extrapolation (Kimotsuki, 1967).

as 'clumsiness'. 32.1% of subjects reported some kind of changes of finger skill in their life, for the reasons of aging, hobby, relocation, lack of physical activity, reduced visual acuity etc.

## 4 DISCUSSION

The finger test was chosen for the reasons of neurological mechanism on handedness as spinalization (Kikuchi,1969) and for the restriction of the time when subjects carried out the physical strength tests.

Low correlation coefficients were observed between age and the tapping test scores. There were large individual differences in scores. Tapping scores were obviously decreasing with aging. Means of tapping scores were reducing with increasing standard deviations. Tapping scores from under-40s to 60s were decreasing with aging.

Shimoyama (1990) reported that tapping frequency lowered with advancing age, which supported this result. The finger tapping test is generally using as a rehabilitation indicator (Noro, 1992), however, it is possible to estimate the finger skills for aging workers.

Generally, tapping scores are reducing in the time test, which is the orthodox type (Figure 6). But some subjects had paradoxical data, which showed inconstancy or increase in time. All of the subjects were ordered to tap as fast as possible before the time test. Figure 7 is the constant score type in time. Figure 8 is the increase score type in time. It is possible to explain this phenomenon, which is influenced by the wrong image as non-maximum speed of subject's tapping strokes and/or by creating the image of a finger movement for tapping. Data volume was not sufficient to describe finger speed with aging; data have continued to be collected, in view of the above.

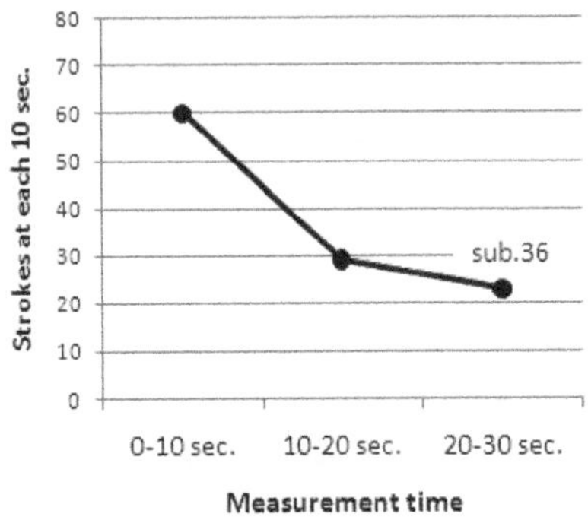


Figure 6. Orthodox type of tapping test.

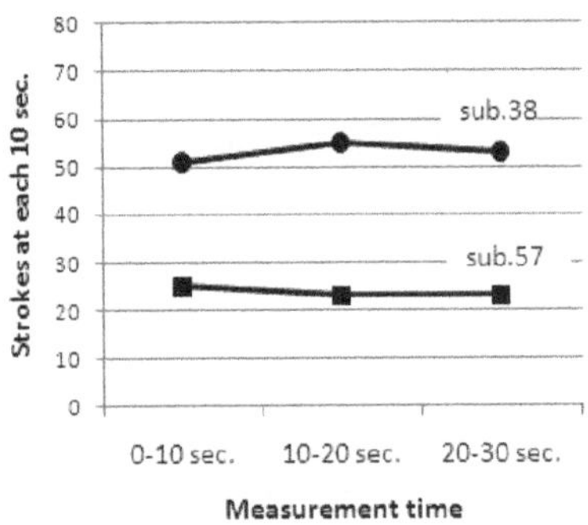


Figure 7. Constant score type of tapping test.

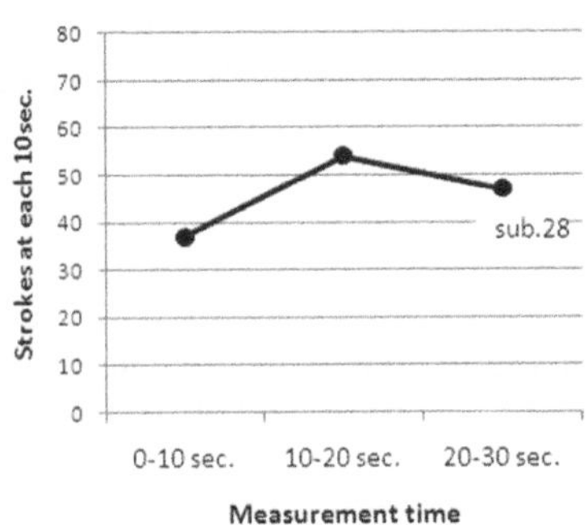


Figure 8. Increase score type of tapping test.

## REFERENCES

Kikuchi, Y. (1969). The neurological mechanism on handedness, *Jap. J. of Ergonomics,* Vol. 5, No. 5, pp. 333–339.

Kimotsuki, K. (1967). On the motion velocity of man (II). *J. Science of Labour,* Vol. 43, No. 4, pp. 585–596.

Noro, H. et al. (1992). Quantitative estimation of finger movement functions by Finger-Tapping Test (part 1). *Jap. Assoc. of Rehabilitation Medicine,* Vol. 29, No. 12, pp. 1183–1184.

Shimoyama, I. et al. (1990). The finger-tapping test. A quantitative analysis. *Arch Neurol*, Vol. 47, No. 6, pp. 681–684.

*Ergonomics in Asia – Shih & Liang (eds)*


# Motion analysis of putting on and taking off short-sleeved shirts

Kanami Tanimizu & Manami Ishiuchi
*Graduate School of Design, Kyushu University, Fukuoka, Japan*

Satoshi Muraki
*Faculty of Design, Kyushu University, Fukuoka, Japan*

Masahiro Yamasaki
*Faculty of Integrated Arts and Sciences, Hiroshima University, Higashi-Hiroshima, Japan*

ABSTRACT: The elderly have functional deterioration and pain in many parts of the body with aging, and have difficulty in putting on and taking off clothes. We analyzed the motion of 56 healthy elderly women (aged 59 to 85 years) and 26 young women (aged 20 to 34 years) putting on and taking off three kinds of short-sleeved shirts (T-shirt, a polo shirt, a tank top), and studied the influence of aging and maximal angles of the shoulder joints on the motion of putting on and taking off their shirts. The methods of taking off shirts were classified into four patterns, and there were significant differences in the pattern distribution among generations and different kinds of clothes. Meanwhile, the maximal angles of the shoulder joints failed to show any relationship with the motion of putting on and taking off shirts.

*Keywords*: the elderly, putting on and taking off, motion analysis, clothes

## 1 INTRODUCTION

The elderly have functional deterioration and pain in many parts of the body with aging, and have difficulty in putting on and taking off clothes. Accordingly, clothes that can be changed easily should play an important role in the independence of the elderly. In addition, elderly women generally showed substantial interest in dressing up and in fashion (Saito and Nakagawa, 2004). In our previous study, we analyzed the motions of putting on and taking off a T-shirt in elderly and young women in order to obtain data to design and make a universally wearable T-shirt; however, various methods of putting on and taking off a T-shirt were identified (Tanimizu et al., 2010). Therefore, the factors that lead to difficulty for elderly women in wearing shirts need to be clarified in order to obtain useful data for designing and making a universally wearable T-shirt.

In the present study, we analyzed the influences of aging, the kind of clothes and the maximal angles of the shoulder joints on the motions of putting on and taking off short-sleeved shirts in elderly and young women in order to clarify the factors that differ in terms of the methods of wearing shirts.

## 2 METHODS

### 2.1 *Participants*

Fifty-six elderly women and twenty-six young women participated in this study (Table 1). No participants had disease of the joints in the upper body. This study was approved by the Research Ethics Committee of the Faculty of Design at Kyushu University. Written informed consent was obtained from all participants.

### 2.2 *Materials*

Three kinds of shirts were used for this study: shirt A was a T-shirt, shirt B was a polo shirt, shirt C was a tank top (Table 2 and Figure 1). We prepared every kind of shirt in three sizes, and these shirts were chosen on the basis of the JIS L4005 sizing systems for women's garments and participants' figures.

### 2.3 *Measurements*

We measured body weight, height and the maximal angles of the shoulder joint. Six motion directions of the shoulder joint were measured: elevation, forward flexion, backward extension, abduction, external rotation and horizontal flexion. Eighteen markers were placed on the head, shoulders, sternoclavicular joints, elbows, wrists, trochanters, right knee and ankles. The participants moved their shoulders voluntarily in six directions. During the experiments, the trajectories of the reflective markers were sampled at a frequency of 60 Hz using a nine-camera Hawk digital motion analysis system (Motion Analysis Corporation), and three-dimensional trajectories were constructed using EVaRT 5.0 software (Motion Analysis Corporation). The maximal ranges of motion were constructed using Kinema Tracer software (Kissei Comtec Corporation).

Participants put on and took off shirts in the usual way. The motion was recorded twice for each shirt using a digital video camera.

### 2.4 *Statistical analysis*

All statistical analyses were performed with statistical software (SPSS, version 17.0). An unpaired t-test analysis was used to compare the physical characteristics and the maximal

Table 1. Physical characteristics of participants.

| | Elderly (n = 56) | | | Young (n = 26) | |
|---|---|---|---|---|---|
| | Mean | S.D. | | Mean | S.D. |
| Age (year) | 70.6 | 6.5 | * | 23.4 | 3.4 |
| Body height (cm) | 151.1 | 4.9 | * | 159.0 | 6.2 |
| Body weight (kg) | 51.4 | 7.9 | | 50.6 | 6.8 |
| BMI (kg/m$^2$) | 22.5 | 3.1 | * | 19.9 | 1.7 |

*$p < 0.05$ vs. Young (unpaired t-test).

Table 2. Textile characteristics of clothes.

| | T-shirt | Polo shirt | Tank top |
|---|---|---|---|
| Quality | 100% cotton | 100% cotton | 100% cotton |
| Construction | Plain weave | Moss stitch | Plain weave |
| Density (g/m$^2$) | 210 | 230 | 200 |

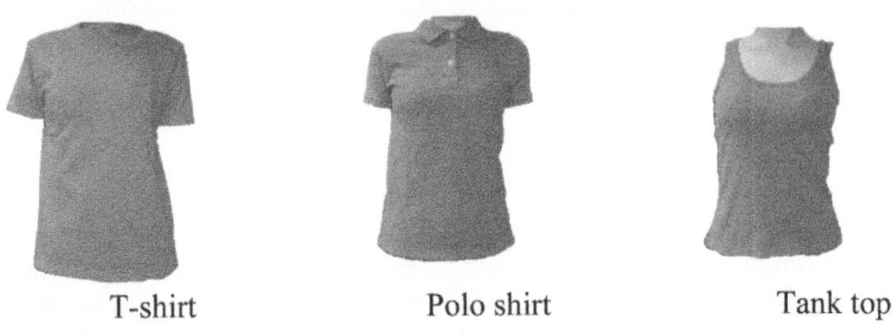


Figure 1. Three kinds of shirts used for this study.

angle of the shoulder joint between elderly and young groups. In addition, a one-way factorial ANOVA analysis was performed to compare the maximal angles of the shoulder joint among patterns of taking off shirts. A p-value < 0.05 was considered to be significant.

## 3 RESULTS AND DISCUSSION

### 3.1 *Maximal angles of the shoulder joint*

There were significant differences in forward flexion, backward extension, abduction and external rotation between generations (Table 3). Allander et al. (1974) reported that the shoulder joint's range of motion decreased significantly with aging, which is consistent with our results.

### 3.2 *Aging and the methods of putting on and taking off shirts*

The methods of putting on and taking off shirts were classified in three and four patterns, respectively. The pattern distribution significantly differed between generations in taking off but not putting on the shirts.

The patterns of putting on shirts were different from the order of putting their arms through the sleeves and their head through the nape (Figure 2). Figures 3 and 4 show the classified patterns of putting on shirts in young and elderly women, respectively. The pattern that was seen the most in both generations was putting their arms through first and their head through the nape last.

The methods of taking off shirts were classified into four patterns (Figure 5). Pattern 1 was pulling their arms through the sleeves first. Pattern 2 was raising the front hem first, and lifting off the shirt. Pattern 3 was raising the front hem first, and pulling their arms through. Pattern 4 was holding the back collar first, and pulling the shirt forward.

Figures 6 and 7 show the classified patterns of taking off shirts in young and elderly women, respectively. Pattern 2 was common in both generations (39.3% elderly women and 38.5% young women); however, Pattern 1 was seen only in young women (0.0% elderly women and 46.2% young women) and Pattern 3 was seen more in elderly women (19.6% elderly women and 7.7% young women). It has been reported that the motion of taking off shirts is more difficult than that of putting them on in the elderly (Okada et al., 2008). These findings indicated that the motion pattern of taking off shirts was affected by some factors related to aging.

### 3.3 *The kind of clothes and the methods of putting on and taking off shirts*

The pattern distribution also significantly differed by the kind of clothes in terms of taking them off, but not putting them on. Although almost all young women did not change the methods of taking them off, irrespective of the kind of shirt, 21.4% of elderly women changed

Table 3. Comparison of the maximal angles of the shoulder joint (degrees).

| | Elderly (n = 56) | | | Young (n = 26) | |
|---|---|---|---|---|---|
| | Mean | S.D. | | Mean | S.D. |
| Elevation | 28.5 | 5.1 | | 28.6 | 6.3 |
| Forward flexion | 149.0 | 8.6 | * | 157.7 | 7.9 |
| Backward extension | 57.6 | 13.6 | * | 66.2 | 24.5 |
| Abduction | 149.2 | 8.8 | * | 156.9 | 7.4 |
| External rotation | 78.0 | 13.6 | * | 93.5 | 11.7 |
| Horizontal flexion | 64.3 | 12.3 | | 63.7 | 10.1 |

*$p < 0.05$ vs. Young (unpaired t-test).

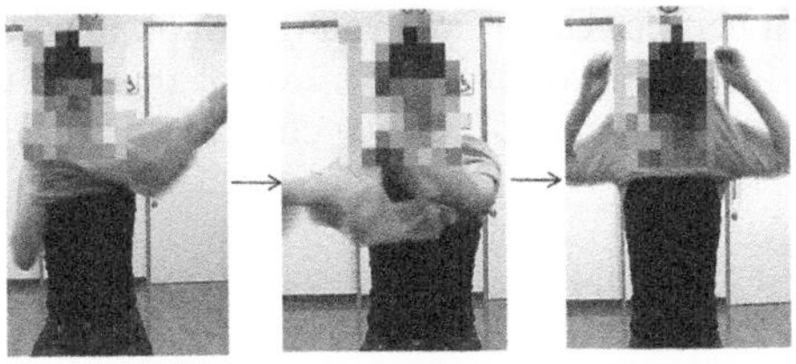

Pattern 1: Putting the arms through shirt first, the head last.

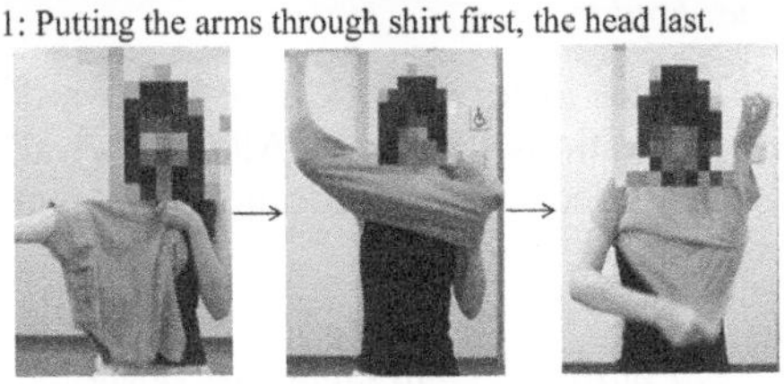

Pattern 2: Putting the arm through shirt, the head, another arm last.

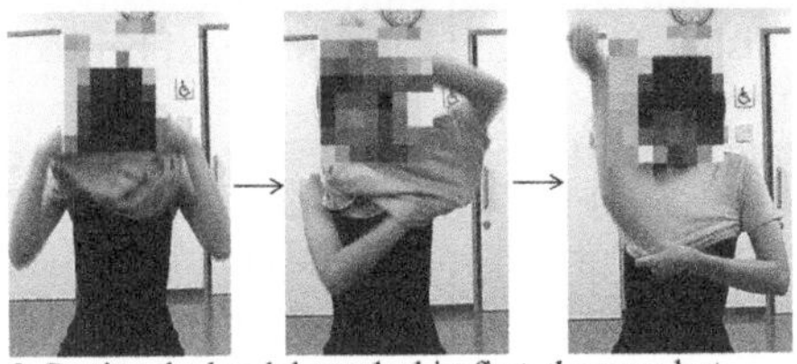

Pattern 3: Putting the head through shirt first, the arms last.

Figure 2. The methods of putting on shirts.

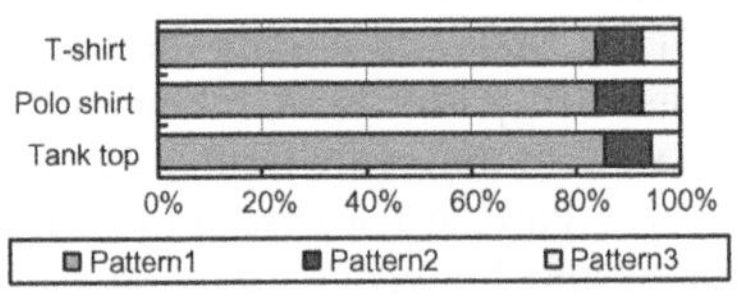


Figure 3. The method of putting on shirts in elderly women (n = 56).

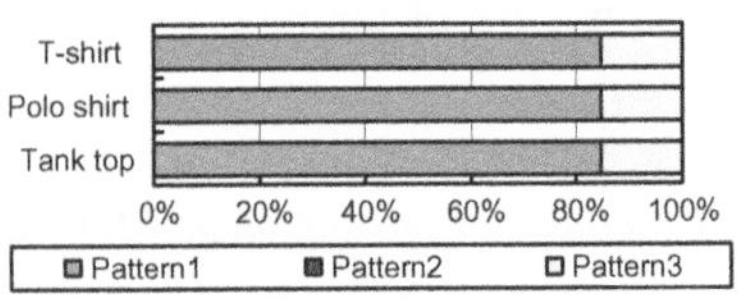


Figure 4. The method of putting on shirts in young women (n = 26).

the method according to the kind of shirt (Figures 6 and 7). In taking off T-shirts in the elderly, the percentage using Pattern 3 increased, which was used instead of Pattern 2. These results indicated that the kind of clothes also influenced the methods of taking them off.

### 3.4 *Maximal angle of the shoulder joint and the methods of putting on and taking off shirts*

There were significant differences in four motion directions of the shoulder joint between generations. However, the present study failed to show any significant relationships between maximal angle of the shoulder joint and the method of putting on and taking off shirts. This was partly because a majority of the elderly participants were in their 60s. Then, we carried out an analysis of variance of maximal angle of the shoulder joint and the methods of taking off shirts in women aged 70 and over, and abduction in Pattern 4 was found to be significantly smaller than that in Pattern 3 in three kinds of shirts (Figures 8–10). In our previous

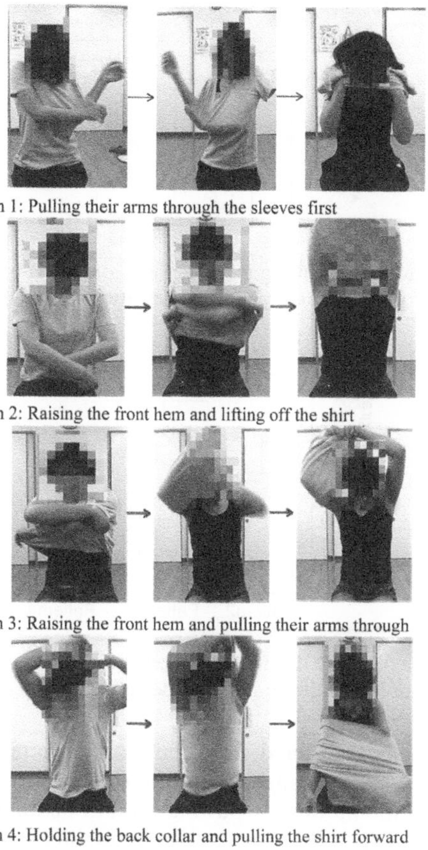

Pattern 1: Pulling their arms through the sleeves first
Pattern 2: Raising the front hem and lifting off the shirt
Pattern 3: Raising the front hem and pulling their arms through
Pattern 4: Holding the back collar and pulling the shirt forward


Figure 5. The methods of taking off shirts.

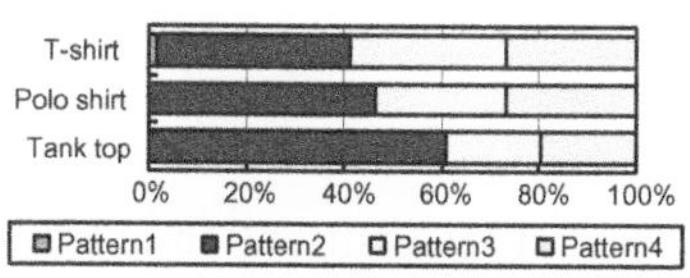


Figure 6. The method of taking off shirts in elderly women (n = 56).

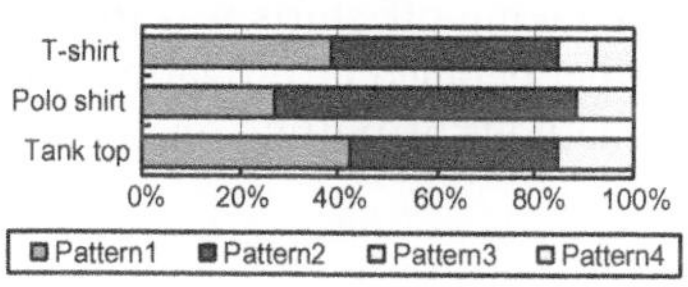


Figure 7. The method of taking off shirts in young women (n = 26).

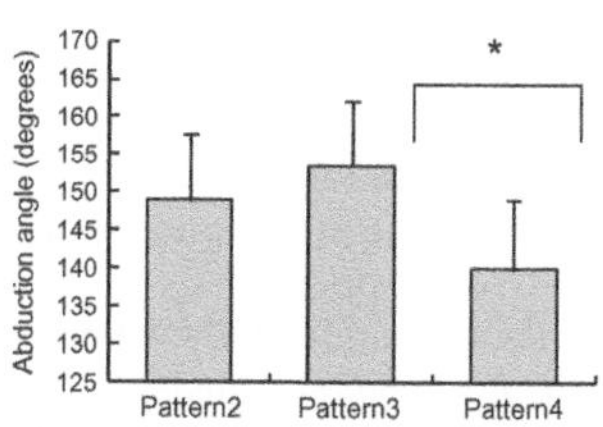


Figure 8. Maximal abduction angle during three methods of taking off T-shirts in elderly women (aged seventy and over). Error bars indicate standard deviation (*$p < 0.05$).

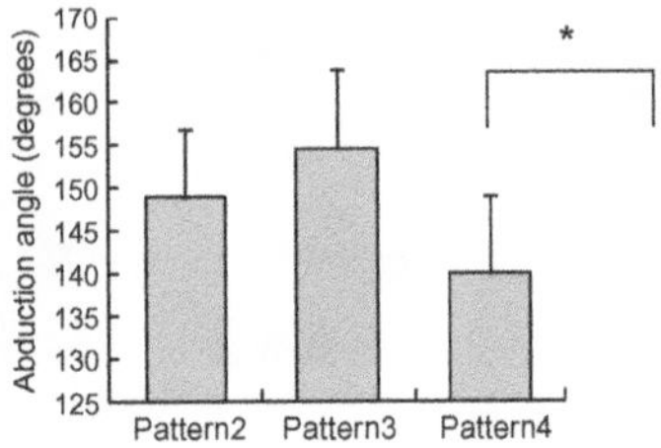


Figure 9. Maximal abduction angle during three methods of taking off Polo shirts in elderly women (aged seventy and over). Error bars indicate standard deviation (*$p < 0.05$).

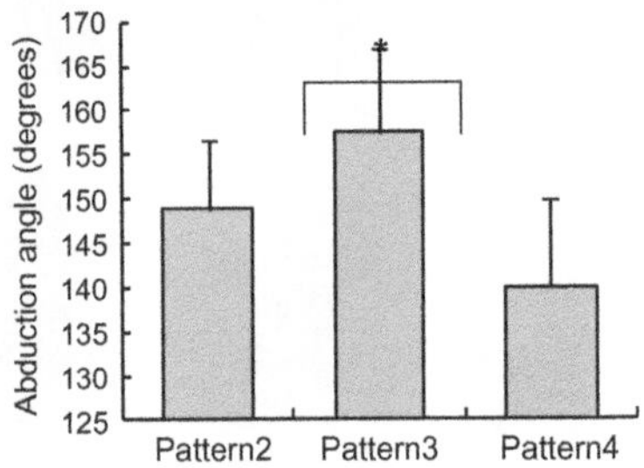


Figure 10. Maximal abduction angle during three methods of taking off Tank top in elderly women (aged seventy and over). Error bars indicate standard deviation (*$p < 0.05$).

study, the height of the elbow differed from the motion patterns of taking off a T-shirt. Accordingly, we expected that the range of motion of shoulder joints was one of the main determinants in selecting the motion of putting on and taking off shirts. The selection of Pattern 4 could be attributable to the restricted joint mobility and the difficulty in raising the arms to a high position. These results suggested that the maximal angle of the shoulder could be a determinant in selecting the method of taking off shirts in women aged 70 and over.

## 4 CONCLUSIONS

The present study showed that the pattern distribution in taking off shirts significantly differed and changed with age, and the kind of clothes had an impact on the methods of taking off shirts. Meanwhile, a significant correlation between the partial maximal angle of the shoulder joint and the method of taking off shirts was found in women aged 70 and over. These findings suggested that further detailed investigation is required in order to clarify the factors behind differences in the methods of putting on and taking off shirts.

## REFERENCES

Allander, E., Bjornsson, O.J., Olafsson, O., Sigfusson, N. & Thorsteinsson, J. (1974). Normal range of joint movements in shoulder, hip, wrist and thumb with special reference to side: a comparison between two populations. *International Journal of Epidemiology,* Vol. 3, pp. 253–261.

JIS L 4005.

Okada, N. & Watabe, J. (2008). Planning clothes for the elderly–clothing designs based on the pushing and pulling movements of the arms while putting on and taking off pull-over upper wear–. *Journal of Home Economics of Japan,* Vol. 59, No. 2, pp. 87–98.

Saito, E. & Nakagawa, S. (2004). Middle-aged and elderly women's consciousness toward dressing up and clothing norms. *Journal of Home Economics of Japan,* Vol. 55, No. 9, pp. 743–751.

Tanimizu, K., Muraki, S. & Yamasaki, M. (2010). The universal designed T-shirt based on motion analysis of putting on and taking off clothes. *The Proceeding of the 3rd International Conference for Universal Design in Hamamatsu,* pp. 0–152.

*Ergonomics in Asia – Shih & Liang (eds)*
*© 2012 Taylor & Francis Group, London, ISBN 978-0-415-68414-9*

# Assessment of work ability of Korean workers in the shipbuilding industry

Seong Rok Chang & Yujeong Lee
*Department of Safety Engineering, Pukyong National University, Busan, Korea*

Dong-Joon Kim
*Department of Naval Architecture and Marine Systems Engineering, Pukyong National University, Busan, Korea*

ABSTRACT: The goal of this study was to assess the work ability of Korean workers in the shipbuilding industry. The Work Ability Index (WAI) is a kind of survey method developed by the Finnish Institute of Occupational Health (FIOH) in 1998 to estimate the work capacity of aged workers. This study surveyed 2,709 persons working in shipbuilding industries in Korea. The average WAI score for all workers was 40.0, denoting a Good level. Current work ability, sick leave during the past year, work ability two years from now, and mental resources were positively correlated with age, but the number of current diseases was negatively correlated with age. Also, workers in the shipbuilding industry had lower work abilities when compared to workers in other industries (housing management 43.6; mechanical industry 41.7; construction industry 41.4). The WAI was analyzed for different age groups (≤ 29; 30–34; 35–39; 40–44; 45–49; 50–54; ≥ 55). The results of the Kruskal-Wallis test showed that a significant difference was identified due to the effects of aging ($p < 0.05$). The mean WAI score of older workers was higher than that of younger workers.

*Keywords*: work ability, WAI, age, shipbuilding industry

## 1 INTRODUCTION

The proportion of people over 65 years old in Korea will increase steadily. It reached the level of 10.7% of the whole population in 2009, denoting Korea as an already aging society. In addition, the Statistics Korea office also expects that Korea would be an aged society and a post-aged society in 2019 and 2026 respectively (Statistics Korea, 2009a). The demographic data shows an increase in the age-dependency ratio, fewer younger workers, and declining workplace participation rates for older workers. Increased rates of employment during the last 10 years in Korea were –9.3%, 0.3%, 40%, 47% and 49% in the ages of twenties, thirties, forties, fifties and sixties, respectively (Statistics Korea, 2009b). In Korea, the increase rate of workers in the shipbuilding industry was 76% for the last 10 years and eight times as compared with that of other industries. In addition, the average age of workers in the shipbuilding industry has steadily increased (Korea Shipbuilders' Association, 2009).

Old age is associated with inevitable time-dependent losses in physical capabilities. However the maintenance of physical capabilities is essential for continuing independence in old age (Bassey 1998). Ilmarinen et al. (2005) suggested that the balance between human health resources and work demands and characteristics are also important dimensions of work ability, and Tuomi et al. (1997) showed that, in addition to the health resources, also competence as well as values, attitude, and motivation could play an important role on work ability. The work ability index (WAI) was constructed to reveal how well a worker is able to perform his or her work (Tuomi et al., 1998). This study focused on the discovery of worker-related and age-related influences on the work ability of Korean workers in the shipbuilding industry using the WAI questionnaire.

## 2 SUBJECTS AND METHODS

### 2.1 *Subjects*

2,709 workers, aged between 18 and 73 years (mean 42.6, S.D. 10.3) and with a work experience in the shipbuilding industry varying from 0.5 to 48 years (mean 10.5, S.D. 9.23) had been examined. The participants were categorized by age, the length of their service in the shipbuilding industry, the type of work, and the type of employment, that is:

- age: under 29 years, 30 ~ 34 years, 35 ~ 39 years, 40 ~ 44 years, 45 ~ 49 years, 50 ~ 54 years, over 55 years
- length of service in the shipbuilding industry: under 1 year, 2 ~ 5 years, 6 ~ 10 years, 11 ~ 15 years, 16 ~ 20 years, 21 ~ 25 years, over 26 years
- type of work: physically demanding, mentally demanding, physically and mentally demanding (mixed)
- type of employment: direct, subcontract.

### 2.2 *Methods*

The WAI questionnaire developed by the Finnish Institute of Occupational Health was used for this study. The work ability measured by the WAI is considered as poor (7 ~ 27 points), moderate (28 ~ 36 points), good (37 ~ 43 points), excellent (44 ~ 49 points). This was a questionnaire study. Before the survey, the WAI questionnaire was translated into Korean. Workers' participation was fully voluntary and questionnaires were sent to workers with the help of safety managers in the Korean shipbuilding industry. Three major shipbuilding companies in the world and two medium companies participated in this study.

To enter the data collected, a database was developed using Microsoft Excel 2000, using the codes provided in the questionnaire. The difference of the WAI between groups in each category was tested using the Kruskal-Wallis test, and the relationship between the WAI and the workers' ages was tested by the Correlation test. Statistically significant were results with p below 0.05. All statistical analysis was performed by SPSS 12.0 K.

## 3 RESULTS

The WAI results of workers in the shipbuilding industry varied from 15 to 49 points, and the mean WAI score was 40.0. As regards work ability categories excellent WAI was found in 26.8% of workers, good in 50.7%, moderate in 21.3% and finally poor only in 1.1% (See Table 1).

Table 2 shows WAI results categorized by age, length of service, type of work, and type of employment. Analysis of WAI score by age, type of job and type of employment showed statistically significant differences ($p < 0.05$). Difference of WAI score by length of service in the shipbuilding industry, however, was statistically insignificant ($p > 0.05$). According to age groups, the WAI results were as follows: the best mean has been noted in the

Table 1. Distribution of work ability.

| | Excellent | Good | Moderate | Poor | Total |
|---|---|---|---|---|---|
| No. | 726 | 1374 | 578 | 31 | 2709 |
| % | 26.8 | 50.7 | 21.3 | 1.1 | 100.0 |

Table 2. Distribution of WAI scores.

| Category | Group | No. | Work Ability Index | | | | P-value |
|---|---|---|---|---|---|---|---|
| | | | Mean | Min. | Max. | Avg. Rank | |
| Age(year) | Under 29 | 306 | 39.4 | 26 | 49 | 1253.1 | 0.000 |
| | 30~34 | 408 | 39.5 | 23 | 49 | 1268.9 | |
| | 35~39 | 374 | 39.8 | 24 | 49 | 1315.9 | |
| | 40~44 | 339 | 40.7 | 24 | 49 | 1462.3 | |
| | 45~49 | 472 | 40.6 | 18 | 49 | 1448.6 | |
| | 50~54 | 409 | 40.1 | 18 | 49 | 1376.9 | |
| | Over 55 | 301 | 40.6 | 15 | 49 | 1451.5 | |
| | missed data | 100 | 37.0 | 8 | 48 | 978.5 | |
| Year of service (year) | Under 1 | 322 | 40.1 | 27 | 49 | 1353.0 | 0.826 |
| | 2~5 | 785 | 40.0 | 18 | 49 | 1350.2 | |
| | 6~10 | 487 | 40.1 | 24 | 49 | 1365.2 | |
| | 11~15 | 254 | 40.3 | 27 | 49 | 1389.5 | |
| | 16~20 | 222 | 40.3 | 20 | 49 | 1393.4 | |
| | 21~25 | 260 | 39.6 | 15 | 48 | 1317.2 | |
| | Over 26 | 249 | 40.5 | 28 | 49 | 1414.2 | |
| | missed data | 130 | 38.4 | 8 | 49 | 1180.1 | |
| Type of work | Physical | 1568 | 40.6 | 18 | 49 | 1451.4 | 0.000 |
| | Mental | 420 | 38.3 | 15 | 49 | 1071.0 | |
| | Mixed | 695 | 39.9 | 24 | 49 | 1317.3 | |
| | missed data | 26 | 35.1 | 8 | 48 | 1140.0 | |
| Type of Employment | Direct | 889 | 39.5 | 18 | 49 | 1263.8 | 0.000 |
| | Subcontract | 1234 | 41.1 | 20 | 49 | 1537.0 | |
| | missed data | 586 | 38.4 | 8 | 49 | 1110.9 | |

Table 3. Relationship between age and WAI items.

| WAI item | Score | r range | Correlation p-value |
|---|---|---|---|
| Total (1 ~ 7) | 7 ~ 49 | +0.084 | 0.000 |
| 1. Current work ability compared with the lifetime best | 0 ~ 10 | +0.181 | 0.000 |
| 2. Work ability in relation to the demands of the job | 2 ~ 10 | +0.019 | 0.341 |
| 3. Number of current diseases diagnosed by a physician | 1 ~ 5, 7 | −0.128 | 0.000 |
| 4. Estimated work impairment due to diseases | 1 ~ 6 | −0.032 | 0.109 |
| 5. Sick leave during the past year (12 months) | 1 ~ 5 | +0.094 | 0.000 |
| 6. Own prognosis of work ability two years from now | 1, 4, 7 | +0.063 | 0.001 |
| 7. Mental resources | 1 ~ 4 | +0.172 | 0.000 |

group of workers from 40 to 44 years old, and the worst one in the group of workers under 29 years old.

When groups had been divided by the type of work, the mean WAI score of the group of physically demanding work was higher than that of the other groups. When groups had been divided by the type of employment, the mean WAI score of the group of subcontractors was higher than that of the group of direct employment.

The relationship between WAI score and worker's age has been evaluated in Table 3. A positive correlation between WAI total score and age is evident ($p < 0.05$). WAI item's score and age are statistically significantly correlated with age except in item no. 2 (job demand) and item no. 4 (estimated work impairment due to disease) ($p > 0.05$). Item no. 1 (current work ability), item no. 5 (sick leave during the past year), item no. 6 (work ability two years from now), and item no. 7 (mental resources) were positively correlated with age, but item no. 3 (number of current diseases) was negatively correlated with age.

In the previous studies in the housing management company, machinery industry, and construction industry in Korea, the mean WAI scores were 43.6, 41.7, 41.4 respectively (Chang 2008, Kang 2005, Chang 2010). (See Table 4) The difference of WAI scores between the shipbuilding industry and the other industries was statistically significant ($p < 0.05$).

Table 4. WAI scores and working field.

| | | | Index | p-value |
|---|---|---|---|---|
| Category | No. | Work ability | Mean | Avg. Rank |
| Housing Management | 736 | 43.6 | 2703.3 | |
| Machinery | 450 | 41.7 | 2208.1 | 0.000 |
| Construction | 214 | 41.4 | 2560.6 | |
| Shipbuilding | 2709 | 40.0 | 1813.5 | |

The WAI score of workers in shipbuilding industry was lower than that of the other industries. One could deduce that mental stress and the physical demand of work in shipbuilding industry explain the low WAI score.

## 4 CONCLUSIONS AND DISCUSSION

The goal of this study was to assess work ability of Korean workers in the shipbuilding industry using the WAI questionnaire.

The WAI result of workers in the shipbuilding industry varied from 15 to 49 points, the mean WAI score was 40.0. As regards work ability categories, excellent WAI was found in 26.8% of workers, good in 50.7%, moderate in 21.3% and finally poor only in 1.1%. According to age groups, the WAI results were as follows: the best mean has been noted in the group of workers from 40 to 44 years old and the worst one in the group of workers under 29 years old. When groups were divided by the type of work, the mean WAI score of the group with physically demanding work was higher than that of the other groups. When groups were divided by the type of employment, the mean WAI score of the group of the subcontract was higher than that of the group of direct employment.

This study focused on the discovery of worker-related and age-related factors on the work ability of Korean workers. Current work ability, sick leave during the past year, work ability two years from now, and mental resources were positively correlated with age, but the number of current diseases was negatively correlated with age. These factors show that a management plan to improve the work ability of older Korean workers must contain a program to lessen mental stress and to promote physical health.

The WAI total score of Korean workers in the shipbuilding industry increased as age increased, particularly in the forties. Researches in Eastern Asia (Chang 2008, Duong 2007, Chumchai 2007) have shown that the change of WAI score by age is not statistically significant. This means that although ageing is associated with inevitable time-dependent losses in physical capabilities, workers in Asia want to work continuously to provide for their family.

The WAI score of workers in the shipbuilding industry was lower than that of the other industries in Korea. One could deduce that mental stress and the physical demands of work in the shipbuilding industry explain the low WAI score. The promotion plan to maintain good work ability in the Korean shipbuilding industry focuses on a decrease in repetitive movements, improving the supervisors' attitudes and increasing physical exercises (Tuomi et al., 1997).

## ACKNOWLEDGEMENTS

This research was financially supported by the Ministry of Education, Science and Technology (MEST) and the National Research Foundation of Korea through Human Resource Training Project for Regional Innovation.

## REFERENCES


Bassey, E.J. (1998). Longitudinal changes in selected physical capabilities: muscle strength, flexibility and body size. *Age Ageing*, Vol. 27, No. 3, pp. 12–16.

Chang, S.R., Nam, C. & Lee, Y. (2008). A Study on the Work Ability Assessment of Housing Manager. *Autumn Conference of the KOSOS*, pp. 233–236.

Chang, S.R. & Lee, Y. (2010). Survey Analysis on Work Ability in Construction Industry, *Pukyong National University,* unpublished.

Chumchai, P., Silpasuwan, P., Viwatwongkasem, C. & Wongsuvan, T. (2007). Work Ability among Truck Drivers in Thailand, *3rd International Symposium on Work Ability*.

Duong, K.V., Nguyen, N.N., Ta, Q.B. & Khuc, X. (2007). Primary Study on Work Ability of Vietnamese Workers, *3rd International Symposium on Work Ability*.

Ilmarinen, J., Tuomi, K. & Seitsamo, J. (2005). New dimensions of work ability, *International Congress Series*, 1280, pp. 3–7.

Kang, J.C., Baek, S.Y. & Chang, S.R. (2005). Investigating the Work Ability of Employees in the Korean Machinery Industry, *Journal of the KOSOS*, Vol. 20, No. 3, pp. 197–201.

Statistics Korea, 2009a. (2009). Statistics on the Aged.

Statistics Korea, 2009b. (2009). Economically Active Population Survey.

The Korea Shipbuilders' Association, (2009). Korea shipbuilding workforce.

Tuomi, K., Ilmarinen, J., Martiainen R. & Klockars M. (1997). Aging, work, life style and work ability among Finnish municipal workers in 1981–1992, *Scand. J. Work Environ. Health*, Vol. 23, No. 1, pp. 58–65.

Tuomi, K., Ilmarinen, J., Jahkola, A., Katajarinne, L. & Tulkki, A. (1998). Work Ability Index. 2nd ed. *Finnish Institute of Occupational Health, Helsinki*.

*Ergonomics in Asia – Shih & Liang (eds)*
*© 2012 Taylor & Francis Group, London, ISBN 978-0-415-68414-9*

# Exercise habits and work ability in Japanese manufacturing industry workers

Yoshiyuki Higuchi, Hiroyuki Izumi, Tadayuki Yokota & Masaharu Kumashiro
*Department of Ergonomics, Institute of Industrial Ecological Sciences, University of Occupational and Environmental Health, Japan*

Takeshi Ebara
*Department of Occupational and Environment Health, Graduate School of Medical Sciences, Nagoya City University, Japan*

ABSTRACT: The purpose of this study is to verify the relationship between work ability and exercise habits among Japanese manufacturing industry workers. The study design was a cross-sectional questionnaire survey. Work ability was assessed by the Work Ability Index (WAI), and exercise habit was obtained by questionnaire including exercise intensity and frequency. The total number of participants was 5,155. Of this total, we analyzed the data of 3,754 participants who had valid responses of WAI and exercise habits. The result of the generalized linear model (GLM) which predicts WAI score by exercise intensity and frequency, showed that the effect of exercise intensity was significant. Also GLM analysis was done in each age group. The result implied that the effect of the combination between intensity and frequency in daily exercise varies depending on age group.

*Keywords*: exercise habit, Work Ability Index, exercise intensity, exercise frequency, generalized liner model

## 1 INTRODUCTION

Exercise habits are an important factor for health and work ability promotion in an occupational setting. Performance of vigorous exercise by adults has some cardiovascular or orthopedic risks, but these risks are acceptably small if the participants initiate an exercise program following appropriate guidelines as to medical clearance and exercise performance (Haskell & Blair). According to previous studies, regular exercise may contribute to maintenance of work ability and enhancement of mental health (Ilmaninen et al., 1991, Tuomi et al., 1997). Especially for the elderly, a regular exercise habit is one of the most important factors for health promotion (Ilmarinen 2001, Chan et al., 2000, Krause et al., 1993).

The purpose of this study is to verify the relationship between work ability and exercise habits among Japanese manufacturing industry workers. In particular, we examine the effect of the combination of intensity and frequency of daily exercise habits.

## 2 METHOD

In the present study we used a cross-sectional questionnaire survey. A set of questions including Work Ability Index (Ilmarinen & Tuomi 2004), exercise habits, age, sex, height, weight, and work condition was prepared. A questionnaire was filled in by 5,155 participants from 13 manufacturing companies in Japan. The survey was conducted in the autumn of 2009.

## 2.1 *Work ability*

Work ability was assessed by the work ability index (WAI). The WAI consists of seven dimensions. The range of the summative index is 7 to 49, which is classified into four categories, "Poor (7–27)", "Moderate (28–36)", "Good (37–43)", and "Excellent (44–49)".

## 2.2 *Exercise habits*

Exercise habits were composed of exercise intensity and frequency. The intensity of daily exercise habits was surveyed by asking respondents to rank the following items: "Almost none", "Easy exercise (e.g., light stretching)", "Mild exercise (e.g., walking at a slow pace over 30 minutes)", "Moderate exercise (e.g., walking at a fast pace over 30 minutes)", and "Hard exercise (e.g., Jogging over 30 minutes)". For the frequency of exercise, we used the following items: "once or twice per month", "once per week", "two or three times per week", and "almost every day".

## 2.3 *Statistical analysis*

Proportions were compared using the chi-square test. Post hoc analysis of the chi-square test was based on the standardized adjusted residuals. The standardized adjusted residuals of a cross table cell indicate whether its observed frequency is significantly different from the expected frequency. We used a significance criteria of 1.96 in the standardized adjusted residuals, and we used the generalized linear model (GLM) to analyze the relationship between exercise habits and work ability. In GLM analysis, we tested the main effect model and interaction effect model. When deciding on an appropriate model, we considered the likelihood ratio test, –2 log likelihood and AIC for goodness of fit. All data were analyzed with SPSS 18.0 j.

# 3 RESULT

The number of total participations was 5,155. Of this total, we analyzed the data of 3,754 participants who had valid responses of WAI and exercise habits.

Demographic characteristics of the participants are shown in Table 1. Regarding percentage of gender, males were 89.6% (n = 3,363), females were 10.4% (n = 391). The mean age was 44.7 years (SD 11.9). The mean of Body Mass Index was 23.0 (SD 3.1).

The mean of WAI scores was 37.9 (SD 5.4). The percentage of WAI categories was: "Excellent" 16.1% (n = 603), "Good" 48.7% (n = 1,827), "Moderate" 31.5% (n = 1,181), and "Poor" 3.8% (n = 143).

The percentage of participants who have some exercise habits in daily life was 52.0% (1,953/3,754). For intensity of exercise habits in daily life, "Almost none" was 48.0% (n = 1,801), "Easy exercise" was 26.5% (n = 994), "Mild exercise" was 11.7% (n = 439), "Moderate exercise" was 7.6% (n = 284), and "Hard exercise" was 6.3% (n = 236). Concerning the frequency of exercise, "once or twice per month" was 5.9% (222/3,754), "once per week" was 13.3% (501/3,754), "two or three times per week" was 16.8% (631/3,754), and "almost every day" 16.0% (599/3,754).

## 3.1 *Proportion of exercise habits by sex, age-group*

The proportion of exercise intensity and frequency by sex is shown in Figure 1 and Figure 2.

For exercise intensity, the proportion of "almost none" was 47.4% with male and 53.2% with female. The proportion of "Hard" was 6.4% with male and 5.4% with female. The result of the chi-square test was not significant (chi-square value = 5.502, df = 4, p = 0.240).

For exercise frequency, the proportion of "once or twice per month" was 5.9% for males and 5.9% for females. The proportion of "almost every day" was 16.2% for males and 14.1% for females. The result of the chi-square test was not significant (chi-square value = 5.497, df = 4, p = 0.240).

Table 1. Summary of variables for this survey.

| Mean (SD) | |
|---|---|
| Sex | |
| male | 3363 (89.6) |
| female | 391 (10.4) |
| Age (years) | 44.7 (11.9) |
| BMI | 23.0 (3.1) |
| WAI score | 37.9 (5.4) |
| WAI category | |
| Excellent | 603 (16.1) |
| Good | 1827 (48.7) |
| Moderate | 1181 (31.5) |
| Poor | 143 (3.8) |
| Exercise Intensity | |
| Almost none | 1801 (48.0) |
| Easy exercise | 994 (26.5) |
| Mild exercise | 439 (11.7) |
| Moderate exercise | 284 (7.6) |
| Hard exercise | 236 (6.3) |
| Exercise Frequency | |
| Once or twice per month | 223 (5.9) |
| Once per week | 501 (13.3) |
| Two or three times per week | 631 (16.8) |
| Almost every day | 599 (16.0) |

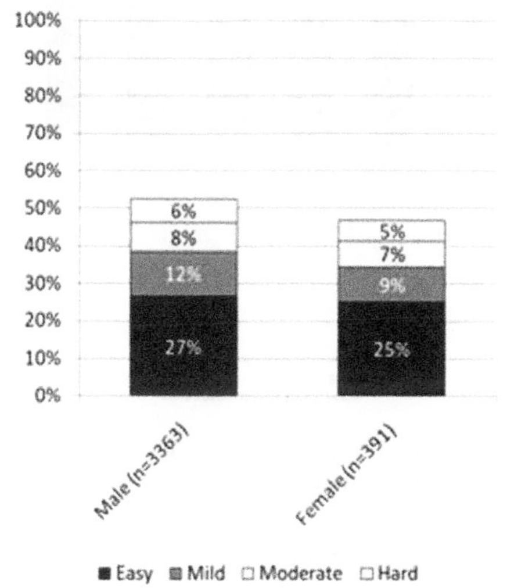


Figure 1. Proportion of exercise intensity by sex.

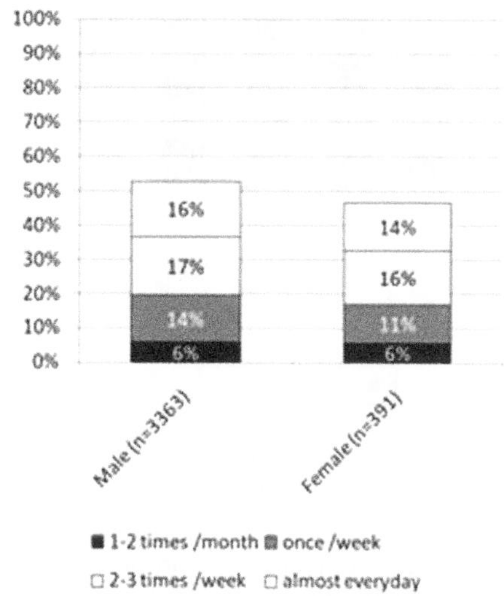


Figure 2. Proportion of exercise frequency by sex.

The proportion of exercise intensity and frequency by age-group is shown in Figure 3 and Figure 4. For exercise intensity, the proportion of "almost none" was 38.0% with 20–29 years, 50.9% with 30–39 years, 49.6% with 40–49 years, 52.3% with 50–59 years and 34.4% with 60 years over. The proportion of "Hard" was 16.8% with 20–29 years, 7.2% with 30–39 years, 5.2% with 40–49 years, 2.6% with 50–59 years and 5.4% with 60 years over. The result of the chi-square test was statistically significant (chi-square value = 224.127, df = 16, $p < 0.001$). According to standardized adjusted residuals, the percentage of "almost none" with the 50–59 years age group was higher than the expected value, and the percentage of "Hard exercise" with 20–29 years age group was higher than the expected value.

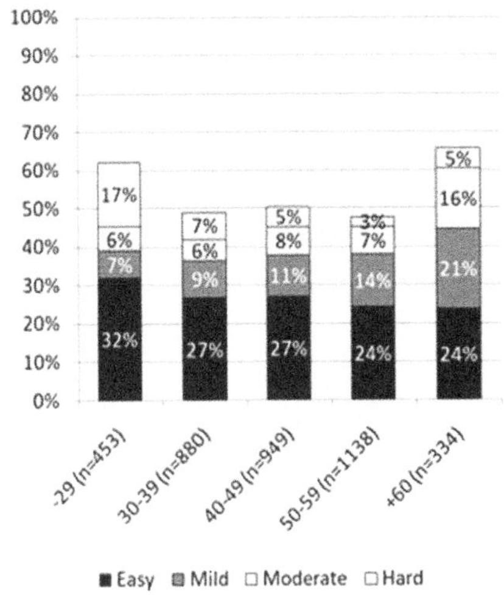


Figure 3. Proportion of exercise intensity by age-group.

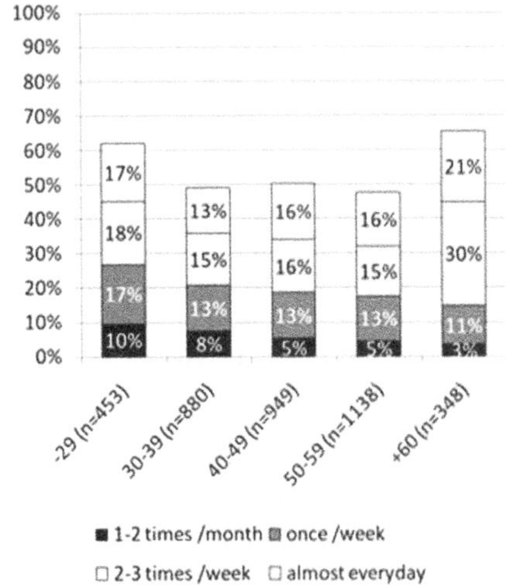


Figure 4. Proportion of exercise frequency by age-group.

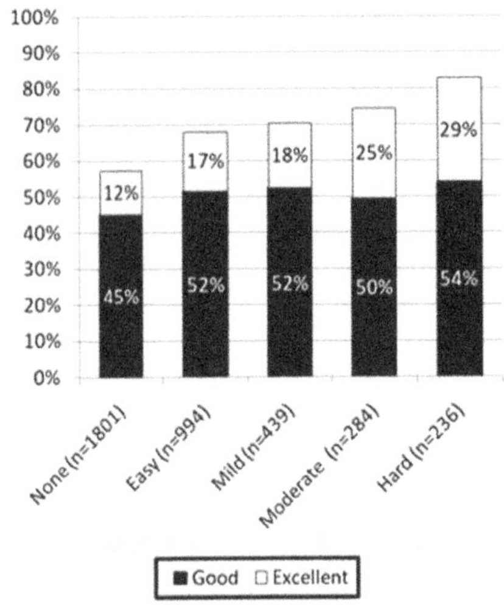


Figure 5. Proportion of "Excellent" and "Good" in each exercise intensity.

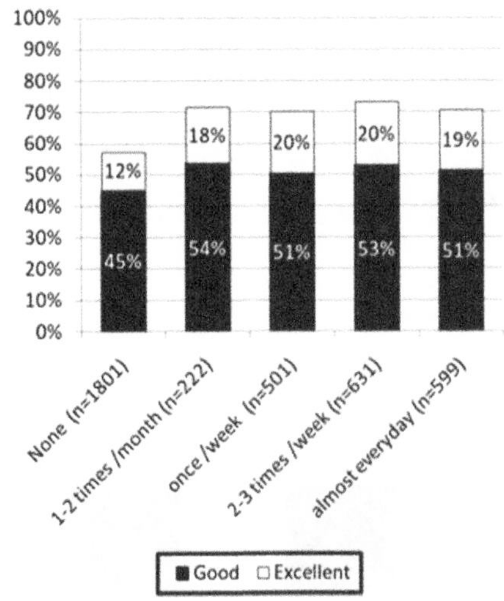


Figure 6. Proportion of "Excellent" and "Good" in each exercise frequency.

For exercise frequency, the proportion of "once or twice per month" was 9.7% with 20–29 years, 7.5% with 30–39 years, 5.3% with 40–49 years, 4.5% with 50–59 years and 3.3% with 60 years over. The proportion of "almost every day" was 16.8% with 20–29 years, 13.3% with 30–39 years, 16.4% with 40–49 years, 15.7% with 50–59 years and 21.3% with 60 years over. The result of the chi-square test was statistically significant (chi-square value = 113.452, df = 16, $p < 0.001$). According to standardized adjusted residuals, the percentage of "once or twice in per month" with 20–29 years and 30–39 age group were higher than the expected values, and the percentage of "almost every day" with 60–69 years age group was higher than the expected value.

### 3.2 *Relationship between exercise habits and work ability index*

The proportions of "Good" and "Excellent" in WAI category for each exercise intensity and frequency are shown in Figure 5 and Figure 6.

In exercise intensity, the proportion of "Excellent" was 12.3% with "Almost none", 16.5% with "Easy exercise", 18.0% with "Mild exercise", 25.0% with "Moderate exercise" and 28.8% with "Hard exercise". The proportion of "Good" was 45.2% with "Almost none", 51.7% with "Easy exercise", 52.4% with "Mild exercise", 49.6% with "Moderate exercise" and 54.2% with "Hard exercise". The result of the chi-square test was statistically significant (chi-square value = 139.413, df = 12, $p < 0.001$). According to standardized adjusted residuals, the percentage of "Excellent" with "Moderate exercise" and "Hard exercise" were higher than the expected value, and the percentage of "Excellent" with "Almost none" was statistically lower than the expected value.

In exercise frequency, the proportion of "Excellent" was 12.3% with "once or twice per month", 18.0% with "Once per week", 19.6% with "Two or three times per week" and 19.4% with "Almost every day". The proportion of "Good" was 53.6% with "Once or twice per month", 50.5% with "Once per week", 52.9% with "Two or three times per week" and 51.3% with "Almost every day". The result of the chi-square test was statistically significant (chi-square value = 101.496, df = 12, $p < 0.001$). According to standardized adjusted residuals, the percentage of "Excellent" with "Once per week", "Two or three times per week" and "Almost every day" were higher than the expected value.

### 3.3 *GLM analysis*

We conducted GLM analysis to verify the relationship between work ability and exercise habits, using WAI score as a dependent variable, and exercise intensity and frequency as independent variables (factors), respectively. In GLM analysis, the "Almost none" data were excluded from the analysis because "Almost none" affects both intensity and frequency.

Table 2. Goodness of fit in GLM analysis.

| | Main effect | Interaction effect |
|---|---|---|
| ALL | | |
| Likelihood ratio test (p) | <0.001 | <0.001 |
| –2 log likelihood | 2503.398 | 2495.434 |
| AIC | 2521.397 | 2531.434 |
| Age group | | |
| 20–29 | 0.240 | 0.009 |
| Likelihood ratio test (p) | 392.578 | 369.666 |
| –2 log likelihood | 410.577 | 405.666 |
| AIC | | |
| 30–39 | 0.209 | 0.120 |
| Likelihood ratio test (p) | 528.662 | 515.508 |
| –2 log likelihood | 546.662 | 551.508 |
| AIC | | |
| 40–49 | 0.022 | <0.001 |
| Likelihood ratio test (p) | 585.298 | 557.680 |
| –2 log likelihood | 603.297 | 593.681 |
| AIC | | |
| 50–59 | 0.159 | 0.028 |
| Likelihood ratio test (p) | 577.334 | 559.450 |
| –2 log likelihood | 595.334 | 595.451 |
| AIC | | |
| 60 over | 0.298 | 0.118 |
| Likelihood ratio test (p) | 326.786 | 317.628 |
| –2 log likelihood | 344.785 | 347.628 |
| AIC | | |

Table 3. Significant regression coefficient in each age group GLM analysis.

| | B | p |
|---|---|---|
| Significant model | | |
| Interaction effect model with 20–29 years | | |
| Moderate * almost every day | 3.277 | 0.007 |
| Moderate * 2–3 times per week | 2.348 | 0.021 |
| Hard * once per week | 1.377 | 0.023 |
| Easy * almost every day | 1.176 | 0.034 |
| Main effect model with 40–49 years | | |
| Hard | 0.815 | 0.007 |
| Mild | 0.459 | 0.040 |
| Interaction effect model with 40–49 years | | |
| Hard * 1–2 times per month | 3.443 | 0.004 |
| Moderate * 1–2 times per month | 2.068 | 0.025 |
| Mild * once per week | 1.754 | 0.001 |
| Hard * 2–3 times per week | 1.683 | 0.004 |
| Moderate * 2–3 times per week | 1.483 | 0.004 |
| Interaction effect model with 50–59 years | | |
| Moderate * once per week | 2.037 | <0.001 |
| Mild * 1–2 times per month | 1.502 | 0.024 |
| Mild * 2–3 times per week | 1.213 | 0.008 |
| Moderate * almost every day | 1.183 | 0.046 |
| Moderate * 2–3times per week | 1.113 | 0.024 |
| Easy * almost every day | 1.081 | 0.008 |

In the result of GLM analysis with the main effect model, a likelihood ratio test is statistically significant (likelihood ratio chi-square = 153.650, df = 7, $p < 0.001$), –2 restricted log likelihood was 2503.398 and AIC was 2521.397. In interaction effect model, likelihood ratio test is statistically significant (likelihood ratio chi-square = 43.396, df = 15, $p < 0.001$), –2 restricted log likelihood was 2495.434 and AIC was 2531.434. According to goodness of fit, the main effect model is better than the interaction effect model. In the main effect model, the effect of exercise intensity was significant (Wald chi-square = 33.985, df = 3, $p < 0.001$), but the effect of exercise frequency was not significant (Wald chi-square = 1.892, df = 3, p = 0.595). For level of exercise intensity, regression coefficient of "Hard exercise (B = 0.763, $p < 0.001$)" and "Moderate exercise (B = 0.414, p = 0.002)" were significant. However, in this model "Easy exercise" is the reference level. Therefore the regression coefficient (B value) means the intensity of effect of work ability when compared with "Easy exercise".

We also conducted GLM analysis in each age group. The result of goodness of fit in each age group is shown Table 2.

The interaction effect model with 20–29 years age group, the main effect model and interaction effect model with 40–49 years age group, the interaction effect model with 50–59 years age group were all significant in a likelihood ratio test. Also a significant regression coefficient in each age group GLM analysis is shown Table 3. In the interaction effect model with 20–29 years age group, the most effective combination between intensity and frequency is "Moderate exercise with Almost every day" (B = 3.277, p = 0.007). In the main effect model with 40–49 years age group, the most effective level is "Hard exercise" (B = 0.815, p = 0.007), and in the interaction effect model the most effective combination between intensity and frequency is "Hard exercise with Once or twice per week" (B = 3.443, p = 0.004). In the interaction effect model with 50–59 years age group, the most effective combination between intensity and frequency is "Moderate exercise with Once per week" (B = 2.037, $p < 0.001$).

## 4 DISCUSSION

The result of WAI-related studies based on the questionnaire data from 13 manufacturing companies showed that exercise habits were strongly related to WAI score. The current study demonstrated that the benefits of daily exercise had a large positive effect on WAI score. Our study shows that the benefits of individual exercise habits and support of those habits by the company occupational health team will be an important key to increasing work ability (Kumashiro et al., 2010). Especially, based on the result of GLM analysis, young workers may need a high frequency of exercise habits, while middle-age workers may need moderate frequency of exercise habits. These results may imply that it is not necessary to have hard exercise every day for promoting work ability in an occupational setting. If so, age-appropriate strategies are required to develop programs containing a suitable amount of exercise to increase work ability. The relationship between frequency of exercise and WAI was not observed in GLM analysis. However, it is difficult to identify a causal association, because this result was based on a cross-sectional survey. In the future, it will be necessary to conduct intervention studies to demonstrate the effect of exercise habits in the working population to promote work ability.

## 5 CONCLUSION

This study focused on the relationship between exercise habits and WAI. WAI in workers with exercise habits was higher than in workers without exercise habits among Japanese manufacturing workers. Exercise intensity had a positive affect on WAI. The effect of combination between intensity and frequency in daily exercise may vary depending on age group. Age-appropriate strategies are important to promote work ability with an exercise program in an occupational setting.

## REFERENCES


Chan, G., Tan, V. & Koh, D. (2000). Ageing and fitness to work. *Occup Med (Lond)* Vol. 50, pp. 483–491.

Haskell, W.L. & Blair, S.N. (1980). The physical activity component of health promotion in occupational settings. *Public Health Rep* Vol. 95, pp. 109–118.

Ilmarinen, J.E. (2001). Aging workers. *Occup Environ Med* Vol. 58, pp. 546–552.

Ilmarinen, J. & Tuomi, K. (2004). Past, present and future of work ability, *Proceedings of the 1st international Symposium on Work Ability*. Helsinki: FIOH.

Ilmarinen, J., Tuomi, K., Eskelinen, L., Nygard, C.H., Huuhtanen, P. & Klockars, M. (1991). Summary and recommendations of a project involving cross-sectional and follow-up studies on the aging worker in Finnish municipal occupations (1981–1985). *Scand J Work Environ Health* 17 Suppl 1, pp. 135–141.

Krause, N., Goldenhar, L., Liang, J., Jay, G. & Maeda, D. (1993). Stress and exercise among the Japanese elderly. *Soc Sci Med* Vol. 36, pp. 1429–1441.

Kumashiro, M., Kadoya, M., Kubota, M., Yamashita, T., Higuchi, Y. & Izumi, H. (2011). The Relationship between WAI, Exercise Habits, and Occupational Stress Employees with Good Exercise Habits Have Greater Work Ability. *4th Symposium on Work Ability*. in press.

Tuomi, K., Ilmarinen, J., Martikainen, R., Aalto, L. & Klockars, M. (1997). Aging, work, life-style and work ability among Finnish municipal workers in 1981–1992. *Scand J Work Environ Health* 23 Suppl 1, pp. 58–65.

*Ergonomics in Asia – Shih & Liang (eds)*
*© 2012 Taylor & Francis Group, London, ISBN 978-0-415-68414-9*

# Investigation of eye movements in battlefield target search

SL. Hwang, MC. Hsieh & ST. Huang
*Department of Industrial Engineering and Engineering Management, National Tsing Hua University, Hsinchu, Taiwan*

TJ. Yeh, FK. Wu, CF. Huang & CH. Yeh
*Department of Power Mechanical Engineering, National Tsing Hua University, Hsinchu, Taiwan*

CC. Chang
*Chung-Shan Institute of Science and Technology, Taoyuan, Taiwan*

ABSTRACT: The purpose of military camouflage is to prevent the enemy from finding out the target the first time so that they may lose opportunities to attack the target. Therefore, the locations of camouflage need to be studied to achieve the best camouflage effect. The purpose of this study is to find out human visual search strategies and performance in battlefields through the eye tracking experiment, and then to suggest hidden locations for anti-camouflage and camouflage. Besides conducting an experiment in a laboratory, the results are further verified in the field, and a standard evaluation method and procedure for field study is derived.

## 1 INTRODUCTION

In nature, creatures usually developed their own ways to identify their target based on experience, and humans are no exception. In the pressure of war, visual searching performance would be decreased significantly. Moreover, if the complexity of the battlefield were greater, then the searching performance would be decreased more obviously (Morelli & Burton, 2004). Knowing how people search the target, one may get some insights into a better camouflage design. Thus, it is important to find out human searching rules on the battlefield.

There were a lot of designs of military camouflage learned from the natural world. Cott (1940) found that any animal camouflage pattern was a result of long-term natural selection, and such a pattern should be the best design. Over the past few years, several studies have been done on camouflage efficiency (North Atlantic Treaty Organization, 2006; Hu, 2007). Nyberg & Bohman (2001) analyzed the characteristics of man-made targets and natural backgrounds. They indicated that camouflage patterns with more natural characteristic features such as irregularity and roughness had a better efficiency on camouflage. However, these studies have evaluated camouflage efficiency in the laboratory, and little is known about camouflage efficiency in the field. Therefore, this study was designed to compare the results of visual search strategies in laboratory with the results in field, and the results of this study could be used as reference for anti-camouflage and camouflage design.

## 2 METHODOLOGY

### 2.1 *Laboratory experiment*

The experiment was designed to analyze human searching behavior by an eye tracker. Subjects would be asked to search for camouflage in different scenes and the eye tracker would record their eye movements. In this phase, the Experiment Builder (SR Research) was used to simulate the experiment, and the eye tracker was used to collect the data of eye movement in the laboratory.

2.1.1 *Experimental design*

A total of thirty subjects were recruited in this experiment, and ten of them had served in the army. All subjects had visual acuity >0.8 (as tested with the Landolt C test) and passed the color discrimination test.

The eye tracker (SR Research) and the experiment builder (SR Research) were used in this experiment. Subjects had to wear the eye tracker (SR Research Eye Link II) during the experiment for the eye movement data collection.

The experimental scenario was showed in a 19-inch monitor. Subjects were asked to search for a target that was hidden in the scene and point out the target by a pointer on the screen. The person wearing the camouflage dressing was a target (Figure 1) and hid in any position in the scene.

Scenes were an independent variable in this experiment including two levels, woodland and urban. In order to reduce the influence of any learning effect, each scene level had three types of background (Figure 2). The dependent variables included searching performance (hit rate, miss rate, false alarm), searching time, and scan path.

The battlefield was simulated by computer in a laboratory. There were a total of thirty photos, including three types of background of each scene, there were four photos with hidden soldiers and one photo without in each type of backgrounds. Besides, there were fifteen inverse photos with the same conditions added to the experiment in each scene. In other words, there were twenty-four photos containing primitive (unaltered) photos and inverse photos with hidden soldiers and six without soldiers in each scene.

2.1.2 *Procedure*

The subjects wore eye trackers to search for a target in each photo, which was shown for fifteen seconds. The subjects had to circle on the screen if they found suspicious targets. During the experiment, the data of visual search path, fixation time, search performance, and searching time would be recorded.

### 2.2 *Field experiment*

In this phase, experimental conditions were carried out in the field and the data of visual search path, fixation time, search performance, and searching time were recorded by the eye tracker. The scenes in this phase were selected in National Tsing Hua University in Taiwan.

Figure 1. The target wears camouflage dressing.

Figure 2. Three types of background of woodland.

2.2.1 *Experimental design*

Another thirty male subjects participated in the field experiment, All subjects had visual acuity > 0.8 and passed the color discrimination test. The subjects had to wear the eye tracker (ASL Mobile Eye) during the experiment. Due to the ASL Mobile Eye System being more efficient than SR Research when worn outdoors, the ASL Mobile eye system was used to collect the data of eye movement in the field.

In the field experiment, the person who hid in the woodland and urban scene wore the same camouflage clothing and hid in any position in each scene. Scenes are an independent variable in this experiment including two levels, woodland and urban. Each scene had three different kinds of backgrounds as laboratory experiment. The dependent variables include searching performance (hit rate, miss rate, false alarm), searching time, and scan path, and there were four trials with hidden soldiers and one trial without in each background. Therefore, there was a total of fifteen trials in each scene.

2.2.2 *Procedure*

The procedure of the field experiment is the same as that in laboratory experiment. The participants wore the eye tracker to search for a target in each scene, and each of them had fifteen seconds to find the person who wore the camouflage dressing hidden in the scenes. During the experiment, the data of visual search path, fixation time, search performance, and searching time would be recorded by the ASL mobile eye system.

## 3 RESULTS AND DISCUSSION

The result of the laboratory experiment indicated that most of the fixation points located on trees, Awn grasses and meadow in woodland scenes, and the main searched objects in urban scenes were walls, windows, doors and trees. In further analysis, the time to gain fixation on trees was significantly longer than that on meadow or Awn grasses in the woodlands (Table 1), and meadow was missed easily. The error rate for trees was higher than the others (Table 2), so the dummy soldier is best to be put by trees in woodland scenes.

In urban scenes, the fixation time on windows or doors was significantly longer than walls and trees (Table 3), and the hit rate was lower and false alarm rate was higher when targets were hidden in windows or doors (Table 4). Subjects spend a shorter fixation time on walls, but they found targets hidden in walls more easily. A dummy soldier is suggested to be best placed by walls in urban scenes. In addition, there was a significant difference between scenes in the time for finding the target. The searching time in the woodland was longer than that in the urban scene.

In the results of the field experiment, the fixation time on Awn grasses is significantly longer than that on meadow or trees in the woodlands, and for searching performance in woodland, the average of the miss ratio is 56%, and the average searching time is 8.024 sec.

Table 1. Average fixation time on woodland.

| Fixation time (ms) | Primitive photos | Inverse photos |
|---|---|---|
| Trees | 7144.7 | 6826.3 |
| Awn grasses | 2800.9 | 3130.5 |
| Meadow | 3307.6 | 3670.4 |

Table 2. Searching performance on woodland.

| | Trees | Awn grasses | Meadow |
|---|---|---|---|
| Hit rate | 0.361 | 0.393 | 0.283 |
| Miss rate | 0.528 | 0.607 | 0.683 |
| False alarm | 0.111 | 0.083 | 0.05 |

Table 3. Average fixation time on urban scenes.

| Fixation time (ms) | Primitive photos | Inverse photos |
|---|---|---|
| Windows/doors | 7083.7 | 7844.7 |
| Walls | 4424 | 4516.3 |
| Trees | 1741.3 | 1173 |

Table 4. Searching performance on urban scenes.

| | Walls | Windows/doors |
|---|---|---|
| Hit rate | 0.413 | 0.395 |
| Miss rate | 0.587 | 0.604 |
| False alarm | 0.042 | 0.106 |

Table 5. Searching performance.

| | Woodland | | Urban | |
|---|---|---|---|---|
| | Lab | Field | Lab | Field |
| Miss rate | 0.56 | 0.606 | 0.5 | 0.596 |
| Hit rate | 0.44 | 0.346 | 0.5 | 0.404 |

In the urban, scenes, fixation time on pillars is significantly longer than that on walls or trees, the average of the miss ratio is 50%, and the average searching time is 9.22 sec.

Comparing the results between the laboratory and the field experiment, there were no significant differences between the average searching time for lab and that for field, neither was there any significant difference between searching performance for lab and that for field study (Table 5). Moreover, in comparison with the field experiment, the experiment in the laboratory needed to consider more factors which could influence the result of the experiment in the urban, such as sunlight, shadow, and weather.

## 4 CONCLUSION

The purpose of this study was to compare the results of visual searching strategies in the laboratory with the results in field, and the result of this study could be used as reference for anti-camouflage and camouflage design. According to the results of both experiments, the best hiding position in the woodland is suggested in the Awn grasses and meadow, and the hiding position in the urban scenes is suggested in the windows, trees, and meadow. This research explains the general searching strategy of humans. It may provide a reference for suitable camouflage areas for military purposes.

## REFERENCES

Cott, H.B. (1940). Adaptive coloration in animals. London: Methuen & Co. Ltd.

Morelli, F. & Burton, P.A. (2004). The impact of induced stress upon multiple—object tracking: research in support of the cognitive readiness initiative. Human Research & Engineering Directorate.

Nyberg, S. & Bohman, L. (2001). Assessing camouflage methods using textural features. *Optical Engineering*, Vol. 40, No. 9, pp. 1869–1876.

Peak, J.E. et al. (2006). Guidelines for Camouflage Assessment Using Observers, North Atlantic Treaty Organization.

*Ergonomics in Asia – Shih & Liang (eds)*


# The effects of acute stress on working memory

V. Lai, C. Theppitak, M. Movahed, D.V.G. Kumudini, Y. Higuchi, H. Izumi & M. Kumashiro
*Department of Ergonomics, University of Occupational and Environmental Health, Japan*

T. Makizuka
*Department of Management and Business, Kinki University, Japan*

ABSTRACT: Stress is a strong modulator of memory function, which is very important and can affect work effectiveness, productivity and safety in the workplace. The aim of this study is to investigate the effects of stress on working memory in young healthy males. A group of 12 young healthy males was submitted to a stressful condition in one day and control condition in another day. After each condition subjects performed the modified version of Sternberg task with complex visual scenes. The results showed that stressful conditions did not affect the performances between variable loads as well as variable delays. Further analyses separating performance into hits (correctly recognized) and CRs (correctly rejected) in term of recognition revealed that the CRs rate increased in the stress group with significant difference. These results indicate that acute stress enhances working memory performance related to the recognition of recollection process but not for familiarity process.

*Keywords*: working memory, stress, recognition

## 1 INTRODUCTION

Stress is a psycho-physiological process, which results from the interaction of the individual with the environment (Luthans, 1998) and results in disturbances caused to the physiological, psychological and social systems, depending upon individual characteristics and psychological processes. Stress is a popular topic these days. In many jobs, perceived work stress is an important part of the overall job demands. It may briefly be defined as "the psychological and physiological effects caused by work performance." If work stress is becoming too high, this may result in negative consequences for job performance, absenteeism, work ability and possibly the development of diseases. However, stress is not always necessarily harmful. Stress can be therefore negative, positive or neutral. For many decades, changes in cognitive function have been considered to be important consequences of stress and anxiety (Maher et al., 1966).

As to the physiological stress, cold pressure stress is an experimental stress paradigm based on a short term painful stimulation by immersing the hand into ice-cold water. It is known to be associated with substantial activation of the autonomic nervous system as well as mild to moderate activation of the hypothalamic pituitary adrenocortical (HPA) axis (McRae et al., 2006; Schwabe et al., 2008).

Psychosocial stressors typically include performance components that employ cognitive resources (e.g., mathematics tasks) and/or include a social- evaluative component. One often used psychosocial stressor is the Trier Social Stress Test (TSST; Kirschbaum et al.,1993). It is known to reliably activate the sympathetic nervous system (SNS) and the HPA axis.

However, stress effects are multi-faceted, with numerous neuromodulator and hormonal effects on different neural and peripheral systems. So in this study we employed the stressor, which combined also a physiological factor (socially evaluated cold pressure test) (SECPT)

(Schwabe et al., 2008) and psychosocial factors (modified version of Trier Social Stress Test - TSST. (Kirschbaum et al., (1993).

Stressful events are generally better remembered than unemotional ones. However, memory is not a unitary process and stress seems to exert different effects depending on the memory type under study. Previous studies (Akirav et al., 2004; Lupien et al., 2002), have suggested an inverted-U relationship between stress and memory, with midrange doses generally enhancing memory, and higher or lower doses not enhancing, or even impairing memory.

Working memory (WM) is a storage system that holds a limited amount of information for a brief time. It is an important cognitive system that is thought to underlie many higher order cognitive functions, including learning, planning and reasoning (Baddeley A. 1992). The Sternberg Item Recognition Task (SIRT) (Sternberg, 1966) is a WM task that emphasizes the maintenance of information with minimal manipulation. It consists of an encoding phase during which subjects see a string of letters that they are to remember, followed by a short unfilled delay or maintenance phase during which subjects attempt to maintain the information, and finally a retrieval phase during which subjects are shown a probe and asked to indicate whether or not the probe is a subset of the encoded string.

Therefore, the current study was designed to examine the effects of complex stress including physiological and psychosocial stressors on the working memory task. An additional issue in this literature is that even when alterations in working memory are found, it is not always clear which stage of working memory is most affected.

## 2 METHODS

### 2.1 *Participants*

Twelve right-handed young healthy male students from Medical University (age 22.1 ± 1.0 years), with a body mass index ranging from 19.57–23.94 (mean 22.16 ± 0.15), volunteered to attend the experiment. One subject was excluded from data analysis because the WM performance was <50%. All subjects provided written informed consent prior to participation.

### 2.2 *Experimental procedure*

All experimental sessions were from 13:00–17:00. Each subject came in on 2 days and was randomly assigned which day for stressful or for control condition. After arrival in the laboratory, subjects were given a resting phase of 15 min. Subjects were required to attend to the stressful or control condition in 3 min. They rested for 3 min and then did 30 trials of working memory tasks for training, and after following a short rest period, subjects engaged in the real WM task with the total time of testing at 22 min.

#### 2.2.1 *Stress test*

*For stressful condition*: The participants were asked to self-introduce in English (foreign language) in 2 min in front of a woman and camera (based on the TSST) and then do the SECPT (Socially Evaluated Cold Pressure Test) in 1 min.

*For control condition*: they put their hand in warm water (~30°C) in 3 min with no camera, not viewed by a woman and no self-introduction.

At the end of the stress test, subjects were asked to rate the level of discomfort, pain and stress they experienced during the stress test from 0 (no pain, stress or discomfort) to 100 (the worst painful, stressful or discomforting). Heart rate and blood pressure were measured before and after the stress test. Salivary cortisol also was assessed before and 2 times after the treatment to validate stress effects.

#### 2.2.2 *Sternberg task*

Fifteen minutes after the stress exposure, working memory performance was tested with the Sternberg item recognition task (Sternberg 1966) (Figure 1) with 2 levels of cognitive load (load 2 and load 5) and 2 levels of maintenance (2 s and 8 s).

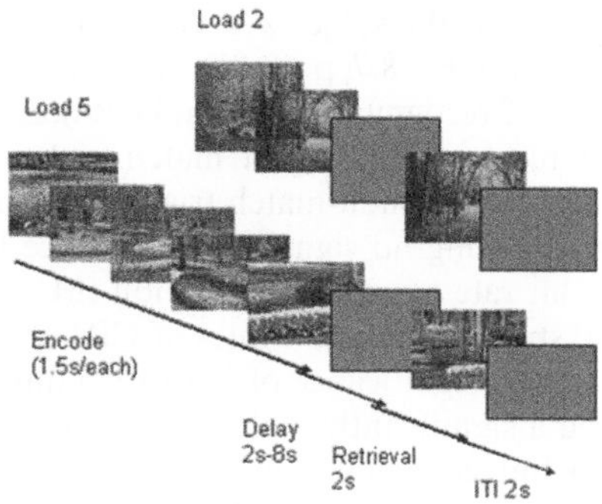


Figure 1. Sternberg task. Subjects performed 40 trials Load 2 (top) and 40 trials Load 5 (bottom).

Each trial consisted of an initial encoding phase during which the subjects saw either 2 or 5 sequentially presented pictures. The pictures in each trial were the same topic and were very similar to increase the difficulty of the task. The encoding period was then followed by a variable length delay period which, in turn, was followed by a retrieval phase (2 s), and lastly was internal interval (1500–2500 ms, fixation/ITI). During the encoding phase, each scene was presented for 1500 ms and was followed by a variable temporal jitter during which the screen was black (400 ms).

During the retrieval phase, subjects had to press "Y" (right arrow on the keyboard) if the probe picture was identical to one of the sample pictures seen during that trial (match) or press the "N" (left arrow) if it was a new picture (non-match). Subjects were asked to respond as quickly and as accurately as possible. In total they had to do 80 trials (40 match and 40 non-match trials) in each condition.

Reaction time and accuracy were calculated to compare the differences of the stress effects on these processes of working memory.

### 2.3 *Stress measures*

Heart rate and systolic blood pressure (SBP) and diastolic blood pressure (DBP) were measured before and after the stress test.

### 2.4 *Electrophysiological recordings*

The electroencephalogram (EEG) was continuously recorded (sample rate = 1000 Hz) from 6 electrodes located at standard electrode positions (F3, F4, Cz, P3, P4 and Oz) of the 10–20 International System. All impedances were kept below 5 kΩ.

Event Related Potential: EEG epochs extended from 200 ms pre- to 1000 ms post-stimulus. The Parietal EM effect was measured by comparing the averaged voltages for Hits and CRs at P3 and F4.

### 2.5 *Statistical analyses*

Data were analyzed by means of ANOVAs, paired t-tests and t-tests for independent samples by using SPSS 16.0.

## 3 RESULTS

### 3.1 *Behavioral results*

*Accuracy:* All subjects performed significantly above chance (>62.5%) at each load (Table 1). There was no significant difference in accuracy as a function of short or long delays ($F = 0.3$, $p > 0.05$).

The 2 ways ANOVA revealed that the subjects' accuracy level decreased with increasing load in both stress and control group ($F = 8.7$, $p < 0.01$). Additionally, we tested specific effects of the stressor on different types of recognition (Hits and CRs). As Figure 2 shows, stress did not increase rates of correctly recognized (Hits) of match trials but enhanced the accuracy rate of items correctly rejected (CRs) of non-match trials. This observation is supported by 2-way analysis ANOVA demonstrating no significant difference between stress condition's hit rate (M = 78.6%) and the hit rate of control condition (M = 85%) at both Load 2 and Load 5 ($p > 0.05$). In contrast, stress did increase rates of CRs compare to control condition ($F = 5.4$; $p < 0.05$) but there was no significance of load x condition interaction. Subjects in the stress condition recognized a significantly higher proportion of CRs (M = 83.6%) than control condition (M = 73.2%). Subjects showed higher accuracy of Hits than CRs ($p < 0.05$) in control condition but no difference in stress condition.

*Reaction Time:* Subjects responded significantly more slowly with increasing working memory loads ($F = 9.9$, $p < 0.01$) in both stress and control group. There was no significant difference in reaction time as a function of short or long delays ($F = 0.005$, $p > 0.05$).

Stress condition showed the trend in faster reaction time with the CRs than the control condition ($F = 3.7$; $p = 0.06$) and no significance of load x condition interaction. And also we saw that the Hits trials had faster reaction time than CRs in control condition ($p < 0.05$) but no difference in stress condition.

### 3.2 *Subjective and physiological stress response*

Participants' subjective assessments, heart rate and blood pressure changes indicated the successful stress induction.

*Emotional response to the stress test:* As expected, subjects in stress condition reported far higher rating of discomfort ($59.1 \pm 10.8$ vs. control condition $2.2 \pm 1.51$, $p < 0.001$) as well as being stressful and painful more than did subjects in the control condition.

*Physiological response:* The stress test elicited significant elevations in SBP ($117.6 \pm 6.7$ vs. control condition $111.4 \pm 9.2$, $p < 0.001$) and DBP ($73.3.6 \pm 6.6$ vs. control condition $66.2 \pm 10.6$, $p < 0.01$), while the control condition did not.

Table 1. Performance data.

| | Control | | Stress | |
|---|---|---|---|---|
| | Load 2 mean (SD) | Load 5 mean (SD) | Load 2 mean (SD) | Load 5 mean (SD) |
| Short delay | | | | |
| RT (ms) | 908 (103) | 1030 (91.3) | 894 (174.8) | 1029 (167.5) |
| Accuracy | 89.5 (11.4) | 72 (13) | 85 (11.3) | 77 (9.8) |
| Long delay | | | | |
| RT (ms) | 920 (77) | 1019 (123) | 890 (167) | 1023 (152) |
| Accuracy | 87.5 (10.1) | 80 (10) | 82.5 (13.4) | 79 (7.4) |

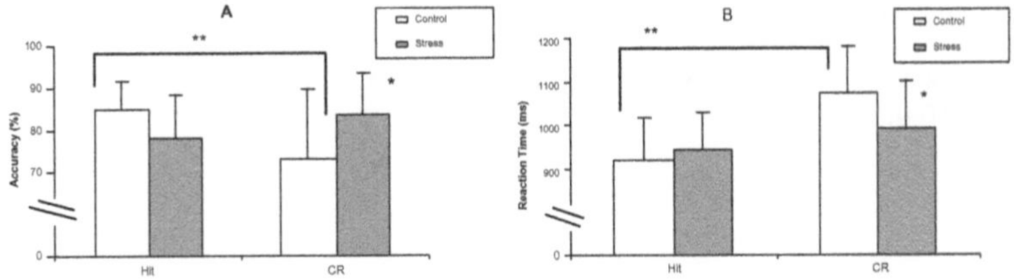


Figure 2. Accuracy (A) and Reaction Time (B) of recognition for retrieval phase of recollection (CRs: items correctly rejected) and familiarity (Hits: items correctly recognized) between stress and control condition. * indicates a significant difference between stress and control groups on correctly rejected ($p < 0.01$). ** indicates significant differences between Hits and CRs in control condition ($p < 0.05$).

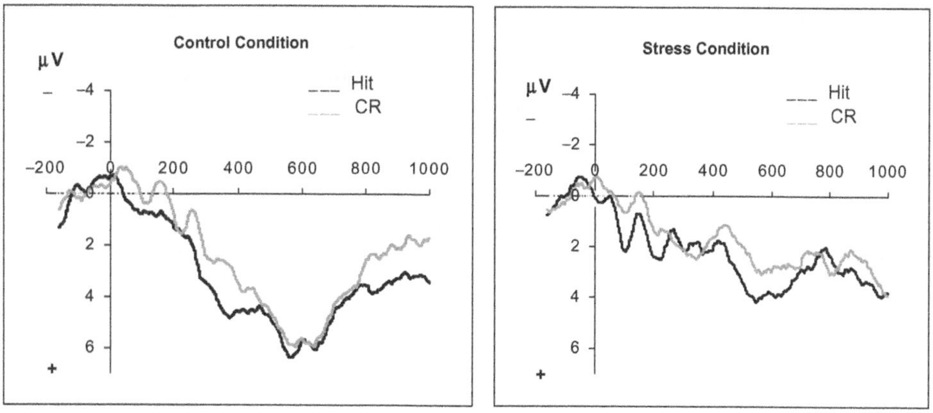


Figure 3. Recognition: old/new comparison. ERP for Hits (old items) and correct rejections (new items).

### 3.3 *ERPs at recognition (old/new effect)*

This figure shows the ERPs elicited at retrieval by old pictures and new pictures that were correctly identified as such (hits and correct rejection) for control condition and stress condition at left parietal (P3). Between 400–800 ms, stress condition's ERP responses have a trend in differentiating new from old trials much more than do those of control condition ($p = 0.06$).

## 4 DISCUSSION

According to dual process models, recognition memory is supported by distinct retrieval processes known as familiarity and recollection. Most researchers would ascribe to the notion that familiarity-based recognition is fast-acting, relatively automatic and does not provide qualitative information about the study episode. Recollection, by contrast, is conceived as a slower, more effortful process that gives rise to consciously accessible information about both the prior occurrence of the test item and the context of that occurrence. (Rugg M.D., 2007). Familiarity is considered to be an automatic (unconscious) process and recollection to be a controlled (conscious) process (Mandler, 1980), and associated with implicit and explicit memory processes, respectively (Jacoby, 1991). According to this conceptualization, recollection, but not familiarity, should depend on medial temporal lobe structures (i.e., hippocampal), as they are key brain regions in the explicit memory system (Squire, 1992).

Hippocampus-independent memory is frequently facilitated by stress (Sandi et al., 2007; Shors, 2006), and this facilitation seems to be dependent on gluocorticoids (Shors, 2003). Glucocorticoids seem to play a key role in stress effects, since an inverted U-shaped function has also been reported for the relationship between glucocorticoid levels and memory and plascity (Abrari et al., 2008; Joels, 2006).

The results of the current study show that acute stress caused an improvement on performance of the Sternberg task. This effect may be due to the relatively moderate increase in stress caused by both psychosocial and physiological factors. These findings add further support to the theory of inverted U shape.

In accordance with other previous studies, the impact of stress on memory cannot be explained as a general memory performance but rather reflects a specific impact on aspects of retrieval (Payne J.D. et al., 2002). In this study we saw that stress enhanced the recollection but not familiarity at retrieval. Despite the fact that familiarity and recollection sometimes activate the same brain regions, they are typically found to be quite distinct functionally (Rugg et al., 2003). Also recollection perhaps requires a higher degree response of matching than familiarity.

Although the present findings resemble the results of another study (Yonelinas A.P. et al., 2011) that moderate stress can enhance the performance of WM, however in contrast that study found that stress increased familiarity-based recognition rather than recollection. The reason may be in the design of that study, in which they gave stress to subjects after the encoding.

According to one view (the high-threshold/signal-detection model), high confidence in a recognition decision strongly implies that the decision was based on recollection, whereas lower confidence necessarily implies that the decision was based on familiarity (Yonelinas, 1994). Another explanation for our results is that perhaps stress at moderate level makes subjects feel a higher confidence than in the control condition so they had better performance at recollection-based recognition (with greater accuracy rate and higher reaction time).

Our results indicated that stress didn't affect the maintenance process but enhanced encoding and retrieval processes. In support of this finding, one study found that stress increases activity in the prefrontal cortex and posterior parietal cortex, specifically during maintenance of items in WM, whereas effects on hippocampal activity are restricted to encoding and retrieval (Weerda R. et al., 2010). Also, possibly in this study the hippocampal of subjects in stress condition was more affected by stressors so the retrieval process was enhanced.

In some situations, memory is enhanced by stressful experience, while in others, it is impaired. Furthermore, gaining a better understanding of how these factors, in addition to emotional arousal, the nature of stressor, attention, motivation, sex and memory practice, interaction will be essential for understanding the role of stress on working memory.

## REFERENCES

Akirav, I.M. & Kozenicky, M. 2004. A facilitative role for corticosterone in the acquisition of a spatial task under moderate stress. Learning and Memory, 11, 188–195.

Baddeley, A. 1992. Working memory, Science, 255 (5044), 556–559.

Kirschbaum, C., Pirke, K.M. & Hellhammer, D.H. 1993. The 'Trier Social Stress Test'—a tool for investigating psychobiological stress responses in a laboratory setting. Neuropsychobiology, 28 (1–2), 76–81.

Luthans, F. 1998. Organizational Behavior. Boston: McGraw-Hill.

Lupien, S.J. & Wilkinson, C.W. 2002. The modulatory effects of corticosteroids on cognition: Studies in young human populations. Psychoneuroendocrinology, 27, 401–416.

Maher, B. 1966. Principles of psychopathology: an experimental approach. New York: McGraw-Hill.

McRae, B.H. 2006. Isolation by resistance. Evolution, 60:1551–1561.

Payne, J.D. & Nadel, L. 2002. The effects of experimentally induced stress on false recognition. *Memory,* 10 (1).

Rugg, M.D. & Curran, T. 2007. Event-related potentials and recognition memory. *Trends in Cognitive Sciences*, 11:251–257.

Schwabe, L. & Haddad, L. 2008. HPA axis activation by a socially evaluated cold pressor test. Psychoneuroendocrinology, 33, 890–895.

Sternberg, 1966. High-speed scanning in human memory, Science, 153, 652–654.

Weerda, R. & Wolf, O.T. 2010. Effects of acute psychosocial stress on working memory related brain activity in men. Hum Brain Mapp, 31 (9), 1418–29.

Wolf, O.T. & Schommer, N.C. 2001. The relationship between stress induced cortisol levels and memory differs between men and women. Psychoneuroendocrinology, 26, 711–720.

Yonelinas, A. 1994. Receiver operating characteristics in recognition memory: Evidence for a dual process model. Journal of Experimental Psychology: Learning, Memory, and Cognition, 20:1341–1354.

Yonelinas, A. & Colleen, M. 2011. The effects of post-encoding stress on recognition memory: Examining the impact of skydiving in young men and women, 14 (2), 136–144.

*Ergonomics in Asia – Shih & Liang (eds)*
*© 2012 Taylor & Francis Group, London, ISBN 978-0-415-68414-9*

# Examination of the mental model construction process of a FLASH game

Shinichiro Kitaoka
*Wakayama University, Graduate School of System Engineering, Japan*

Toshiki Yamaoka
*Wakayama University, Faculty of System Engineering, Japan*

ABSTRACT

*Objective*: The mental model construction process when users operate 328 different electrical products was examined.
*Methods*: First, the functions of the products were investigated. Next they were classified under 82 categories according to their functions. 132 subjects were selected from the categories. After the selection, 2 researchers described their thoughts in the format of three-point task analysis. Next their thoughts were classified, and the items were analyzed by the mental model construction process. To verify the items, 2 experiments were performed.
*Results*: The thoughts were classified into 5 items and 13 subordinate items. As a result of the experiments, all items were examined when the participants were operating them. All participants omitted to operate the iPod shuffle.
*Discussion*: It is considered that there are 5 stages in the mental model construction process and several of 13 subordinate items are relevant to each stage. If people missed an operation in the middle of the mental model construction process, they couldn't achieve the operation and couldn't construct the mental model.

*Keywords*: mental model, mental model construction process, three-point task analysis

## 1 INTRODUCTION

### 1.1 *Importance of metal model in operation*

In recent years, electrical products have become more difficult to use. Therefore, many users can't understand the operation and so they do not operate them correctly. When a user operates an electrical product, the, user's mental model is important in understanding their operation. D.A. Norman wrote, "People form internal, mental models of themselves and of the things and people with whom they interact. These models provide predictive and explanatory power for understanding the interaction [1]."

When a user has already had a mental model of a product, he/she operates the product according to his/her mental model. On the other hand, if the mental model has differed from the product's system image, he/she can't understand the operation and might fail to operate it correctly.

### 1.2 *Purpose of this study*

The purpose of this study is to investigate the mental model construction process.

## 2 METHOD

### 2.1 *Selection of subjects*

First, 328 products were picked from the "Similar commodity and labor screening criterion" and "Census of Manufactures Product Classification in 2009." These describe all products which are sold and produced in whole of Japan. Second, their functions were investigated using the functional approach of value engineering [2]. The products were classified into 82 categories according to their functions. Finally, several subjects were selected from each category. As a result 132 subjects were selected. Examples of categories and subjects are showed as below.

### 2.2 *Setting up the hypothesis of the mental model construction process*

Two researchers who have majored in ergonomics investigated their thoughts while they operated the objects. They described their thoughts on the three-point task analysis [3] format (Table 2). Three-point task analysis is an evaluation method to find problems in a system. It was evaluated using the 3 viewpoints "effective acquirement of information," "ease of understanding and judgment" and "comfortable operation." The objects of investigation are the basic task of each subject. The task they thought about was written in the "Task" column, the information they got was written in the "effective acquirement of Information" column, their judgement while they operated the subject was written in the "ease of understanding and judgment" column, the operation they did was written in the "comfortable operation" column. The example of a medical thermometer is shown below.

After the investigation, the thoughts written in "ease of understanding and judgment" were classified using card sorting because it is considered that a mental model is constructed to understand the operation in this manner.

### 2.3 *Verification of the items*

To verify the items which are conducted while users operate the product, a verbal protocol analysis was used to compare the items and verbal protocol while operating the objects. Two experiments were conducted.

Table 1. Groups and subjects.

| Category | Products |
|---|---|
| Warm | Heater<br>Electrically heated carpet<br>Portable body warmer |
| Gather | Broom |

Table 2. Three-point task analysis.

| Task | Effective acquirement of information | Ease of understanding and judgement | Comfortable operation |
|---|---|---|---|
| We turn on power | A power button is on the point | I'll take the power<br>If I push the button, the power could be taken. | I push the power button |
| | A screen lights<br>A sound was rung | I pushed the button and a sound was rung<br>I'll insert it | |
| We insert it under the armpit | Shape of the thermometer<br>The position of the thermometer | I'll insert the tip of the thermometer under the left armpit<br>I understand the thermometer was inserted<br>I'll wait for the temperature reading | I insert it under the left armpit |

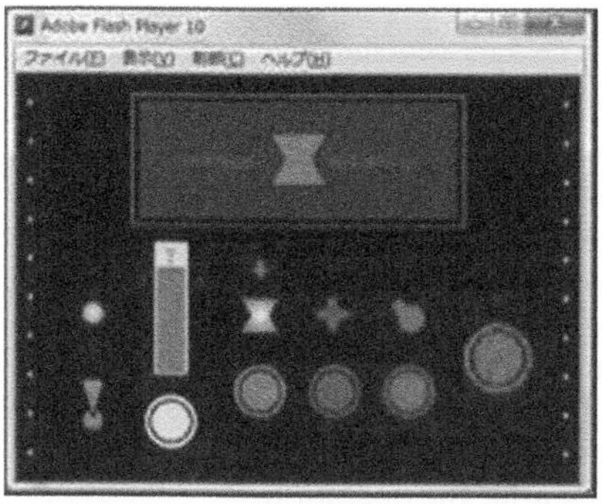

Figure 1. Spaceship interface made by FLASH.

The appliance of Experiment 1 was the spaceship interface made by FLASH (Figure 1). The participants were 4 students and their average age was 22. They said what they thought while they operated it. They weren't informed how it is operated. Task was to destroy the target on the display.

The appliance of Experiment 2 was the iPod shuffle fourth generation. The participants were 5 students and the average age was 22. In a similar way, they were not informed how it works. They had never used it before, however they have used a music player. Two tasks were conducted in Experiment 2. Task 1 was to change the playlist. Task 2 was to set it to random. After the experiments researchers interviewed them about the appliance.

## 3 RESULTS

### 3.1 *Results of the card sorting*

The thoughts were classified into 5 items and 13 subordinate items. They are shown below.

1. Determining the objectives.
   To determine the objective in using the product.
   1. Grasping the objectives
      This is to grasp the objective of operating the objects.
   2. Recalling the objects
      To recall where the objects to use are, and shape of the objects.
2. Gathering necessary information for operation.
   To gather information for operation, the product's status and the position of the parts to use, etc.
3. Searching the objects.
   To search product's parts and its position according to mental model.
4. Reasoning about the objects.
   To reason about the product's parts, characters, and illustrations etc.
5. Reasoning about the object's status.
   To reason about the object's status.
   3. Determining the procedure of operation
      To determine the procedure when operating the objects.
6. Reasoning using the script.
   To reason out the next operation using the user's instructions.
7. Understanding from the object's status.
   To understand the next operation according the object's situation.
   4. Reasoning how to operate the objects
      To reason how to operate the object from its status and information about it.

8. Ascertaining the operation.
   To ascertain the operation.
9. Reasoning about the relation between the objects.
   To reason about relations between the objects.
10. Predicting object's status after the operation.
    To predict object's status after user operates the objects.
11. Reasoning about the procedure in detail.
    To reason about the operation procedure in detail.
    5. Ascertaining the operation was achieved
       To ascertain whether the operation was achieved after the operation.
12. Ascertaining the feedback.
    To ascertain the feedback after the operation.
13. Compare the forecast and the object's status.
    To compare the forecast before operation and the object's situation and evaluate the operation is achieved.

### 3.2 *Results of Experiment 1*

All participants achieved the task. However, the verbal protocols that were applied to "Recollection of object" and "Inference of detailed operation's procedure" weren't performed.

### 3.3 *Results of Experiment 2*

The verbal protocols were applied to all items. All participants achieved task 1 but they couldn't achieve task 2.

During the interview, one of them said 'I couldn't understand how to operate the button to change the playlist.' 'After pushing button, I listened for a response. But I couldn't understand the meaning.'

## 4 DISCUSSION

### 4.1 *Verification of items*

The result of Experiment 2 showed that 5 items and 13 subordinate position items are performed while users operate products.

The participants couldn't achieve task 2. It considered that it is caused by the object's operation being different from another music player and there is no clue representing the playlist has been changed.

### 4.2 *Relationship between items and failing operation*

We consider the relations between cause of failure and 13 subordinate items.

From the results of the interview, the reasons they fail the task seem to be 'the participants couldn't understand how to operate the button' and 'the participants couldn't understand the meaning of the response.'

It is considered that 'They couldn't understand how to operate the button' is applied to 'Ascertaining the operation' and 'the participants couldn't understand the meaning of response' is applied to 'Ascertaining the feedback.' It is considered that the participants might fail at 'Ascertaining the operation' and 'Ascertaining the feedback.'

Due to these failures, they couldn't understand the function of the button and construct the mental model.

This showed that if users failed to perform items, they couldn't achieve the objective and construct the mental model.

### 4.3 *Comparison of the 5 items and Gulf model*

We compare the 5 items and the Gulf model (Norman, 1983). First, users determine the purpose of operation at 'Determining the objective' stage. Next, they gather the information to achieve the objective at 'Gathering necessary information for operation' stage and they decide the next operation at 'Determining the procedure of operation' stage. Next, they reason the operation at 'Reasoning how to operate the object' stage. After they operate the object, they ascertain the status of the object and evaluate the operation is achieved at 'Ascertaining the operation was achieved' stage. Next, we compare the items and the 3 layers model [4]. It is showed that mental model is constructed through the three layers.

Table 3. Compare 5 items and Gulf model.

| 5 Items | Gulf model |
|---|---|
| Determining the objective | Establishing the goal |
| Gathering necessary information for operation | Forming the intention |
| Determining the procedure of operation | |
| Reasoning how to operate the object | Specifying the action sequence |
| Ascertaining the operation was achieved | Interpreting the state<br>Evaluating the system state with respect to the goals and intentions |

Table 4. Correspondence between 5 items and three stages of constructing a model.

| 5 Items | 3 layers model of mental model construction process |
|---|---|
| Determining the objective | Remembrance level |
| Gathering necessary information for operation | |
| Determining the procedure of operation | Structuring network level |
| Reasoning how to operate the object | |
| Ascertaining the operation was achieved | Mental model construction level |

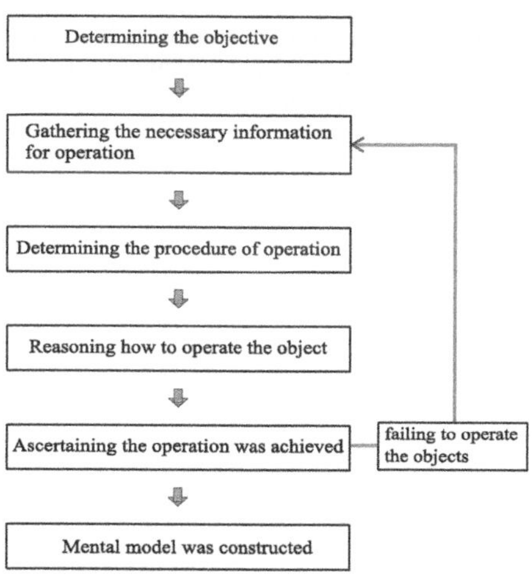


Figure 2. Mental model construction process.

'Remembrance level' is remembering the concept and knowledge. 'Structuring network level', is understanding the relation between concept and knowledge. 'Mental model constructing level' is where the network is completed and is buried in the schema. We now consider the correspondence between the 5 items and 3 layers of the 3 layers model of a mental model (Table 3).

First, users remember the concepts and knowledge at 'Determining the objective' and 'Gathering necessary information for operation.' Second, they understand the relation between concept and knowledge at 'Determining the procedure of operation' and 'Reasoning how to operate the object.'

After the operation, they complete the network and construct a mental model at 'Ascertaining the operation was achieved.'

Figure 2 showed the mental model structural process. First, users determine the objective to operate the object at 'Determining the objective.' Next, they gather necessary information to operate the object at 'Gathering necessary information for operation.' Next, they determine the procedure of the operation at 'Determining the procedure of operation.' Next, they reason the operation of the objects at 'Reasoning how to operate the object.' After the operation, they ascertain the operation was achieved at 'Ascertaining the operation was achieved.' If they ascertain the operation was achieved, a mental model is constructed. However, if they ascertain the operation is not achieved or the operation is failed, they return to 'Gathering necessary information for operation' and gather the information to modify their mental model.

## 5 CONCLUSION

It is considered that mental model construction process has 5 stages, and several subordinate items are related to each stage. If a user fails to perform the items in the middle of process, they couldn't understand the operation and performed the next stage.

## REFERENCES


[1] Norman, D.A.: *The Psychology of Everyday Things*, Basic Books, 1988.

[2] Naoaki Tejima: *Value Engineering Practice—technology to increase customer satisfaction*, pp. 48–55, 1993.

[3] Toshiki Yamaoka: *Lecture on Ergonomics for Hardware and Software Design* (in Japanese), Musashino Art University Press, pp. 154–161, 2003.

[4] Masatoshi Rin, Toshiki Yamaoka, Ryota Mori & Hidetoshi Yoshioka: Grasp of Senior People's Mental Model Construction Process when Using Calculator. In *Proc. of 12th Int. Conf. on Human-Computer-Interaction International, 2007.*

*Ergonomics in Asia – Shih & Liang (eds)*
*© 2012 Taylor & Francis Group, London, ISBN 978-0-415-68414-9*

# Effect of physical exercise on the processes of cognition

Hiroyuki Izumi, Kei Kinugawa, Takehiro Kishigami, Kazukuni Hirabuki, Kana Morii, Yoshiyuki Higuchi & Masaharu Kumashiro
*Department of Ergonomics, IIES, University of Occupational and Environmental Health, Japan*

ABSTRACT: Physical exercises are known not only to have a physical effect but also an enhancement effect on cognitive functions. The aim of this study is to investigate the effect of physical exercise on cognitive processes and discuss its mechanism from the viewpoint of brain activity change. Twelve healthy male students were recruited as subjects from UOEH, Japan. A simple reaction task and a selective reaction task using the computer were used to evaluate cognitive function. Subjects were requested to perform a cycle ergometer exercise for 20 min at 65% exercise intensity of maximal oxygen uptake. The change of blood oxygen concentration at the frontal cortex was measured by NIRS during the experiment as an index of brain activity. Physical exercise shows many effects on cognitive function. Each process, cognition and decision-making, reaction shows a different effect of physical exercise. It seems that the change of blood flow at the frontal cortex by physical exercise affects the cognitive function.

*Keywords*: physical exercise, cognitive function, reaction time, brain activity, NIRS

## 1 INTRODUCTION

Cognitive function is one of the important factors for safety in daily working life. It is known that cognitive function is affected by workers' conditions such as arousal and fatigue level. The physical exercise induces warming up effect also getting fatigue. Warming up is the most popular strategy to decrease the risk of injury in the sports field. Also in the occupational health field, many kinds of physical exercise are introduced as warming up activities. The physical exercises are also known not only to have a physical effect but also an enhancement effect on cognitive functions.

Many studies have examined the effects of acute exercise on cognitive function (Brisswalter et al., 2002; Etnier et al., 1997; McMorris and Graydon 2000; Tomporowski 2003).

McMorris et al. (2008, 2009) reported that cognitive function was impaired during strenuous exercise at 80% maximum aerobic power, and discussed an exercise–cognition interaction from a hormonal perspective. Exercise has been found to induce several physiological changes in the central nervous system, including circulatory, metabolic, and neurohormonal effects (Dalsgaard 2006; Nybo and Secher 2004; Meeusen et al., 2006).

The aim of this study is to investigate the effect of physical exercise on cognitive processes and discuss its mechanism from the viewpoint of brain activity change.

## 2 METHODS

### 2.1 *Subjects*

Twelve healthy male students were recruited as subjects from UOEH, Japan (mean age ± SD: 22.1 ± 1.4 years old, mean weight ± SD: 64.6 ± 6.1 kg). None of the subjects had a history of cardiovascular, cerebrovascular, or respiratory diseases.

## 2.2 *Evaluation of cognitive function*

A simple reaction task and a selective reaction task using the computer were used to evaluate cognitive function. These computer tasks were developed on the Turbo Delphi 2006. Subjects were asked to push a key on the keyboard as soon as possible when a target alphabet appeared on the display. Duration of each trial was set to 3 s to minimize the effect of task on physical exercise. An alphabet appeared within 2 s after the trial start and the appearance duration was 500 ms. The latency of response for each trial was measured as a reaction time. In a simple reaction task, the target is every alphabet and 10 trials were involved in a test. In a selective reaction task, the target is only "s" and 50 trials including 10 trials that showed "s" were required to be done in a test. Total number of trials was 60 and around 3 min is needed for one test.

## 2.3 *Physiological measurements*

Heart rate was continuously monitored during exercise to check the exercise intensity. Blood oxygen saturation level ($SpO_2$) was continuously monitored during exercise from the tip of the forefinger. Both of the indexes were recorded with the digital polygraph system (Polymate AP-1132, DIGITEX LAB. CO., LTD. Tokyo, Japan).Critical flicker frequency (CFF) was measured to evaluate arousal level before and after exercise. Cerebral oxygenation was monitored with a NIRS system (BRAIN OXIMETER TOS96, TOSTEC Co., LTD., Tokyo, Japan). Probe holders were attached at the left and right side of the forehead (over the frontal cortex). Each probe holder contained one light source probe and one detector placed at 40 mm from the source. The depth of monitoring is around 30 mm. The light source generated two wavelengths of near-infrared light (760 and 850 nm). Hemoglobin concentrations were calculated using NIR light received by each detector.

## 2.4 *Experimental procedure*

Maximal oxygen uptake ($VO_2$ max) was measured by using a cycle ergometer to determine physical capacity of each subject. The range of $VO_2$ max was from 38.6 to 57.8 ml/kg/min and the mean $VO_2$ max was 46.5 ± 6.6 ml/kg/min (mean ± SD). Measurements of maximal oxygen uptake were performed the day before the experiment day.

On the experimental day, physical exercise with cycle ergometer were carried out for 5 minutes for recording basal value of physiological indexes. Subjects required warming-up for 4 min at 30% exercise intensity of maximal oxygen uptake at first. Then exercise at 65% of maximal oxygen uptake was performed for 20 min. Simple reaction time and selective reaction time were measured before exercise and at 10, 20 min after starting exercise. CFF was measured before and after exercise for evaluation of arousal level. Heart rate and $SpO_2$ was monitored during exercise to check the exercise intensity. The change of blood oxygen concentration at the frontal cortex was measured by NIRS during the experiment as an index of brain activity.

# 3 RESULT AND DISCUSSION

## 3.1 *Reaction time*

Figure 1 shows the change of simple reaction time of each subject (subjects 1 to 12). Simple reaction time after 20 min exercise became shorter than before exercise (0 min, after warm-up) in 11 of the 12 subjects.

A possible reason for the shortening of reaction time is a learning effect. Figure 2 shows the change of simple reaction time of each subject. A clear trend can't be found in selective reaction time. In addition to the process of the simple reaction task, a decision-making process is included in selective reaction task.

The differences between selective reaction time and simple reaction time were calculated to indicate the change of time for decision-making process and are shown in Figure 3.

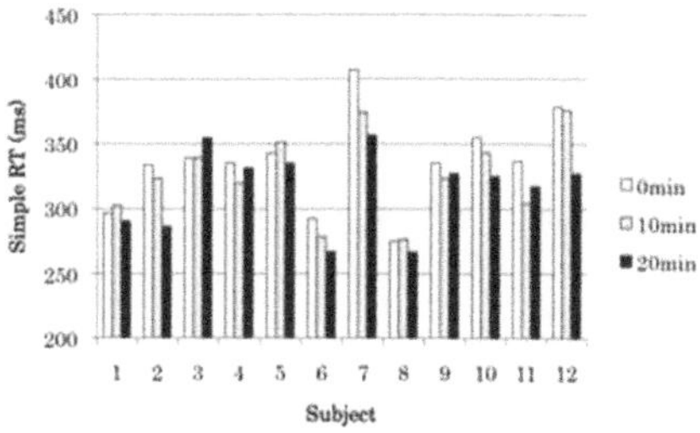


Figure 1. The change of simple reaction time of 12 subjects (before exercise (0 min), after 10 min and 20 min exercise).

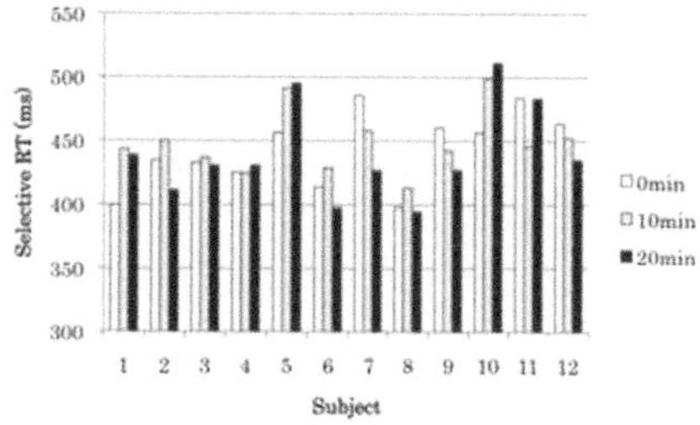


Figure 2. The change of selective reaction time of 12 subjects (before exercise (0 min), after 10 min and 20 min exercise).

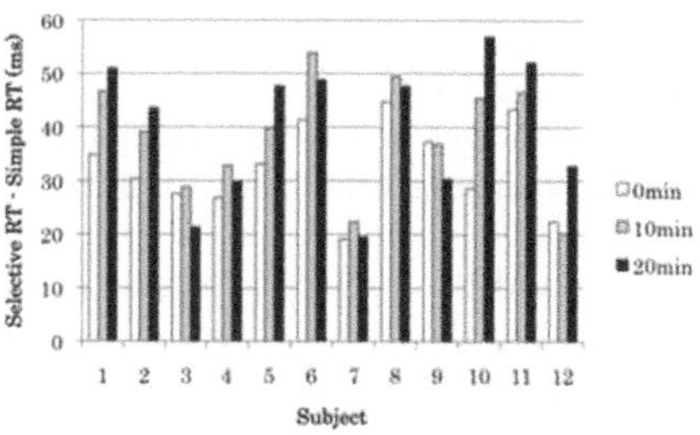


Figure 3. The change of difference between selective reaction time and selective reaction time of 12 subjects (before exercise (0 min), after 10 min and 20 min exercise).

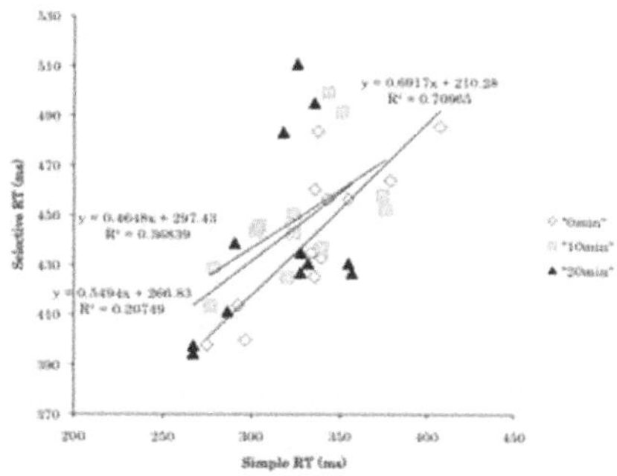


Figure 4. The change of difference between selective reaction time and selective reaction time of 12 subjects (before exercise (0 min), after 10 min and 20 min exercise).

10 of 12 subjects show an extended time for decision-making after 20 min exercise. It seems that the decision-making process is extended by continuous exercise.

Figure 4 shows relationships between simple reaction time and selective reaction time. Before exercise (0 min), simple reaction time and selective reaction time shows a certain level of correlation (R2 = 0.70965). The correlation coefficients after 10 and 20 min exercise are quite small compare to before exercise (R2 = 0.20749 and R2 = 0.36839).

This might mean that physical exercise affected the processes that aren't included in both reaction times. This process might be the decision-making process and the effect of exercise varies between individuals.

### 3.2 *Cerebral oxygenation*

Figure 5 shows the change of oxygen concentration in the blood at the frontal cortex of subject No. 5. The average of oxygen concentration in pre-rest duration (time region 1) set to 0. This graph shows a typical pattern of oxygen concentration. Region 4 is warming-up and region 6 and 8 are physical exercise. Regions 3, 5, 7 and 9 are simple and selective reaction tasks. Oxygen concentration was decreased in warming-up (4) and early stage of the first exercise (6). However, oxygen concentration increases more than rest level in the second half. Decreasing oxygen concentration might be caused by decrease of blood flow due to skeletal muscle activation. Adaptation to exercise intensity occurred in the second half of the first exercise. We think that these changes of oxygen supply might be one of the causes of unstable change in reaction time after 10 min exercise.

The number placed on the x-axis in the graph indicates the time region of each procedure (1: pre-rest, 4: warming-up, 6 and 8: physical exercise, 3, 5, 7 and 9: simple and selective reaction task).

Figure 6 shows the change of oxygen concentration in the blood at the frontal cortex of subject No. 3. No. 3 subject shows different behavior in reaction time. In his case, the

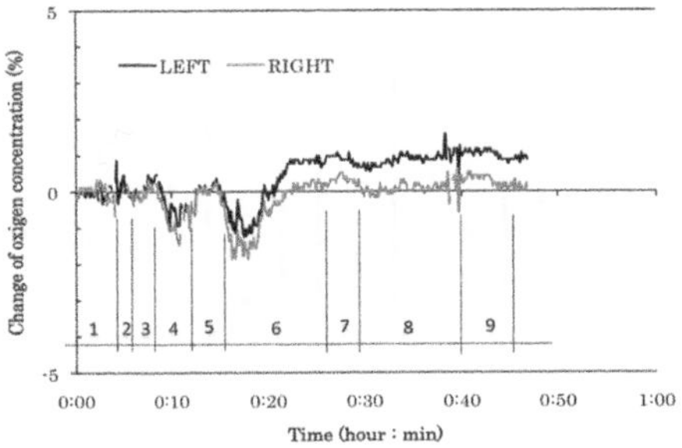


Figure 5. The change of oxygen concentration in the blood at frontal cortex (subject No. 5).

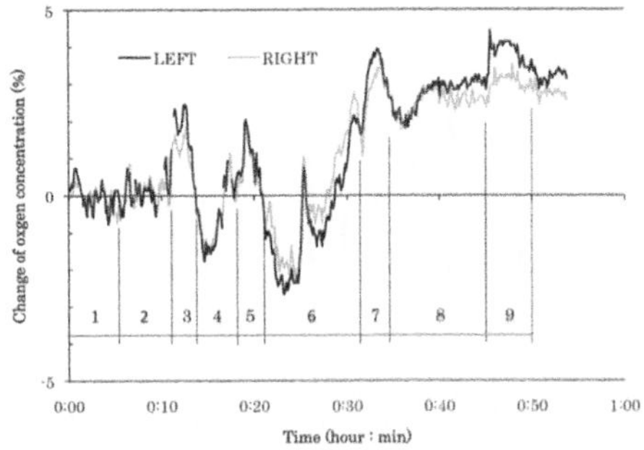


Figure 6. The change of oxygen concentration in the blood at frontal cortex (subject No. 3).

difference between selective reaction time and simple reaction time was shortened by 20 min exercise. The change of oxygen concentration shows great sensitivity to every event such as physical exercise and reaction tasks. Peaks at the event are larger and clearer than subject No. 5. Increasing oxygen concentration during reaction tasks means brain activation by task execution.

Average of $SpO_2$ in exercise are 96.1 ± 0.6% (No. 5) and 97.7 ± 0.6% (No. 3) in the first exercise, 96.0 ± 4.3% (No. 5) and 97.4 ± 0.6% (No. 3) in the second exercise. Subject No. 3 indicated a larger and more stable ability to supply oxygen. This might relate to a larger oxygen concentration change in the subject.

The number placed on the x-axis in the graph means the time region of each procedure (1: pre-rest, 4: warming-up, 6 and 8: physical exercise, 3, 5, 7 and 9: simple and selective reaction task).

## 4 CONCLUSION

Physical exercise shows many effects on cognitive function. Each process, cognition and decision-making, reaction shows a different effect due to physical exercise. It seems that the change of blood flow at frontal cortex by physical exercise affects the cognitive function.

## REFERENCES

Brisswalter, J., Collardeau, M. & Arcelin, R. 2002. Effects of acute physical exercise characteristics on cognitive performance. *Sports Med* 32: 555–566.

Dalsgaard, M.K. 2006. Fuelling cerebral activity in exercising man. *J Cereb Blood Flow Metab* 26: 731–750.

Etnier, J.L., Salazar, W., Landers, D.M., Petruzzello, S.J., Han, M. & Nowell, P. 1997. The influence of physical fitness and exercise upon cognitive functioning: a meta-analysis. *J Sport Exerc Psychol* 19: 249–277.

McMorris, T. & Graydon, J. 2000. The effect of incremental exercise on cognitive performance. Int *J Sport Psychol* 31: 66–81.

McMorris, T., Collard, K., Corbett, J., Dicks, M. & Swain, J.P. 2008. A test of the catecholamines hypothesis for an acute exercise–cognition interaction. *Pharmacol Biochem Behav* 89: 106–115.

McMorris, T., Davranche, K., Jones, G., Hall, B., Corbett, J. & Minter, C. 2009. Acute incremental exercise, performance of a central executive task, and sympathoadrenal system and hypothalamic-pituitary-adrenal axis activity. *Int J Psychophysiol* 73: 334–340.

Meeusen, R., Watson, P., Hasegawa, H., Roelands, B. & Piacentini, M.F. 2006. Central fatigue: the serotonin hypothesis and beyond. *Sports Med* 36: 881–909.

Nybo, L. & Secher, N.H. 2004. Cerebral perturbations provoked by prolonged exercise. *Prog Neurobiol* 72: 223–261.

Tomporowski, P.D. 2003. Effects of acute bouts of exercise on cognition. *Acta Psychol* 112: 297–324.

# *Part III: Usability and interface*

*Ergonomics in Asia – Shih & Liang (eds)*
*© 2012 Taylor & Francis Group, London, ISBN 978-0-415-68414-9*

# A proposal for GUI design index for service machine

Rie Kihara
*Wakayama University Graduate School of Systems Engineering, Wakayama, Japan*

Toshiki Yamaoka
*Wakayama University, Wakayama, Japan*

ABSTRACT:

*Objective*: In this study, a Graphical User Interface (GUI) design index for machines that provide service is proposed.
*Method*: 3 types of GUI for service were constructed based on 3 service types. The present machine's GUI was divided into three groups relating to function by cluster analysis and correspondence analysis. Design items were picked up from research about divided GUI groups by Formal Concept Analysis. The formal concept terms and UI design terms were classified into 3 type of GUI. The GUI design index was composed of design items and divided terms. The index was divided into 3 levels from the viewpoint of a category hierarchy. Finally, usefulness of GUI design index was examined by 3 researches.
*Results*: GUI design index for service machines is useful because it is understood by researchers.
*Conclusion*: The results show that the GUI index is good if the GUI design needs to reflect the service.

*Keywords*: graphical user interface, GUI design, service

## 1 INTRODUCTION

1. *Background*
GUI is used increasingly in service now. However, GUI design has not often been constructed for service.
2. *Objective*
This study reports on proposing GUI design index for service machines. Using this index enables GUI to be used in design for services.

## 2 3 GUI CLASSIFICATIONS FOR SERVICE

Previous studies have divided services into 3 groups. The following list shows these 3 groups.

1. Efficiency oriented group.
   This group's service is efficient, simple and easy.
2. Confidence oriented group.
   This group's service satisfies users' wants and gains users' confidence.
3. Entertainment oriented group.
   This group's service gives users a sense of fulfillment and special experiences.

The followings are things to need for fitting services of these 3 groups.

1. Efficiency oriented GUI.
   This GUI provides service efficiently, simply and easily.

2. Confidence oriented GUI.
   This GUI provides confidence and users' wants.
3. Entertainment oriented group.
   This GUI fits the place of service and gives users special experiences.

These 3 classifications are defined as '3 GUI classifications for service'.

## 3 EXTRACTING CHARACTERISTICS OF GUI

### 3.1 *Method*

GUI collected for service machines was divided into 3 groups from a functional point of view. Functions needed for the 3 classifications were abstracted. Table 1 shows these functions.

A matrix table was composed of function and GUI. If the GUI had these functions, the corresponding cell was checked 1. Conversely, if the GUI did not have these functions, the corresponding cell was checked 0. The data were analyzed by correspondence analysis and cluster analysis.

Design characteristics were abstracted from every cluster. These clusters were divided by correspondence analysis and cluster analysis. Characteristics were obtained for every cluster by items of design and layout. Table 3 shows the items of design and layout.

Table 1. Functions needed by 3 classifications.

| Type | Functions |
|---|---|
| Efficiency | No inactivity, Short cut, Accurate words, Usability |
| Confidence | Unification, Validity of behavior, Feedback, Report of moving |
| Entertainment | Surprising element, Feeling of seeing beauty, Feeling of unreality |

Table 2. Part of matrix table (function & sample).

| | Efficiency | | | |
|---|---|---|---|---|
| Sample | No inactivity | Short cut | Accurate words | Usability |
| ATM1 | 1 | 1 | 1 | 1 |
| ATM2 | 1 | 1 | 1 | 1 |
| ATM3 | 1 | 1 | 1 | 1 |
| ATM4 | 1 | 1 | 1 | 1 |
| Copy1 | 1 | 1 | 0 | 1 |
| Loppi | 0 | 1 | 0 | 0 |
| Ticket 1 | 1 | 1 | 1 | 1 |
| Ticket 2 | 1 | 1 | 0 | 1 |
| Book search 1 | 0 | 0 | 0 | 0 |
| Book search 2 | 0 | 1 | 0 | 0 |

Table 3. Items of design and layout.

| Type | Items of design and layout |
|---|---|
| Configuration | One page, Multiple page |
| Main color | Warm color, Cool color, Achromatic color |
| Color scheme | Analogy, Intermediate, Contrast |
| Image | Picture, Illustration, Not image |
| Character | Used, Not used |
| Line | Straight, Curve |

Table 4. Part of matrix table (Confidence type).

| | Configuration | | Main color | | |
|---|---|---|---|---|---|
| | One page | Multiple page | Warm color | Cool color | Achromatic color |
| Self register | 0 | 1 | 0 | 1 | 0 |
| Copy 2 | 0 | 1 | 0 | 0 | 1 |
| Copy 3 | 1 | 0 | 0 | 0 | 1 |
| Book search 2 | 0 | 1 | 0 | 0 | 1 |

A matrix table composed of design and layout items is shown in Table 3. The sample GUI was made from those clusters. If the GUI had these items of design and layout, the corresponding cell was checked 1. Conversely, if the GUI did not have these items of design and layout, the corresponding cell was checked 0. The data were analyzed by Formal Concept Analysis.

### 3.2 *Results*

Extracted characteristics are as below.

## 4 EXTRACTING CONCEPT

A clear concept is needed for constructing GUI. Therefore, design concepts and images are extracted by formal concept terms. Table 6 shows the design concepts and images.

Following the main requests and sub requests defined by previous studies, the 3 groups' functions and these design concepts, design images were arranged as the following items.

1. Efficiency oriented GUI.
   This GUI needs simple and basic designs.
2. Confidence oriented GUI.
   This GUI needs natural, fresh and unified designs and pages for every task.
3. Entertainment oriented GUI.
   This GUI needs novel and original designs.
   These have to be referred to in design.

## 5 GUI DESIGN INDEX FOR SERVICE MACHINES

UI design items were divided into 3 classifications. In addition to these were added missing items from Table 5 and Table 6. The following items were collected from these items and divided into 3 levels.

A category hierarchy was identified when this index was divided into 3 severities.

In Figure 1, the basic level is equal to the Level 2. The lower level is equal to the Level 1. The higher level is equal to the Level 3.

A GUI design is made easy by adding these levels when the GUI has many classifications.

## 6 DESIGN OF RESEARCH UTILIZING THE GUI INDEX

### 6.1 *Method*

Designs made using the GUI index were examined using the following data. The designs are seen to be better than designs which were not made with GUI index in their service.

Therefore, 3 researches were done.

Table 5. Characteristics of 3 clusters.

| Cluster | Items |
|---|---|
| Efficiency cluster | One page, Straight line, Analogy, No character, Picture and Illustration |
| Confidence cluster | Multiple page, Analogy, Cool color or Achromatic color<br>No character, Illustration only, Percentage of straight lines and curves are the same |
| Entertainment cluster | All multiple page, Many illustrations and some pictures, Contrast, Character |

Table 6. Extracted design concepts.

| Type | Design concept |
|---|---|
| Efficiency | Simple, Basic, Functional, Linear |
| Confidence | Fresh, Clean, Natural, Refreshing, Stability, No discomfort, Unity, Familiarity<br>Curvilinear, Timeless |
| Entertainment | Fun, Classy, Chic, Unique, Symbolic, Fantastic, Novelty |

Table 7. GUI design index for service machines.

| Type | Level | Items |
|---|---|---|
| Efficiency | 3 | Identity, Appropriate terminology & message, Universal Design, Mental model, Affordance |
| | 2 | Efficient access to information, Simplicity, Mapping, Emphasis, Metaphor |
| | 1 | Support of proficiency, Search ease, List, Efficient operation |
| Confidence | 3 | Cross-cultural, Confidence, Consistency |
| | 2 | Multifaceted provided information, Reduce the burden of memory, Operating principle, Feedback |
| | 1 | Operation feeling, Tolerance & flexibility, Help, Ensuring user's independence, Analogy, Natural color |
| Entertainment | 3 | Eye catching, Illustration |
| | 2 | Fun, Sense of accomplishment, Character, Feeling of service image |
| | 1 | Feeling, Novelty |

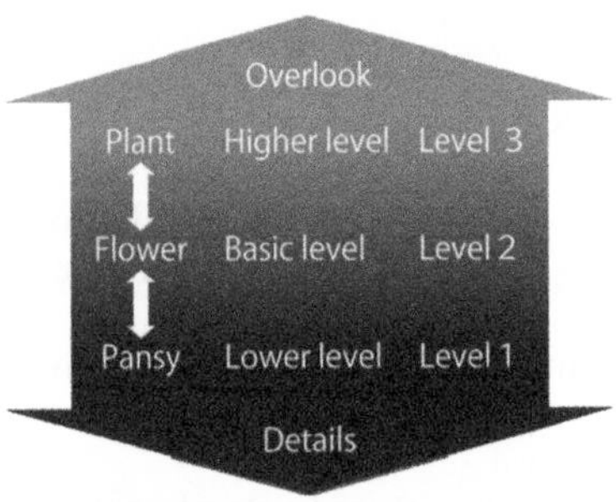


Figure 1. Category hierarchy and severities.

1. Research into design images.
   Participants were asked about images which samples have in the SD method. These data were analyzed using principal component analysis.
2. Research into using and seeing images.
   Participants were asked about using and seeing images which samples have in the Likert scale of 9 gradations. These data were researched in an allowable range measure method.

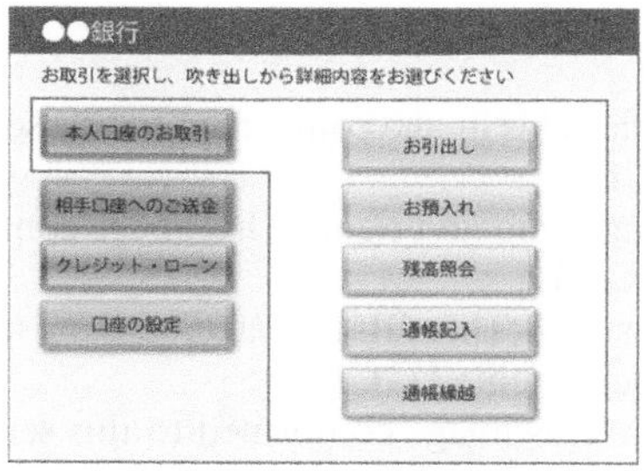


Figure 2. A sample of efficiency and confidence oriented GUI.

Table 8. GUI design index used in Figure 2.

| Type | Severity | Items |
|---|---|---|
| Efficiency | 3 | Identity, Appropriate terminology & message, Universal Design, Mental model, Affordance |
| | 2 | Efficient access to information, Simplicity, Mapping, Emphasis, Metaphor |
| | 1 | Search Ease, List, Efficient operation |
| Confidence | 3 | Confidence, Consistency |
| | 2 | Multifaceted provided Information, Reduce the burden of memory, Operating principle |
| | 1 | Analogy, Natural color |
| Entertainment | 3 | Eye catching |

3. Research of factor giving design images.
   Participants were asked the following items about every sample.
      A sample is 'B' since 'A'.
      The 'B' means a design image.
      The 'A' means a cause
   Participants wrote 'A' and 'B' every sample. These data were arranged under the DEMATEL method.

Figure 2 shows an example of design sample used in the GUI index.

This GUI is designed with the following GUI index. Table 8 shows the index.

Used items are 13 items from 14 Efficiency items, 6 items from 12 Confidence items and 1 item from 8 Confidence items.

### 6.2 *Result*

The results are as follows.

1. Each GUI's image is given in the designs
   1. Efficiency oriented GUI.
      Basic, Simple and Easy, Comfortable, Familiar.
   2. Confidence oriented GUI.
      Clean, Firm, Trustable, Basic, Reassuring.
   3. Entertainment oriented GUI.
      Affinity, Comfortable, Impressive, Simple and Easy, Fast.
   4. Efficiency and Confidence oriented GUI.
      Trustable, Familiar, Positive, Certain.
   5. Not used GUI 1 (Service: Entertainment).
      Far, Concise, Ordinary, Familiar.
   6. Not used GUI 2 (Service: Entertainment).
      Concise, Steady, Ordinary, Slow.

2. Each GUI's using and seeing images.
   1. Efficiency oriented GUI.
      This operation was simple. This image conformed to this service.
   2. Confidence oriented GUI.
      This image was reassuring, calm and clean. This image conformed to this service.
   3. Entertainment oriented GUI.
      This image was fun and active. This GUI created an atmosphere. This has impact.
   4. Efficiency and Confidence-oriented GUI.
      This procedure was easy. This image conformed to this service. Users can achieve the goals quickly.
   5. Not used GUI 1 (Service: Entertainment).
      This image wasn't fun. This didn't have impact.
   6. Not used GUI 2 (Service: Entertainment).
      This image wasn't characteristic.
3. Factor giving design images.
   1. Efficiency oriented GUI.
      Black of back color was fashionable and strong. Many colorful buttons were fashionable and had gaiety.
   2. Confidence oriented GUI.
      Color scheme of green and brown gave image of nature and calm.
   3. Entertainment oriented GUI.
      Black and pink had the image of adult female.
   4. Efficiency and Confidence oriented GUI.
      A fresh, calm and serious image was felt because blue colors were used.
   5. Not used GUI 1 (Service: Entertainment).
      A fresh, boring image was felt and it was hard to see because blue and white were used.
   6. Not used GUI 2 (Service: Entertainment).
      Black of back color was hard and repulsive. Many similar items were confusing.

### 6.3 *Discussion*

It was found that design using the index was better than design not using the index from the results. In other words, efficiency oriented GUI is thought to be simple by users; confidence oriented GUI is thought to be reassuring and calm; in the entertainment oriented GUI it is thought that service image is conformed. GUI didn't give an image conformed to the service if the index wasn't used. Therefore, this index helps in the design of GUI conformed to its service.

## 7 DISCUSSION

GUI design index for service machines is useful.

This was understood from the research. This shows that GUI matching up to service can be designed if you know the service and use this GUI index.

At first, you should know machine's service type if you design a service machine's GUI. After acquiring knowledge of the service, using this index will make your GUI fit the service.

## REFERENCES


Mayuko Yoshida, Toshiki Yamaoka, (2007). A proposal of screen design rule for service concepts, *paper book for 2007 Annual Meeting of Japan Ergonomics Society Kansai Branch*, pp. 85–88.

Seisaku Kawakami, *An Introduction to Cognitive Linguistics*, pp. 32–35, Kenkyusha Press, 1998.

Toshiki Yamaoka, *Form Engineering for Design*, pp. 136–144, Kogyo Chosakai Publishing, 2010.

Toshiki Yamaoka, *Lecture on Ergonomics for hardware and software design*, pp. 277–289, Musashino Art University Press, 2003.

*Ergonomics in Asia – Shih & Liang (eds)*
*© 2012 Taylor & Francis Group, London, ISBN 978-0-415-68414-9*

# A study on construction of a community tool with a social network service for uniting older and younger people

Masahiro Shibuya
*Tokyo Metropolitan University, Tokyo, Japan*

Akiko Yamashita
*Kanagawa University, Yokohama, Japan*

Fumitaro Goto
*Kitami Institute of Technology, Kitami, Japan*

Koki Mikami
*Hokkaido Institute of Technology, Sapporo, Japan*

ABSTRACT: To find out how much social network service (SNS) information older people had, a preparatory investigation was conducted using a questionnaire. The subjects were 1 man and 51 women going to a dance studio in Yokohama City. The results showed that although they had some interest in SNS, they did not have much knowledge of the service. Therefore we conducted another investigation into needs for general SNS functions using a questionnaire. The subjects were dance trainees and staff of a junior high school and a high school, totaling approximately 100 persons. Furthermore we asked 229 students of a junior high school and a high school the same questions.

The analysis of the results revealed that younger people's needs are clearly different from older people's. We found out that to meet the older people's needs it is necessary to communicate using ones' real names like world-famous Facebook, not Japanese mixi type services, and expand the functions according to each person's needs. In this paper are mentioned what kind of community older people (a typical digital divide example) want, whether or not SNS can be a community tool for uniting older and younger people, and also show how an SNS site should be constructed.

*Keywords*: SNS, elderly people, mixi, Facebook, dance

## 1 INTRODUCTION

Young people's basic abilities as full-fledged members of society (basic knowledge for the real world) have been lowering recently. The authors have been studying educational methods and curriculums necessary to improve young people's basic knowledge for the real world (Shibuya M. et al., 2010). Because we think that the real problems which society or companies have give students more motivation than the ones made purposely for learning, we made a practical teaching manual for the management from planning to operation of an open dance course held by a nonprofit foundation.

In this study SNS was intended for an information-sharing learning material among the trainees and the students managing the studio. We thought it very effective to use IT accessible to any student anytime because not all the staff members belonging to different colleges, departments or years could meet frequently.

Construction of an SNS site for elderly people makes it possible to send information to every generation, accumulate and circulate it, which will promote intellectual exchange

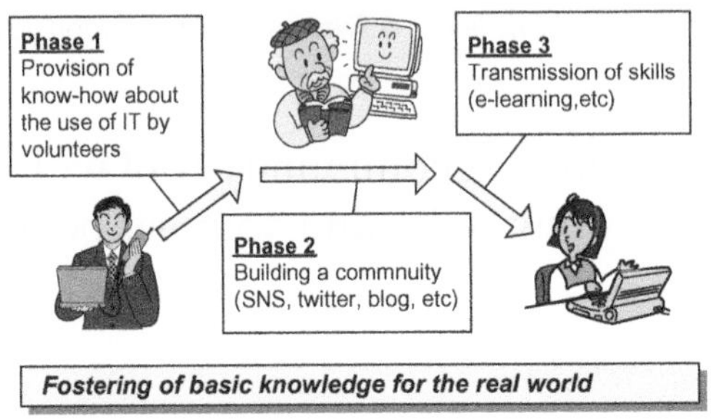


Figure 1. The concept.

among circles or citizens. This will lead to construction of an e-learning system utilizing elderly people and succession of their skills. It is also expected that communicating with other generations with no everyday contact will foster young people's basic knowledge of the real world.

Figure 1 shows the concept of this research. In Phase 1 is investigated how to teach older people the know-how younger people have about the use of IT. Young people greatly concern themselves in the construction and operation of an SNS site and deepen the exchange between generations.

In Phase 2, older people form a community by use of SNS etc. In this phase support systems are investigated to encourage older people to use IT. In Phase 3 older people who have acquired IT skills are supported so that they can open their knowledge to the public. If other people can use the knowledge with e-learning, their skills can be transmitted smoothly. In this paper is described how to construct a social network service site used in all the phases.

## 2 THE METHODS AND THE BACKGROUND

### 2.1 *The background*

With information communication technology (ICT) developing, new types of communication tools have been born recently. Younger Japanese people tend to set more importance on the mailing function than the phoning function as the main function of the cell phone, and the appearance of "Twitter" has increased the importance of literal communication on the network. SNS includes world-famous Facebook and mixi of Japanese make. According to Wikipedia, the number of the total Facebook members is 350,000,000 as of January 2010 and that of mixi is 30,120,000 as of July 2010. Both numbers are increasing now. However, these types of ICT are not accepted by every generation. As of September 2008, the population of elderly people was 28,190,000, which was 22.1% of the total population. This speed of social aging indicates that about one out of four persons will be 65 years old or older in six years from now.

Moreover, there have been more and more nuclear families in Japan and those households have reached about 60% of the total households. Daiwa Institute of Research Holdings Ltd. says that in the USA 39% of the people of 55 years old or older use SNS, while in Japan the percentage is only 4%. The four percent seems to include not a few people who have registered and do not use the SNS, not knowing what it is like.

Recently SNS sites for some particular areas (Local SNS) have been created by local self-governing bodies or nonprofit organizations to bring the people living there together and to send information peculiar to their areas. There are many studies in which activating plans of local areas were modeled and examined using the regional social networking service. For example, Toriumi et al. proposed an SNS communication model for the purpose of clarifying the conditions to make the best of a small-scale SNS (Toriumi F. et al., 2010). However, the conditions to make the best of the regional social networking service have not been verified.

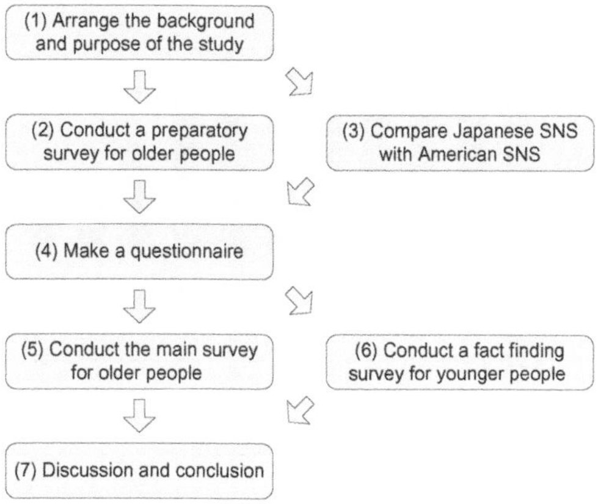


Figure 2. The procedure of the study.

In Japan Yashiro City in Kumamoto Prefecture became its pioneer in 2003. The effect of the Ministry of Internal Affairs and Communications' demonstrative experiment helped the establishment of many local social network services throughout Japan, but not a few of them have been closed or left as they are with little renewal because of the problems of maintenance and continuation. There are many papers and reports on SNS, but some sites in use are not properly linked or have not been renewed for more then one year, and they are not worth reference.

In Japan two of the three major companies running SNS for commercial use focus on games. There is a background that in SNS the intensive management of personally identifiable information compels the service provider to take an exclusive position in respect of information access, which leads to few new attractive functions. In spite of the above, new ICT to replace SNS has not been put to practical use so far, and we have no choice but to use SNS as a communication tool.

However high-tech cell phones or e-books have been introduced one after another, which is making the network familiar to all of us. SNS will probably develop to be easy to use for everyone. The purposes of this study are to find out what kind of community elderly people, a typical digital divide example, want, whether or not SNS can be a community tool, and whether or not creation of a human network, that is, the goal of SNS is possible by making it easy to use.

### 2.2 *Methods*

In Japan SNS is widespread chiefly among younger people, and the main users are in their teens, twenties and thirties. On the other hand, overseas SNS is used by many different generations, young and old. In order to find out how this happened, a questionnaire was conducted on younger and older people, and the needs for SNS were analyzed. Since Facebook has not spread into Japan, the seeds of domestic and foreign SNS were analyzed. Because Japanese older people might have a negative feeling toward using the Internet terminals or no interest in it, a preparatory survey of subconsciousness was conducted for the next questionnaire. The procedure of the study is shown in Fig. 2.

## 3 RESULT

### 3.1 *The preparatory survey*

According to a survey by the Ministry of Internal Affairs and Communications, in the personal use of the Internet by different generations, the use of the Internet by people of 60 years

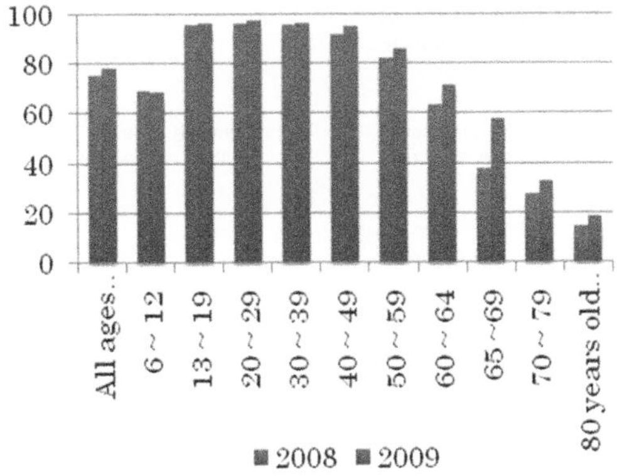


Figure 3. Use of the Internet by generations.

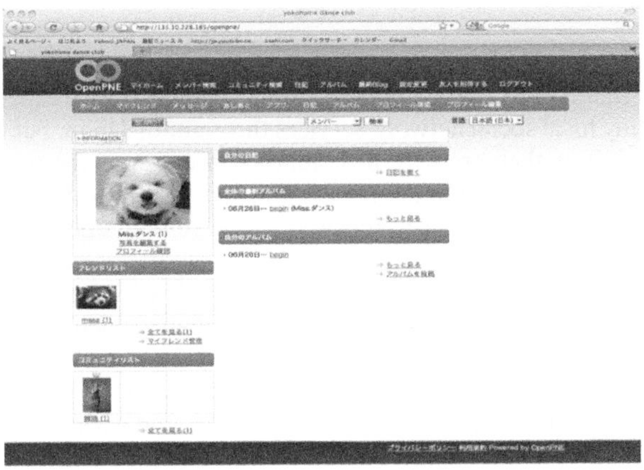


Figure 4. A usage example of the dance SNS.

old or over has grown, and the rate for the people of 65 to 69 years old is 58%, which shows a sharp increase. In the use of the personal computer, more than 80% of people in their teens, twenties, thirties and forties use personal computers, while about 40% of people in their late sixties use them. There is a generation gap, but the growths are significant. Figure 3 shows the use of the Internet by generations.

In order to find out how much knowledge of SNS older people had and whether or not they were interested in SNS, a preparatory survey was conducted. The method was a questionnaire survey. The subjects were one man and 51 women going to a dance studio in Yokohama, Japan.

In this survey how to use the Internet was investigated. The rate of Internet use of the subjects was 40%, and 5 subjects used personal computers, 12 subjects cell phones and 4 subjects used both. The rate of the subjects who knew about SNS was 42%.

The questionnaire consisted of 12 questions. The results show that about 60% of the people wanted to use it as a learning system and that less than 20% wanted functions to create means or a meeting place for promoting communication between friends or acquaintances. It is thought that this is because older people are weak in the activity of dispatching information themselves compared with younger people. This made it clear that for older people the main SNS functions such as "profile" or "diary" should be low-keyed. Considering the results of the preparatory investigation, an SNS site for the dance studio was built using an open source SNS, but no good evaluation was obtained (Figure 4).

### 3.2 *Comparison of SNS between Japan and the USA*

The results of the preparatory survey showed that older people did not consider SNS unnecessary, so we decided to find out why SNS was not widespread from the viewpoint of the mechanism of SNS. A comparison was made between worldwide Facebook with the largest number of members and Japan's most famous SNS, "mixi". The results are shown in Table 1. In Japan world-famous Facebook is inferior in the market share. It is thought that this is caused by the fact that many Japanese people are afraid of information leak and have the hard habit of exchanging information anonymously on the net. Some experts doubt whether in Japan Facebook will exceed "mixi" in members. Others do not.

There is a great difference in the use mode between "mixi" and Facebook. There used to be no custom of inputting letters on the keyboard in Japan. Not a few Japanese are poor at keyboard-inputting, and there is a cell phone service called "i-mode". It is one of NTT docomo's cell phone services which enables the user to send or receive e-mail and to read Web pages on the Internet and so on. With "i-mode" widespread in Japan, there is much more e-mail communication by cell phone than in other countries, and every generation uses the cell phone to access the Internet. However, older people are not good at fingertip movement, and do not like letter-inputting using "i-mode". NTT once developed "L-mode" to replace "i-mode", but it did not become popular. These use modes are thought to be the reason why many "mixi" users are younger people.

### 3.3 *Execution of a questionnaire*

The subjects for older people were 75 men and women including the people in the preparatory investigation, and 22 male and female teachers belonging to a private junior high school and a senior high school in Tokyo. The subjects for younger people were 229 students of 9th-graders to 12th-graders going to the same schools. Two kinds of questionnaires were made, and the survey was conducted in November and December, 2010.

### 3.4 *The contents and results of the questionnaire*

In this passage are described question items to which respondents gave characteristic answers. Considering the fact that many older people do not know about SNS, the questionnaire contents for older people were devised so that they could answer the questions in the alternative judgment, such as "Isn't it convenient for you to be able to do such-and-such using the Internet SNS?"

Table 1. Comparison of the social network services representing Japan and the USA.

| | Mixi | Facebook |
|---|---|---|
| Number of members | 30,120,000 (2010/07) | 350,000,000 (2010/01) |
| Registration | Anonymous | Not anonymous |
| Users | Younger generation | Many different generations |
| Use mode | Many accesses by cell phone | Many accesses by personal computer |
| Functions | Abundant | Simple |
| Contents | Plenty of amusements like games and fortunetelling | Abundant functions such as communication tools |
| Communication | Suited for communing within company, with old friends, current classmates or colleagues | Suited for communing with foreign friends, or school-and-business-related people |
| Disclosure | A large choice of information items in publicity or nonpublicity | A small choice of information items in publicity or nonpublicity |
| Applications | Provided by "mixi" | Provided by third parties |

To find out what older and younger people want using the Internet, the question, "which of the following five items do you have interest in or think convenient?" was asked. The results are shown in Figure 5.

A. Can get important information anytime
B. Can make contact with friends easily
C. Can get emergency news like an earthquake promptly
D. Can keep a diary or read friends' diaries
E. Can share favorite information with friends.

To find out what older people thought about sending information anonymously on the Internet, they were asked the question, "Do you feel a sense of incompatibility toward communing with other people under cover of a false name?". As shown in Table 2, more than half of them answered in the affirmative. There was a difference between the dance group and the group of teachers.

Younger people were asked which they entered their SNS, anonymously or under their real names. There were 90 persons using their real names, and 95 persons using false names. And also the question, "What are you anxious about in communing on the Internet?" was asked. The results are shown in Table 3.

To find how younger people used SNS, the purposes of their SNS use were asked. One hundred and twenty-nine of the subjects used SNS, and they were asked what contents of the SNS they used (Figure 6).

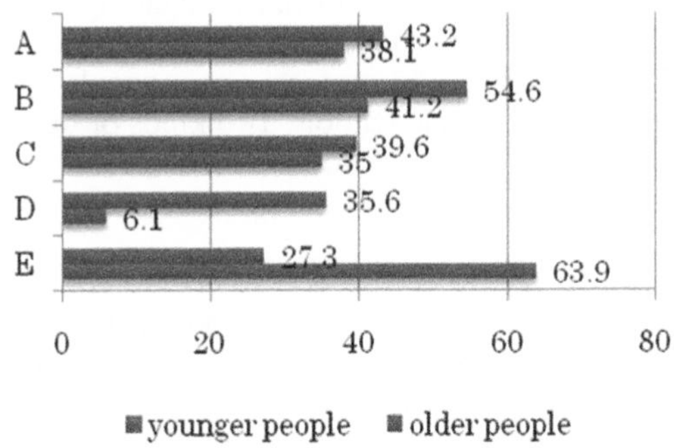


Figure 5. The needs of younger and older people for the internet (for key see text).

Table 2. The sense of incompatibility toward anonymous communication (older people).

| | Dance group | Teacher group |
|---|---|---|
| Yes | 57% | 82% |
| No | 43% | 18% |

Table 3. Anxieties about communing on the Internet (younger people).

| Anxieties | Number of respondents |
|---|---|
| Can't see the other person | 85 |
| Can't hear the other person | 34 |
| To be watched by many and unspecified persons | 67 |
| Misuse of personal information | 166 |
| Lies or deceptions | 151 |

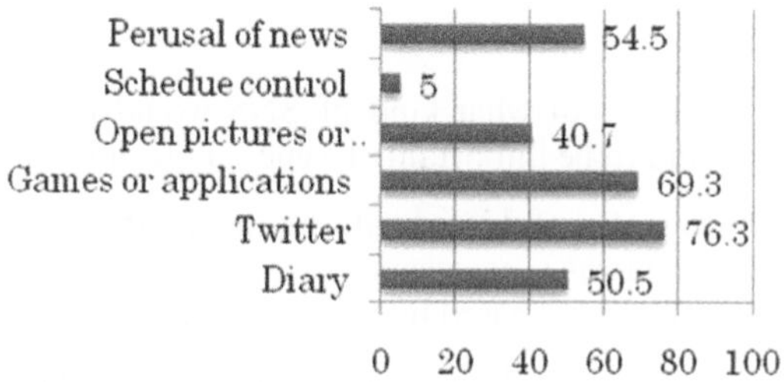


Figure 6. Purposes of SNS use (young people).

Lastly, both generations were asked to write down what SNS functions they wanted. Many of the older people answered, "Learning system and sharing of useful information", while many of the younger people answered, "Handy means for communication and games".

## 4 DISCUSSION

Based on the results of both the questionnaire and Japan's statistics, the difference between younger people's needs and older people's, difference between Japan's SNS seeds and America's, and matching of the needs and the seeds were examined, and then an ideal social network service was sought out.

Difference of the SNS needs between younger generation and older generation clearly appears in the items D and E of Figure 4. One generation wants one thing, and the other wants the opposite. Many younger people want to keep a diary or read other people's diaries, while older people hardly want it. The latter dislike to send information, disclosing themselves, and many of them want functions for learning systems and sharing of useful information. These two are thought to be reflected in the items D and E. On the other hand, younger people want SNS to provide means for self-presentation or communication.

Secondly, the differences of Japan's SNS seeds and American ones were examined. In Japanese SNS both the real-name system and the anonymity one exist at the same time, and this is a good way for younger people who want to commune anonymously. In the USA, SNS communication is considered to be formed by bringing real human relationships onto the network, which produces contents. The users disclose their names, addresses or social positions, which causes social responsibility in use, with the result that serious troubles can be avoided. This way of thinking is connected with a small choice of disclosure. The authors think it is a sign that each person takes responsibility for sending information. In Japan, people tend to use "twitter" for sending information, and want SNS to give amusements. Therefore, many people join SNS anonymously, but it is thought that older people who are reluctant to commune anonymously find difficulty in accepting that.

Lastly, both the generation differences of SNS needs and the differences between American SNS seeds and Japanese ones were jointly examined. Judging synthetically from the diffences of SNS functions and contents which younger and older people want, it is thought that Japanese SNS suits younger people and that American SNS suits older people. However, statistics by the Japanese government show that very few older people use personal computers, and that many older people use cell phones. Encouraging them to use Facebook does not lead to a solution. In the answers to the descriptive questions, there were several complaints about how to operate it. To encourage older people with no habit of inputting on the keyboard to use SNS, the use of touch-panel type terminals, which have been popular in recent years, is thought to be good. However, instead of using the touch-panel type terminal as it is, it is necessary to devise the interface for the benefit of older people who are weak in fine finger movement and have weakening eyesight. In the West, research in construction of SNS sites for older people is in progress, and I would like to use that as reference data (Gibson L. et al., 2010; Cornejo R., 2010).

## 5 CONCLUSIONS

Research was conducted to find out what kind of SNS would be good for older people. The following three were found to be important. To encourage older people to join SNS:

- a mechanism for the sharing of useful information is necessary.
- participation by use of real names is necessary.
- use of information terminals with devised interfaces is necessary.

Construction of an SNS site for elderly people makes it possible to send information to every generation, accumulate and circulate it, which will promote intellectual exchange among circles or citizens. This will lead to construction of an e-learning system utilizing elderly people and succession of their skills. It is also expected that holding communication with other generations with no everyday contact will foster younger people's basic knowledge of the real world. We think it most important to construct an SNS site making it possible to connect the elderly to the young and create a win-win relation between both generations.

## REFERENCES


Cornejo, R. (2010). Integrating Older Adults into Social Networking Sites through Ambient Intelligence. *Proceedings of the 16th ACM International Conference on Supporting Group Work (GROUP 2010),* Sanibel Island, Florida, USA.

Gibson, L., Moncur, W., Forbes, P., Arnott, J., Martin, C. & Bhachu, A. (2010). Designing Social Networking Sites for Older Adults. *Proceedings of BCS HCI 2010*, Dundee, UK BCS, September 6–10, 2010.

Shibuya, M., Yamashita, A., Goto, F. & Mikami, K. Construction of a SNS site for elderly people and its problems, *The 9th Pan-Pacific Conference on Ergonomics,* November 7–11, 2010.

Toriumi, F., Ishida, K. & Ishii, K. (2008). Proposal of Small-SNS Model and Activation Simulation. *IEICE Trans,* B, Vol. J91-B, No. 4, pp. 397–406.

*Ergonomics in Asia – Shih & Liang (eds)*
 *ISBN 978-0-415-68414-9*

# The effects of display elements on information retrieval in geographical intelligence systems

Feng-Yi Tseng, Chin-Jung Chao & Wen-Yang Feng
*Chung-Shan Institute of Science and Technology, Tao-Yuan, Taiwan*

Sheue-Ling Hwang
*Department of Industrial Engineering and Engineering Management, National Tsing Hua University, Hsinchu, Taiwan*

ABSTRACT: Visual search tasks carried out in a state of alertness with multiple sources might be affected by display elements. This study investigates the effect of screen dimension and icon size, and aims to contribute towards task performance for the use of the interface in the geographical intelligence systems. Thirty-six participants carried out a visual search task using a mock geographical intelligence system. The results showed that screen dimension and icon size had a significant effect on task performance and cognition load. The findings of this study can help designers to develop more efficient and comfortable interfaces to fit user needs, and design recommendations are given.

*Keywords*: visual search, screen dimension, icon size, galvanic skin response, completion time, accuracy

## 1 INTRODUCTION

When people perform a visual search task on the alert with multiple sources, an integrated visual display terminal (VDT) is the most effective tool for human-computer interaction (HCI). In modern systems, many people have shifted their role from manual labor to a supervisory control. They monitor complex interfaces through the computer screen (Liu & Su, 2006). In order to acquire useful information, humans who use computers at work must inevitably be exposed to occupational stressors. Research has identified the use of computers at work as a contribution factor for occupational stress (Smith et al., 1999). Hence, the HCI design must have the potential to minimize human cognitive load, especially when they are performing complex tasks.

The typical graphical user interface (GUI) is often better for reducing cognitive load than traditional hardware equipment. GUI can overcome position restriction in space and present meaning in an outline form. Icons are used in a wide variety of ways in GUIs, one of the more common of which is the representation of objects recognized by the operating system (Byrne, 1993). One application of GUI that is growing rapidly is performing geographical intelligence. Geographical intelligence systems, which can be used in battlefield management, sonar detection tasks and the main control room of a control tower, is a set of data which relates to natural resources, water resources, vehicle routing, travel problems and traffic planning and so on. Computerized systems integrate the above-mentioned information from various sources. Operators have to carry out geographical data analysis, assessment, and decision-making in a very short time. However, military or civilian forces often encounter great difficulties of orientation in unknown environments in geographic intelligence (Pfendler & Schlick, 2007). This makes the interface of the geographical intelligence system to be one of the most important factors for reducing the human's cognitive load and preventing safety-related accidents.

Visual search tasks involve detecting relatively few signals that occur at unpredictable times. The operators usually may lose vigilance when there is less external stimulus to keep them alert. Psycho-physiological measures were widely used to understand the cognitive load of humans in response to an external stimulus. In the past, physiologists used these measures of cognitive load to evaluate human emotional response and behavioral stimulus (Min et al., 2003, Dickson & McginnieS, 1966). Experimenters rely on these measures through the internal states of people, so as to help understand the actual cognitive load of subjects. In the modern human-computer interfaces, cognitive load has gradually become an important factor in people's stress and arousal levels, especially under high intensity conditions and complex tasks of work (Shi, Choi, Ruiz, Chen & Taib, 2007; Liang, Lin, Hwang, Huang, Yenn & Hsu, 2009). Actually, cognitive load of all kinds may be described as perceived level of effort associated with reasoning, learning and thinking (Paas et al., 2003). For this reason, cognitive load can be used as an indicator of users' work efficiency.

Cognitive load of behaviors involves changes in peripheral autonomic activity. The measures of physiological responses such as heart rate variability (HRV), blood pressure (BP), electroencephalogram (EEG) and so on, can help experimenters to obtain information on human cognitive load. However, these measures often left subjects feeling uneasy, because of the complex wires which were attached to the experimental apparatus obstructed subject's actions, and the experimental calibration always consumed much time. A psycho-physiological measure called galvanic skin response (GSR) is a measure of conductivity of human skin and is often used as a measure of cognitive effort (Critchley et al., 2000). GSR measures changes in electrical resistance across two regions of skin. The simple way to measure GSR was with two wires connect to the forefinger and middle finger of one hand. Also, the calibration of GSR is very fast. For the duration of the experiments, the electrodermal activity is easily detected with a DC amplifier and experimenters can also monitor the response of subjects in real time.

GSR is a psycho-physiological phenomenon which consists of an increase or a decrease in skin resistance. GSR indicates the activity of sweat glands, which reflects the human action of the sympathetic nervous system. GSR is regarded as an indicator revealing the arousal level accurately. There are also some researchers who have indicated that GSR represents perception and attention. Nevertheless, GSR can detect the change in skin resistance which can be used as an index of response from the autonomic nervous systems. For this purpose, GSR can indicate a change of human potential response to an attention-getting or alerting stimulus.

The objective of this study is to examine the influence of geographical intelligence systems on several display parameters in the GSR of operators. A lab experiment was conducted to explore the human cognitive load on the visual process, and the result may provide design guidelines for battlefield management, sonar detection tasks and main control room of control tower designed to increase arousal level.

## 2 METHOD

### 2.1 *Participants*

Thirty-six volunteers were chosed (21 men and 15 women) with at least 3 years experience of using computers, aged between 22 and 33 years ($M = 26.7$ yr, $SD = 3.04$). They were paid for their participation in this study. All participants had a corrected visual acuity of 0.8 or better, as well as normal color vision. Each subject had to sign a letter of consent before the experiment. Participants adapted to the ambient illumination for 20 min prior to the experiment starting. The participants were instructed to avoid VDT work for at least 3 hours before the experiment, so as to prevent visual fatigue.

### 2.2 *Apparatus*

An Optec 2000 tester which can show six Ishihara plates was used to test the color vision of each participant. Another instrument, a Topcon-acp.8 vision tester, was used to test all participants' visual acuity. To investigate the responses of the autonomic nervous systems, a GSR test was conducted. A PowerLab/16 sp series system was used for measuring the responses of the autonomic nervous system. These physiological signals were measured for the GSR, one channel on the index finger and middle finger of the left hand (in units of microsiemens; μS). GSR data were analyzed by "Chart" software, which uses the averaged amplitude of signals at intervals of experiment, contrasted with the averaged amplitude of base line signals before the experiment. An Intel Pentium (R) M processor 1500 MHz notebook was connected to PowerLab for recording GSR data. The type specification of the stimuli presentation screen was a CHIMEI CMV T38A 21-inch color Liquid Crystal Display (LCD) monitor with a display resolution of 1280 × 800 pixels at a refresh rate of 60 Hz. The monitor was calibrated with a Minolta CS-100 chroma meter. Before the experiments began, the monitor was warmed up for 2 hours.

### 2.3 *Experimental design*

An experiment was designed to evaluate the effects of potential cognitive load indicators on visual search performance. This experiment employed a two-factor design. The independent variables were screen dimensions with 3 levels (7 inches, 15 inches and 21 inches) and icon sizes with 5 levels (40, 50, 60, 70 and 80 minutes of the visual angle with an observation distance of 50 cm). There were three dependent variables in this experiment. One of the dependent variables was GSR which was used as a measurement to validate the effect of the cognitive load levels on the participants. The other dependent variables were accuracy and completion time to estimate task performance of geographical intelligence system.

A third of the participants were randomly assigned to each screen dimension, which was a between-subject factor. Icon sizes were within-subject factors. All the participants were exposed to five icon sizes in the experiment. Each interface was completed with the same icon size, but with random icon arrangements. In short, each participant had to execute 80 tasks totally. After participants had finished the experiment, the dependent variables were analyzed using analysis of variance (ANOVA) for statistical testing.

### 2.4 *Experimental workplace conditions*

The experiment was implemented in a windowless laboratory. The wall of the laboratory sheltered participants from the greater part of the external light and sound sources. So there was almost no interference by external factors. A simulated geographical intelligence workstation was set up in the laboratory. The environmental illuminant was a diffuse light source from a fluorescent tube. The illumination of the laboratory simulated the conditions of geographical intelligence in battlefield management, sonar detection tasks and the main control room of a control tower. The ambient illumination of the laboratory was about 50 lux. The design of the light source was such that it would not result in glare or reflection illumination which would influence participants in this experiment.

### 2.5 *Stimuli*

Common symbology is the basis of geographical intelligence systems. In the real world, the chain of command system that combines Command, Control, Communications, Computers, Intelligence, Surveillance and Reconnaissance (C4ISR) is a key supervisory control apparatus. The C4ISR interfaces collect and integrate all kinds of information retrieved, and then translate that information into common and meaningful symbols for real-time feedback to all users in the geographical intelligence system (Ahlberg et al., 2007). In order to avoid misunderstanding symbols, the symbology of the geographical intelligence system must be unanimous, adequate and identifiable. The common warfighting symbology of "Military Standard 2525b," which represents tactical situations in war and other dangerous situations, was adopted for the icons used in the simulated geographical intelligence system of this experiment.

The experimental independent variable of screen dimension was set at 7, 15 and 21 inches which simulated the small, middle and large display devices. In this study, the stimuli presentation screen of CHIMEI CMV T38A 21-inch color LCD monitor was set by the test program at 7, 15 and 21 inches.

### 2.6 *Procedure*

Firstly, participants had to adjust the workstation suitable for their body. Then they took the visual and color vision tests. Before the experiment, the participants had adapted to the ambient illumination for 10 minutes. Simultaneously, the PowerLab/16 sp series system was calibrated for each participant. After the calibration, the average amplitude was measured for 5 minutes as the base line of each participant. Then each participant could practice the experimental interface operation for approximately 5 minutes, and they could request more practice if they wished.

In the beginning of the experiment, each trial was initiated by a random icon shown on the screen. During that period, participants needed to memorize the pattern of the icon. When the participants were ready, they pressed the "Space" button to enter into the test program of simulated geographical intelligence interface, which included the "Military Standard 2525b" icons, presented on the maps. Participants searched for the designated icon on the test interface. When they found targets, they pressed the "Enter" button and reported the icon location. When participants finished the experiment, the data were calculated by computer.

## 3 RESULTS

This study determined the relative importance of the various display factors that could affect the human alertness performance. The results were analyzed by ANOVA assessing effects of screen dimensions and icon sizes on alertness performance in terms of cognition load (GSR measured changes in electrical resistance). The ANOVA significance level was set at $p < 0.05$). The results of the experiments were listed below.

### 3.1 *Completion time*

The average completion time for each level of factors was shown in Figure 1. Completion time as dependent variable demonstrated a significant main effect on completion time due to screen dimension ($F_{2,239} = 25.57$, $p < 0.01$) and icon size ($F_{4,79} = 52.88$, $p < 0.01$). The interaction effect between screen dimension and icon size ($F_{8,239} = 2.09$, $p < 0.05$) was significant. The best level for a particular factor was the level that gives the lowest value of completion time. Based on the results of experiment, icon sizes (40–80 minutes) were negative related to completion time. In addition there was a positive correlation between screen dimension and completion time.

### 3.2 *Accuracy*

The average accuracy for each level of four factors is shown in Figure 2. With accuracy as the dependent variable, this showed a significant main effect on accuracy for screen dimensions

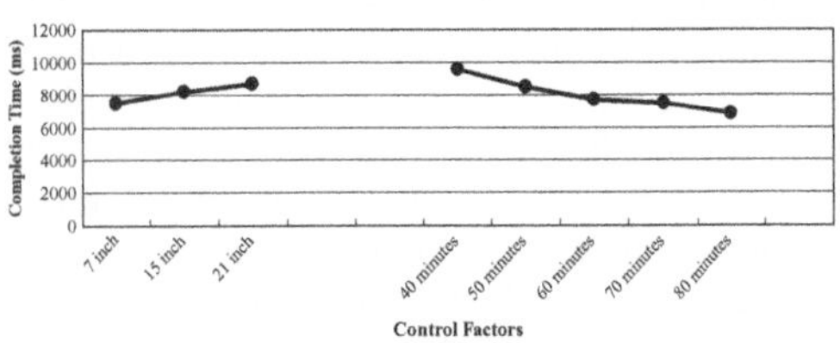


Figure 1. Plots of factor effects in terms of completion time.

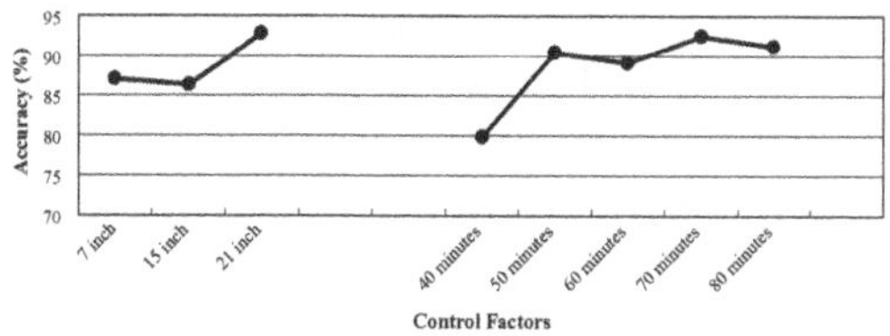


Figure 2. Plots of factor effects in terms of mean accuracy.

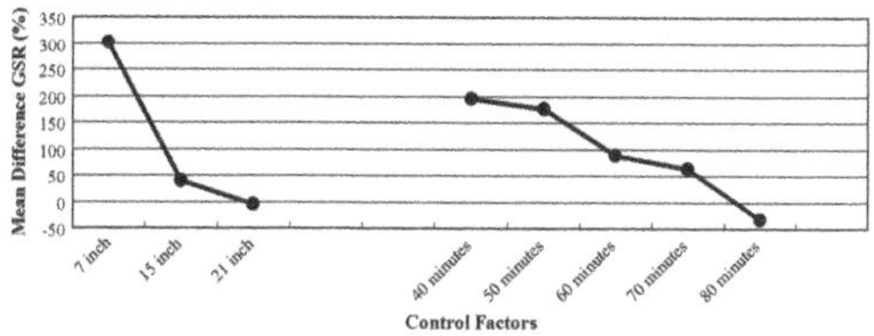


Figure 3. Plots of factor effects in terms of mean difference of GSR.

($F_{2,239} = 13.57$, $p < 0.01$) and icon size ($F_{4,79} = 15.11$, $p < 0.01$). The best level for a particular factor was the level that gives the highest value of accuracy. The experimental results indicated that the screen dimension of 21 inches yielded relatively better accuracy compared to 7 and 15 inches. In addition, icon sizes (40–80 minutes) were positively related to accuracy, with 40 minutes associated with lowest accuracy and 70 minutes associated with highest accuracy.

### 3.3 *Mean difference of GSR*

The physiological data of GSR across this study were analyzed. Because there was no "frame of reference" as baselines for people, the extremely individual essence of GSR data made it nearly impossible to directly compare GSR measurements across individual participants (Lin, Imamiya, Omata & Hu, 2006). In order to minimize the influence of individuals, the experiment adopted normalization of the GSR data by using the formula: (signal-baseline)/baseline. The normalization results of the mean difference in GSR for each independent factor are shown in Figure 3. There were significant differences in normalized GSR by an ANOVA analysis. The results showed significant main effects of screen dimensions ($F_{2,239} = 92.39$, $p < 0.01$) and icon size ($F_{4,79} = 16.27$, $p < 0.01$) on GSR. The result of interaction effects between screen dimension and icon size ($F_{8,239} = 8.40$, $p < 0.01$) was significant. The best level for a particular factor was the level that causes the lowest mean difference of GSR value. The optimum setting in the experiment was shown in Figure 3. The results of GSR reflected the participants, stress levels. The experimental results indicated that the screen dimension of 21 inches resulted in the lowest stress level, followed by 15 and 7 inches. In addition, the factor of icon sizes (40–80 minutes) was positively related to GSR, with 40 minutes associated with the highest stress level and 80 minutes associated with the lowest stress level.

## 4 DISCUSSIONS

Through the assessment in the display parameters of a geographical intelligence system, it was found that screen dimensions and icon sizes were the important factors from the analysis of result data. In brief, this experimental study was designed to examine the effects of display parameters on geographical intelligence systems. The results are discussed as follows.

Across the experiment in this research, the effects of screen dimension were significant on all the dependent variables. The findings indicated that higher screen dimensions yielded better accuracy and lower stress level, but took a longer completion time. The findings of screen dimension effect on accuracy and mean difference of GSR with consistent with previous results (Sweeney and Crestani, 2006; Reeves, Lang, Kim & Tatar, 1999). The reason for longer completion time may be that operators are searching for the targets in a geographical intelligence system, they spend a longer time to find all the targets. To search between designated icons and other MIL-Std 2525b icons on the display in a large screen required more saccades (rapid jerky movements of the eye) in systems on the 21 inches screen than that on the 15 and 7 inch screens. However, a larger screen dimension can also reduce the stress level of operators and increase search accuracy. The results can be used in control room design to promote a person's operational performance.

The effects of icon size were significant on all the dependent variables. Clearly, small icons were consistently worse than large icons on the completion time and accuracy. But small icons yielded lower stress levels than large icons. Lin et al. (2006) found that physiological data of GSR were mirrored in subjective reports assessing stress level. The results of the present study demonstrate that larger screen dimensions and small icons reduce GSR stress. It might be inferred that the least spatial density under the combination of the 15-inch screen dimension and 40-minute icon size may reduce human stress level. Lin also reported that GSR data correlate with task performance data in a video game (the more success participants became, the higher task performance level was). With a decrease of the task performance level, the normalized GSR increases.

## 5 CONCLUSIONS

Based on the experimental design results which had been investigated in a geographical intelligence system, findings from this study demonstrated that minor design differences of interface could have a significant impact on users' performance and cognition load. Changing icon sizes might allow users to easily glean valuable information with minor effort. But the change of icon size was not limitless. Available display capacity should match a suitable icon size. Thus, an appropriate combination of the screen dimension and icon size could improve users' task performance.

In the study, GSR data was used to measure cognition load of the body. The findings showed that when an individual came under suitable stress, it might have a negative effect on performance. Results reported that stress produced negative consequences on target search performance. For example, completion time and accuracy of icon sizes were negatively related to GSR, with 40 minutes associated with the highest stress level (poor task performance) and 80 minutes associated with the lowest stress level (better task performance). However, maintaining a highly stressed state requires an intensive level of cognition load, and consequently leads to users being in a state of exhaustion.

Lastly, this study employed a promising approach for experimental investigation of the effects of many design parameters on task performance. This study found some support for GSR that is consistent with the task performance responses. The findings could help the designers to develop more efficient and comfortable interfaces to fit user needs. In addition to this, it may helpfully be applied to a real geographical intelligence system. Although this study has been able to indicate some of the variables that affect search performance of geographical intelligence systems, it is also necessary in future researches to establish which combinations of display parameters are the most effective in minimizing fatigue.

## REFERENCES

Ahlberg, S., Horling, P., Johansson, K., Jored, K., Kjellstrom, H., Martenson, C., Neider, G., Schubert, J. & Svenson, P. (2007). An information fusion demonstrator for tactical intelligence processing in network-based defense. *Information Fusion*, Vol. 8, pp. 84–107.

Byrne, M.D. (1993). Using icons to find documents: simplicity is critical, *CHI'93 Proceedings of the INTERACT'93 and CHI'93 conference on Human factors in computing systems*.
Critchley, H.D., Elliott, R., Mathias, C.J. & Dolan, R.J. (2000). Neural activity relating to generation and representation of galvanic skin conductance responses: a functional magnetic resonance imaging study. *The Journal of Neuroscience*, Vol. 20, pp. 3033–3040.
Dickson, H.W. & McginnieS, E. (1966). Affectivity in the arousal of attitudes as measured by galvanic skin response. *The American Journal of Psychology*, Vol. 79, pp. 584–589.
DoD, Military Standard–Interoperability Standard for Interface Standard. (1999). Common Warfighting Symbology, MIL-STD 2525b.
Liang, G.F., Lin, J.T., Hwang, S.L., Huang, F.H., Yenn, T.C. & Hsu, C.C. (2009). Evaluation and Prediction of On-Line Maintenance Workload in Nuclear Power Plants. *Human Factors and Ergonomics in Manufacturing and Service*, Vol. 19, pp. 64–77.
Lin, T., Imamiya, A., Omata, M. & Hu, W. (2006). An empirical study of relationships between traditional usability indexes and physiological data. *Australasian Journal of Information Systems*, Vol. 13, pp. 105–117.
Liu, C.L. & Su, K.W. (2006). A Fuzzy Logical Vigilance Alarm System for Improving Situation Awareness and Trust in Supervisory Control. *Human Factors and Ergonomics in Manufacturing and Service*, Vol. 16, pp. 409–426.
Min, B.C., Chung, S.C., Park, S.J., Kim, C.J., Sim, M.K. & Sakamoto, K. (2002). Autonomic responses of young passengers contingent to the speed and driving mode of a vehicle. *Internal Journal of Industrial Ergonomics*, Vol. 29, pp. 187–198.
Paas, F., Tuovinen, J.E., Tabbers, H. & Van Gerven, P.W.M. (2003). Cognitive load measurement as a means to advance cognitive load theory. *Educational Psychologist*, Vol. 38, pp. 63–71.
Pfendler, C. & Schlick, C. (2007). A comparative study of mobile map displays in a geographic orientation task. Behaviour & Information Technology, Vol. 26, pp. 455–463.
Reeves, B., Lang, A., Kim, E.Y. & Tatar, D. (1999). The effects of screen size and message content on attention and arousal. *Media Psychology*, Vol. 1, pp. 49–67.
Shi, Y., Choi, E.H.C., Ruiz, N., Chen, F. & Taib, R. (2007). Galvanic Skin Response (GSR) as an Index of Cognitive Load. *Computer Human Interaction Conference,* San Jose, CA, USA.
Smith, M.J., Conway, F.T. & Karsh, B.T. (1999). Occupational Stress in Human Computer Interaction. *Industrial Health*, Vol. 37, pp. 157–173.

*Ergonomics in Asia – Shih & Liang (eds)*
*© 2012 Taylor & Francis Group, London, ISBN 978-0-415-68414-9*

# Ergonomic evaluation of icons for dashboard

Chia-Fen Chi, Ratna Sari Dewi & Shin-Cheng Chen
*Department of Industrial Management, National Taiwan University of Science and Technology, Taipei, Taiwan*

ABSTRACT: A computer program has been developed to conduct grouping and matching test of icons in a car dashboard and control panel. The tests are designed in order to understand the perceived grouping and comprehension of the icons. In the grouping test, each participant was asked to group 88 icons shown on the screen according her/his own comprehension about the similarity of the icons. For the matching test module, 44 icons were divided into dashboard and control panel groups and shown to the participants to evaluate their comprehension rate. Grouping test results show that shape and function are the most frequently used coding dimensions in grouping. From the confusion matrix of matching test results, most of the mismatches between icons and their referent were asymmetric confusion. This research also analyzed icons with poor recognition rate based on language barrier, shape features, and relationship between shape and features. Based on this analysis, recommendations for better icon design are proposed.

*Keywords*: car icons, grouping test, matching test, icon design

## 1 INTRODUCTION

The use of specific icons or symbols in a car should consider the icons' understandability level. To be easily understood by the driver, the icons (symbols) should follow ergonomic design criteria. Those criteria are legibility, high recognition rate, and interpretability (Campbell, et al., 2004); detectability, discriminable from others, and meaningful to the drivers (Green, 1993); and continuity, closure, symmetry, simplicity, and unity in structural properties (Easterby, 1970).

ISO Standard 2575 (International Organization for Standardization, 2004) has become a *de facto* world-wide standard where car manufacturers tend to acquire any symbols suitable for their need (Green, 1993). After reviewing various researches in this field, the ISO working group also proposed four different evaluation tests for the standardization of symbols (Easterby and Zwaga, 1974 in Brugger, 1999 and Campbell, et al., 2004). Three kinds of tests, production test, appropriateness ranking test, and comprehension/recognition test are used for limiting the number of candidate symbols. Meanwhile the matching test is used to ensure comprehension, recognition, and no confusion between symbols within a set (Campbell, et al., 2004; Brugger 1999). Other than the ISO procedure described above, several other tests had been used for the systematic evaluation of symbols such as familiarity task, association task, reaction time, preference, legibility distance, and certainty (Campbell et al., 2004). The choice of an evaluation technique for assessing specific icons depends on the purpose of the symbol and where it will be used (Mackett-Stout and Dewar, 1981). For example, reaction time or glance legibility might also be measured if symbols will require a rapid response or if there is short exposure duration (Campbell, 2004).

The current research intends to evaluate a set of icons for car dashboard and control panel. Computer aided testing can be employed in order to increase the effectiveness and efficiency of icon evaluation processes. Using a computer program, context effect on icon evaluation (Vukelich & Whitaker, 1993; Wolff & Wogalter, 1998) can be easily included in the evaluation protocol. The evaluation result can be used to improve the grouping arrangement and recognition of icons used on a car dashboard.

## 2 METHODS

### 2.1 *The computer program*

The computer program is written in Visual Basic. It was designed to run grouping tests and matching tests to collect the perceived similarity and comprehension of icons to be used on a car dashboard. Fifteen students took part in the matching and grouping test experiments. All participants were aged 19–23 years old and only 2 of them have a driving license.

#### 2.1.1 *Grouping test module*

In the grouping test experiment, each participant is asked to group 88 icons shown on the screen according to her/his own comprehension about the similarity of the icons. Based on their own preference, each participant can arrange the icon group layout freely by pointing and dragging each of the individual icons. The final arrangement is recorded for further analysis. The screen for the grouping test is shown in Figure 1.

#### 2.1.2 *Matching test module*

In matching test experiment, 44 icons are divided into two interfaces, as shown in Figure 2 and Figure 3, along with their referent meanings listed in random order. For each icon, participants would click and drag the most appropriate meaning to the answer box besides the icon and revise the answer until the final icon is answered. All icons are positioned on a car dashboard (Figure 2) or control panel (Figure 3) to provide the context. The chosen referent for each icon by each participant is collected for further analysis.

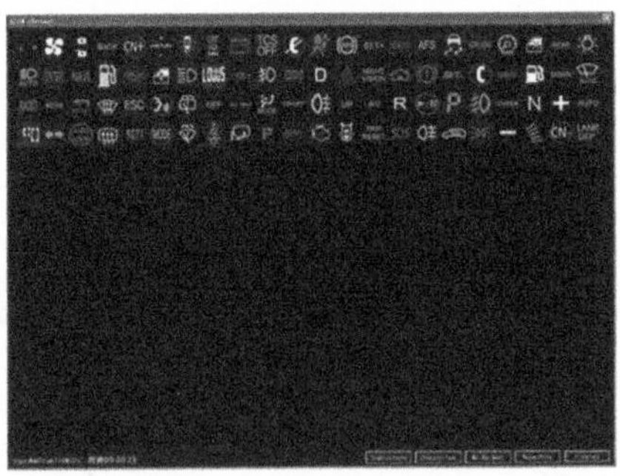

Figure 1. Grouping test interface.

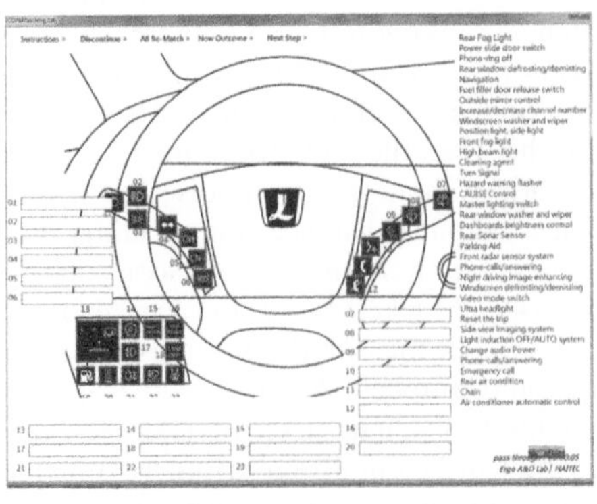

Figure 2. Matching test interface for dashboard.

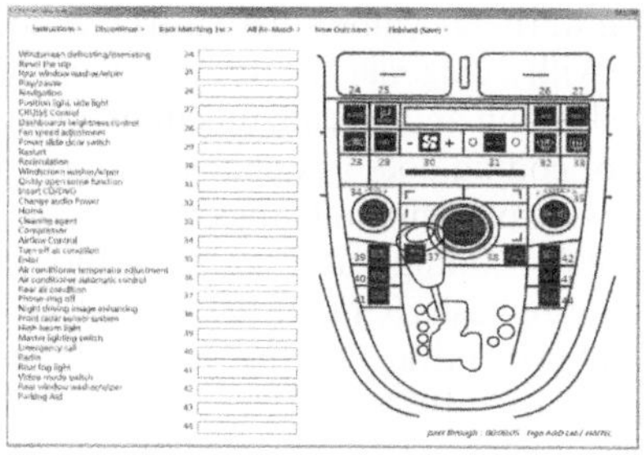

Figure 3. Matching test interface for control panel.

## 3 RESULT AND ANALYSIS

### 3.1 *Icon groupings*

The participants grouped the icons based on any combinations of the following dimensions: color, shape, text, and function. The frequency of these four coding dimensions is shown in Table 1. Participants had classified the icons in the following most frequently used dimensions: shape, function, color, and lastly text existence and format.

### 3.2 *Matching test*

The chosen meaning associated with each icon is tabulated as a confusion matrix as shown in Figure 4 and Figure 5. The percentage of correct responses can be derived from the ratio of correctly answered icons to the number of participants. A correct recognition rate of 66.7%, according to International Organization for Standardization (ISO) Standard 3864 (1984), is adopted to divide the icons tested in the matching test into low-recognition group and high-recognition group. Out of 44 icons, only 17 icons (39%) satisfy the recognition criteria, while 27 out of 44 (61%) have recognition rate below the ISO standard.

The confusion matrix (Figure 4) gives the symmetric and asymmetric confusions (Lin, 2004). Figure 4 indicates that most of the mismatches are asymmetric confusion caused by incorrect matching between the icon and its corresponding referent. On the other hand, symmetric confusion occurs when two icons are confused with each other, and was found between the following pairs of icons:

i. compressor and rear air conditioning
ii. recirculation and air flow control
iii. windscreen defrosting and rear window defrosting
iv. master lighting switch and high beam light
v. master lighting switch and position light, side light
vi. rear air conditioning and rear fog light
vii. rear window washer and wiper and windscreen washer and wiper
viii. outside mirror control and side view imaging system.

Since only a small number of participants took part in this experiment, the above symmetric confusions only came from one to three participants.

## 4 DISCUSSION

Each participant was asked to finish the grouping test then the matching test to prevent any learning effect. Since most (13 out of 15) participants had no prior experience with driving, participants could only categorize icons based on their imagination. Although shape similarity was the most frequently chosen dimension for grouping, based on the chi squared test result, there is no significant difference between all of coding dimensions. In Taiwan, students were taught about traffic symbols including warning symbols have a triangular shape and in yellow color while prohibition sign is a circle with a diagonal line through it and mostly in red color. They are also familiar with text contained within and specific functions embedded in the symbols.

This research finds that color coding does not gain superiority in frequency compared to other coding dimensions. Only one participant (subject 6) used color as the only single basis of the classification. Basically, color coding was found to be the most effective coding dimension for search, locate, and counting tasks (Christ, 1975; Sanders & McCormick, 1998). But, color coding was less powerful for identification that involves the conceptual recognition of the meaningfulness of the icon (Christ, 1975; McCormick, 1998). The low frequency of color coding used in this experiment explained that in grouping icons, subjects not only do the searching task but also do the identification task. This means that the consistency of shape and function is also considered when they identify the exact meaning of the icons.

Table 1. Number of classification/code for each subject.

| Subject # | Color | Shape | Text | Function |
|---|---|---|---|---|
| 1 | 4 | 3 | 1 | 0 |
| 2 | 1 | 1 | 1 | 4 |
| 3 | 3 | 0 | 2 | 6 |
| 4 | 6 | 0 | 0 | 0 |
| 5 | 0 | 4 | 2 | 1 |
| 6 | 0 | 0 | 2 | 6 |
| 7 | 0 | 2 | 2 | 5 |
| 8 | 2 | 1 | 1 | 1 |
| 9 | 2 | 2 | 4 | 0 |
| 10 | 4 | 3 | 3 | 3 |
| 11 | 5 | 4 | 2 | 4 |
| 12 | 0 | 6 | 3 | 1 |
| 13 | 3 | 8 | 3 | 10 |
| 14 | 3 | 7 | 4 | 1 |
| 15 | 2 | 8 | 1 | 3 |
| Total | 35 | 49 | 31 | 45 |

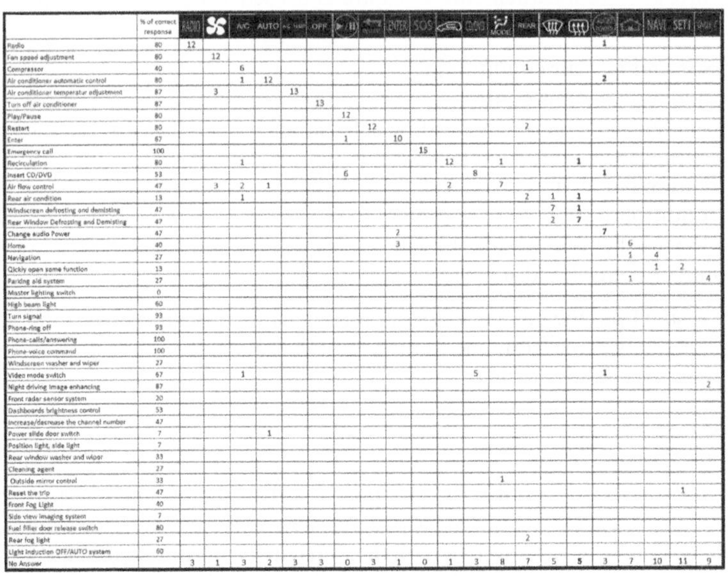

| | % of correct response | RADIO | | A/C | AUTO | | OFF | | | ENTER | SOS | | CD/DVD | MODE | REAR | | | | | NAVI | SET1 | |
|---|---|---|---|---|---|---|---|---|---|---|---|---|---|---|---|---|---|---|---|---|---|---|
| Radio | 80 | 12 | | | | | | | | | | | | | | | | 1 | | | | |
| Fan speed adjustment | 80 | | 12 | | | | | | | | | | | | | | | | | | | |
| Compressor | 40 | | | 6 | | | | | | | | | | | 1 | | | | | | | |
| Air conditioner automatic control | 80 | | | 1 | 12 | | | | | | | | | | | | | 2 | | | | |
| Air conditioner temperatur adjustment | 87 | | 3 | | | 13 | | | | | | | | | | | | | | | | |
| Turn off air conditioner | 87 | | | | | | 13 | | | | | | | | | | | | | | | |
| Play/Pause | 80 | | | | | | | 12 | | | | | | | | | | | | | | |
| Restart | 80 | | | | | | | | 12 | | | | | | 2 | | | | | | | |
| Enter | 67 | | | | | | | 1 | | 10 | | | | | | | | | | | | |
| Emergency call | 100 | | | | | | | | | | 15 | | | | | | | | | | | |
| Recirculation | 80 | | | 1 | | | | | | | | 12 | | 1 | | | 1 | | | | | |
| Insert CD/DVD | 53 | | | | | | | 6 | | | | | 8 | | | | | 1 | | | | |
| Air flow control | 47 | | 3 | 2 | 1 | | | | | | | 2 | | 7 | | | | | | | | |
| Rear air condition | 13 | | | 1 | | | | | | | | | | | 2 | 1 | 1 | | | | | |
| Windscreen defrosting and demisting | 47 | | | | | | | | | | | | | | | 7 | 1 | | | | | |
| Rear Window Defrosting and Demisting | 47 | | | | | | | | | | | | | | | 2 | 7 | | | | | |
| Change audio Power | 47 | | | | | | | | | 2 | | | | | | | | 7 | | | | |
| Home | 40 | | | | | | | | | 3 | | | | | | | | | 6 | | | |
| Navigation | 27 | | | | | | | | | | | | | | | | | | 1 | 4 | | |
| Qickly open some function | 13 | | | | | | | | | | | | | | | | | | | 1 | 2 | |
| Parking aid system | 27 | | | | | | | | | | | | | | | | | | 1 | | | 4 |
| Master lighting switch | 0 | | | | | | | | | | | | | | | | | | | | | |
| High beam light | 60 | | | | | | | | | | | | | | | | | | | | | |
| Turn signal | 93 | | | | | | | | | | | | | | | | | | | | | |
| Phone-ring off | 93 | | | | | | | | | | | | | | | | | | | | | |
| Phone-calls/answering | 100 | | | | | | | | | | | | | | | | | | | | | |
| Phone-voice command | 100 | | | | | | | | | | | | | | | | | | | | | |
| Windscreen washer and wiper | 27 | | | | | | | | | | | | | | | | | | | | | |
| Video mode switch | 67 | | | 1 | | | | | | | | | 5 | | | | | 1 | | | | |
| Night driving image enhancing | 87 | | | | | | | | | | | | | | | | | | | | | 2 |
| Front radar sensor system | 20 | | | | | | | | | | | | | | | | | | | | | |
| Dashboards brightness control | 53 | | | | | | | | | | | | | | | | | | | | | |
| increase/decrease the channel number | 47 | | | | | | | | | | | | | | | | | | | | | |
| Power slide door switch | 7 | | | | 1 | | | | | | | | | | | | | | | | | |
| Position light, side light | 7 | | | | | | | | | | | | | | | | | | | | | |
| Rear window washer and wiper | 33 | | | | | | | | | | | | | | | | | | | | | |
| Cleaning agent | 27 | | | | | | | | | | | | | | | | | | | | | |
| Outside mirror control | 33 | | | | | | | | | | | | | 1 | | | | | | | | |
| Reset the trip | 47 | | | | | | | | | | | | | | | | | | | | 1 | |
| Front Fog Light | 40 | | | | | | | | | | | | | | | | | | | | | |
| Side view imaging system | 7 | | | | | | | | | | | | | | | | | | | | | |
| Fuel filler door release switch | 80 | | | | | | | | | | | | | | | | | | | | | |
| Rear fog light | 27 | | | | | | | | | | | | | | 2 | | | | | | | |
| Light induction OFF/AUTO system | 60 | | | | | | | | | | | | | | | | | | | | | |
| No Answer | | 3 | 1 | 3 | 2 | 3 | 3 | 0 | 3 | 1 | 0 | 1 | 3 | 8 | 7 | 5 | 5 | 3 | 7 | 10 | 11 | 9 |

Figure 4. Confusion matrix.

| | | | | | | | | MODE | NIGHT VISION | | | CN+ | | | | | | TRIP RESET | | LDWS | | | AUTO |
|---|---|---|---|---|---|---|---|---|---|---|---|---|---|---|---|---|---|---|---|---|---|---|---|
| Radio | | | | | | | | | | | | | | | | | | | | | | | |
| Fan speed adjustment | | | | | | | | | | | 1 | | | | | | | | | | | | |
| Compresser | | | | | | | | | | | | | | | | | | | | | | | |
| Air conditioner automatic control | | | | | | | | | | | | | | | | | | | | | | | |
| Air conditioner temperatur adjustment | | | | | | | | | | | | | | | | | | | | | | | |
| Turn off air conditioner | | | | | | | | | | | | | 1 | | | | | | | | | | |
| Play/Pause | | | | | | | | | | | | | | | | | | | | | | | |
| Restart | | | | | | | | | | | | | | | | | | | | | | | |
| Enter | | | | | | | | | | | | | | | | | | | | | | | |
| Emergency call | | | | | 1 | | | | | | | | | | | | | | | | | | |
| Recirculation | | | | | | | | | | | | | | | | | | | | | | | |
| Insert CD/DVD | | | | | | | | | | | | | | | | | | | | | | | |
| Air flow control | | | | | | | | | 1 | | | | | | | | | | 1 | | | 2 | |
| Rear air condition | | | | | | | | | | | | | | | | | | | 2 | | | 1 | |
| Windscreen defroating and demisting | | | | | | | 3 | | | | | | | | 2 | | | | | | | 1 | |
| Rear Window Defrosting and Demisting | | | | | | | 1 | | | | | | | | | 2 | | | 1 | 1 | | | |
| Change audio Power | | | | | | | | | | | | | 3 | | | | | | | | | | |
| Home | | | | | | | | 1 | | | | | | | | | | | | | | | |
| Navigation | | | | | | | | | | 1 | | | | | | | | 2 | | 3 | | | |
| Qickly open some function | | | | | | | | | | | | | | | | | | | | | | | |
| Parking aid system | | | | | 1 | | | | | 2 | | | 1 | | | | 5 | | | | | | |
| Master lighting switch | 0 | 2 | | | | | | | | 1 | | 1 | | 1 | | | | | | | | | 2 |
| High beam light | 1 | 9 | 1 | | | | | | | | | | | | | | | | | | | 2 | |
| Turn signal | | | 14 | | | | | | | | | | | | | | | | | | | | |
| Phone-ring off | | | | 14 | | | | | | | | 1 | | | | | | | | | | | |
| Phone-calls/answering | | | | | 15 | | | | | | | | | | | | | | | | | | |
| Phone-voice command | | | | | | 15 | | | | | | | | | | | | | | | | | |
| Windscreen washer and wiper | | | | | | | 4 | | | | | | | | 3 | | | | | | | | |
| Video mode switch | | | | | | | | 10 | | | | | | | | | | | | | | | |
| Night driving image enhancing | | | | | | | | | 13 | | | | | | | | | | | 1 | | | |
| Front radar sensor system | | | | | | | | | | 3 | | | | | | 1 | | | | | | | |
| Dashboards brightness control | | | | | | | | | | | 8 | 1 | | | | | | | | 1 | | | |
| Increase/decrease the channel number | | | | | | | | 1 | | | | 7 | | | | | | | | | | | |
| Power slide door switch | | | | | | | | | | 3 | | | 1 | | | | | | | | | | |
| Position light, side light | 11 | | 2 | | | | | | | | | | | 1 | | | | | | | | | |
| Rear window washer and wiper | | | | | | | 3 | | | | | | | | 5 | | | | | | | | |
| Cleaning agent | | | | | | | 5 | | | | | | | | 3 | 4 | | | | | | | |
| Outside mirror control | | | | | | | | | | | | | 2 | | 1 | | 5 | | | 3 | | | |
| Reset the trip | | | | | | | | | | | | | | | | | | 7 | | | | | 2 |
| Front Fog Light | | | | | | | | | | | | | | 3 | | 1 | | | 6 | | | | |
| Side view imaging system | | | | | | | | | | | | | | 5 | | | 1 | | | 1 | | | |
| Fuel filler door release switch | | | | | | | | | | | | | | | | | | | | | 12 | | |
| Rear fog light | | | | | | | | | | | | | | | | | | | | | | 4 | |
| Light induction OFF/AUTO system | | | | | | | | | | | | | | | | | | | | | | | 9 |
| No Answer | 2 | 3 | 0 | 1 | 2 | 0 | 0 | 3 | 3 | 6 | 2 | 2 | 8 | 5 | 3 | 6 | 4 | 6 | 5 | 7 | 2 | 5 | 2 |

Figure 5. Confusion matrix (cont).

From the matching test, this study shows that only about 39% of the icons satisfy the 66.7% of correctness criterion of ISO 3864. It is important to understand why so many icons cannot be recognized correctly, hence they can be redesigned. The majority of the mismatches were caused by asymmetric confusion (Figures 4 and 5) which indicated that most errors were caused by a general lack of recognition rather than confusion between similar icons.

Regarding symmetric confusion, the most distinctive symmetric confusion was found between windscreen washer and wiper and rear window washer and wiper due to visual and conceptual similarity as stated in Lin (1994). Since most of participants had no driving experience prior to the test, 9 ISO standard icons (rear window washer and wiper; master lighting switch; rear fog light; position light, side light; windscreen washer and wiper; front fog light; rear window defrosting; windscreen defrosting; and high beam light) have less than 66.7% of correctness. It would be interesting to know whether driving experience can significantly improve the recognition performance of the tested icons (Piamonte et al., 2001).

In order to provide information for redesigning icons, the authors proposed to analyze the causes of low recognition rate icons based on language barrier, shape features, and the relationship between text and shape features. For icons using English text, the low recognition rate can be caused by participants not knowing the vocabulary or acronyms and abbreviations used in the icons. For example, the icon "LDWS" is an acronym of Lane Departure Warning System. Since all of the participants were Taiwanese and non-native English speakers, only 1 among 15 subjects could correctly match this icon. The other reason is word/words used in icons are not complete enough to convey the right meaning. Another example is the icon "SETI" used to indicate quickly open some function. Only 2 subjects could correctly match this icon with its meaning.

There are three issues corresponding to the use of specific shape features in icons. The first one is the similarity of shapes used in two icons made the participants confuse each of the icons and its right referent. For example, the icons for rear fog light and front fog light icons use the similar shape features and they are different only on the direction and angle of the light beam. Next issue is the similarity of functions among icons. For instance, icons for master lighting switch, position light, and high beam light all have similar functions, that is, to convey light control and adjustment in car. The last issue is shape ambiguity; the shape used in an icon can represent more than one meaning. For example in the fan speed adjustment icon, the fan feature used in the icon also can be interpreted as temperature adjustment.

The relationship between text and shape features conveyed in an icon is also an important issue. The shape and the text should match and be compatible to each other. In this study, it is found that only 3 subjects could match the front radar sensor system icon to its right meaning. One possible explanation of this case is the incompatibility between the word "SONAR" and the car shape that contained in this icon. SONAR is commonly regarded as an abbreviation of so (und), na (vigation), and r(anging) and is regularly used in the marine field. Connecting the term within car context is less straightforward for many participants of this study.

## 5 CONCLUSION

Ergonomic evaluation and redesign of icons for car dashboard and control panel is important because better recognition rate of an icon will increase driver understanding and safety. The administered grouping and matching tests revealed that the majority of the icons cannot be recognized by student participants without prior driving experience. From the grouping test, shape, function, color, and text were found to be the frequently used coding dimensions for dashboard and control panel icons. Based on the matching test, confusions/errors found in this study suggest more detailed recommendations regarding text and shape features to be incorporated in icons. The computer program developed in this study facilitates the experiments effectively; and as well as helping in administering the research data, it also made the interface more user-friendly for the subjects.

## REFERENCES


Brugger, Christof. (1999). Public Information Symbols: a Comparison of ISO Testing Procedures. In *Visual Information for Everyday Use: Design and Research Perspective*, by Harm J.G. Zwaga, Theo Boersema and Henriette C.M. Moonhout, pp. 305–313. Taylor and Francis.

Campbell, J.L., Richman, J.B., Carney, C. and Lee, J.D. (2004). *In-Vehicle Display Icons and Other Information Elements Volume I: Guidelines.* U.S. Department of Transportation, Federal Highway Administration.

Christ, Richard E. (1975). Review and Analysis of Color Coding Research for Visual Displays. *Human Factors,* pp. 542–570.

Easterby, RS. (1970). "The perception of symbols and machine displays." *Ergonomics*, pp. 149–158.

Easterby, R.S. and Zwaga, H.C.G. (1976). Evaluation of Public Information Symbols; ISO Tests:1975 Series. *AP Report 60.* Birmingham: University of Aston: Department of Applied Psychology.

Easterby, R.S. and Zwaga, H.J.G. (1974). The Evaluation of Public Information Symbol. *AP Note 50,* Birmingham: University of Aston: Department of Applied Psychology.

Green, Paul. (1993). Design and evaluation of symbols for automobile controls and displays. *In Automotive Ergonomics,* by Brian Peacock and Waldemar Karwowski, pp. 237–268. London: Taylor and Francis Ltd.

International Organization for Standardization. *Road vehicles-Symbols for controls, indicators and telltales*. Geneva: International Standard Organization, 2004.

International Organization for Standardization. *Safety colours and safety signs (ISO 3864).* Geneva: International Organization for Standardization, 1984.

Lin, Rungtai. (1994). A study of visual features for icon design. *Design Studies*, pp. 185–197

Mackett-Stout, Janice and Robert Dewar. (1981). Evaluation of Symbolic Public Information Signs. *Human Factors,* pp. 139–151.

Piamonte, D. Paul T., John D.A. Abeysekera and Kjell Ohlsson. (2001). Understanding small graphical symbols: a cross-cultural study. *International Journal of Industrial Ergonomics*, pp. 399–404.

Sanders, Mark S and Ernest J McCormick. (1992). *Human Factors in Engineering and Design*. Singapore: McGraw-Hill, Inc

Vukelich, Mark and Leslie, A. Whitaker. (1993). The Effects of Context on the Comprehension of Graphic Symbols, *Human Factors and Ergonomics Society Annual Meeting Proceedings,* Forensics Professional, pp. 511–515.

Wolff, Jenifer Snow and Michael, S. Wogalter. (1998). Comprehension of Pictorial Symbols: Effects of Context and Test Method. *Human Factors,* pp. 173–186.

*Ergonomics in Asia – Shih & Liang (eds)*
*© 2012 Taylor & Francis Group, London, ISBN 978-0-415-68414-9*

# *XGameFlow*, extended model for measuring player's enjoyment in playing a game

Jasson Prestiliano & Danny Manongga
*Faculty of Information Technology, Satya Wacana Christian University, Central Java, Indonesia*

Chiuhsiang Joe Lin
*Department of Industrial and System Engineering, Chung-Yuan Christian University, Chung-Li, Taiwan, R.O.C.*

ABSTRACT: People usually expect certain entertainment when they play some games. This can be measured by using a game enjoyment measurement model called *GameFlow* and *EGameFlow*. However, they still need improvement in particular being able to measure how the game can affect the player's life. In this study, an extended model of the player's enjoyment model called *XGameFlow* will be analyzed. This model adds one criterion, that called persuasion. The criterion itself will be tested by using a combination of within-subject design and between-subject design. The test involved three different games and participants from two different countries. The criterion will be validated by using confirmatory factor analysis to recognize whether it should be added to the previous model. The results proved that *XGameFlow* model can be used as a measurement tool of player enjoyment in playing games, and the model does not depend too much to the difference of a player's culture or country.

*Keywords*: *GameFlow*, player's enjoyment, game

## 1 INTRODUCTION

A game is an interactive system that can be controlled and which can give the player certain amount of fun as an aspect of the game itself. The word "computer game" indicates an integrated computer system where one player (or more) takes decisions and controls the game objects to reach the player's aims (Crawford, 2003).

Games are important because they can be an alternative or a breakthrough to entertain people, reduce their stress, and refresh their minds in a certain way that only can be found in a well-designed game (Prasida, 2009). A well-designed game can bring enjoyment and positive emotions to most of the players who spend their time playing games (Korhonnen, Montola & Arrasvuori, 2009). To design a good and enjoyable game, especially for a computer game, a game designer should play many enjoyable games to identify which aspects of the game are enjoyable (Crawford, 2003).

A game that can bring enjoyment to the player is the game that can catch the player's attention to keep playing the game and to acquire something from it. Players usually wish for a certain amount of entertainment when they play games. Some players feel excitement when they complete some tasks that are given in the game. Otherwise, there are some players that are attracted by the storyline, artwork, music and game plays or game mechanism. (Yannakakis & Hallam, 2008).

The enjoyment itself will make the game be played by many people and the messages or the objectives included in the game could be acknowledged by the players.

A study of game enjoyment has been conducted and has resulted in a measurement model called *GameFlow* (Sweetser & Wyeth, 2005). Also, several years later, there was a study to modify *GameFlow* into *EGameFlow* that focused on educational games (Fu, Su & Yu, 2008).

However, both studies only discuss how the game can affect the player when they are playing the game, and they don't discuss how a game can persuade the players, influence their mind or even change their behavior after they have finished playing the game. This issue is important to be discussed because many people are persuaded by some kind of games.

In this study, a discussion about the new criterion in measuring player enjoyment is conducted. The new criterion is called persuasion. This study uses the combination of within- and between-subject designs as the test method and confirmatory factor analysis as the research method to analyze and determine the validity of persuasion in the player enjoyment model.

## 2 PLAYER ENJOYMENT MODEL

### 2.1 *GameFlow*

The player enjoyment model that has been tested and used is the *GameFlow* model that had been made by Sweetser and Wyeth (Sweetser & Wyeth, 2005) and which was modified by the study that focused on e-learning game development, the *EGameFlow* (Fu, Su & Yu, 2008). The model itself has eight criteria to be measured as shown in Figure 1.

1. Games should require concentration and the player should be able to concentrate on the game. The player should give his or her attention to the game activities. Those are the essence of concentration.
2. Games should have clear goals, so the player knows the aim of the game. The clarity of the goal should be informed in the game. The information could be given directly or indirectly. The game usually has some intermediate goals to reach, in order to achieve the main goal of the game.
3. Concerning the challenge, the game should be challenging and match the player's skill level. The difficulty should be adequate and improve gradually as the level is increased but still keep a balance with the player's skill. The challenge itself will make the player keep playing the game without feeling boredom or anxiety (Sweetser & Wyeth, 2005).
4. The player's skill or knowledge improvement is a criterion in which the game can improve the knowledge or skill of the player (Fu, Su & Yu, 2008). In this criterion, players will be taught something they didn't know before.
5. For control or automation, the player should feel a sense of controlling the game. The game will automatically follow the instruction that had been given. The player should find

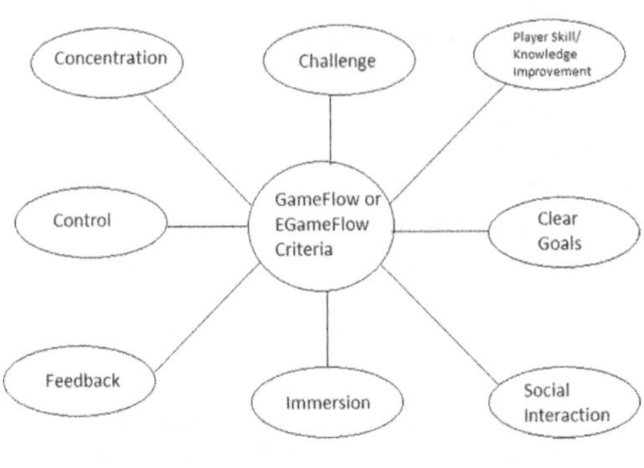


Figure 1. *EGameFlow* Criteria (Sweetser & Wyeth, 2005; Fu, Su & Yu, 2008).

that the game can give them a sense of control through the role, the object and purpose of the game. Moreover, the game can give them some automatic barrier or warning when the player is going too far off track and will cause an error that stops the player from continuing the game.

6. Player should receive appropriate feedback at the appropriate time. They should be given notice about success, failure, events and status immediately. The interface also can give information about something that was or is happening in the game (Sweetser & Wyeth, 2005).
7. Immersion is when a player feels drawn into the game and feels a deep emotion within the game. The player should feel involved inside the game. The player can feel that the music, the artwork and the story can draw the player to enter the game world, and also remember it although they are not currently playing the game (Scheible, Tuulos & Ojala, 2007).
8. Social interaction is where the game could support and create opportunities in social interaction. Communities could be built inside and outside the game. The player can interact with other players from different places to discuss the game (Sweetser & Wyeth, 2005).

### 2.2 *The new criteria: Persuasion*

When somebody plays games, the player receives many inputs from the game world. It can be seen that when the player plays some action game, he starts to jump, scream and express other emotions in his immersion in the game. This kind of immersion can bring the player to a brand new world that can teach him anything and let his knowledge improve. Furthermore, there are also some games that can change the mind, even the behavior. This means not only increasing the player's knowledge, but also improves the skill and behavior. (Berkovsky, Bhandari & Kimani, 2004).

The criterion where a game influences the mind or behavior of the player will be called Persuasion. Persuasion can be an advantage or a disadvantage, depending on the message that is embedded in the game. In many famous games this criterion are often embedded and it becomes advantageous for the player. *America's Army* is a game that is studied to analyze the persuasion inside each game. Because that game can persuade the player to be the real army since the aspect of the game is designed to train real young Americans in the army about battlefield and strategy (Lovlie, 2007). Another study is *World of Warcraft, Doom* and *Madden N.F.L* (*National Football League*), the game that provides an ordinary world in new or different ways, and these games can persuade the player to do something, even in a mobile game, a game still can persuade the player to do what it suggests (Bogost, 2007).

There are also several games with intensive violence that can make people do something outside their normal mind like kill someone else just like in the game events, that's one of the disadvantages of persuasion in a game. However, that kind of game could be said to succeed in persuading the player to do something beyond his or her ordinary behavior.

In this study, for example, in a game called *Saving Kantan*, there is an embedded message in the game that suggests people should start protecting the endangered animals from hunters. The message is not delivered directly to the player in words, but by playing the game, the game tries to persuade people in order to start changing their minds about wildlife preservation.

Persuasion itself has a close relationship with two criteria from the previous studies; they are player skill/knowledge improvement and immersion (Sweetser & Wyeth, 2005). The difference between persuasion and player skill/knowledge improvement is that persuasion can affect the way of thinking, although there is not any skill or knowledge improvement, the player is only persuaded to do something or influenced to think something different. Immersion is how a player can enjoy the world of the game, but after the player exits from the game world, the player will only memorize it or only want to know more about the game, but persuasion makes the player want to do something based on the game. Persuasion is required in this model. It doesn't only improve the knowledge of the player, but also persuades the player to do something. It's more like the player will learn something new and be persuaded to search the relevance with the real world and then will implement it in the real world. That's why this criterion is required to be added to the model. With this new criterion, the model is named *XGameFlow*, or *eXtended-GameFlow*.

## 3 METHODOLOGY

### 3.1 *Research method*

The study combines within-subject design and between-subject design to test whether the model can be implemented to measure a player's enjoyment. Within-subject design is used, which has more power but less variability (Lambdin & Shaffer, 2009) to measure some variable among several items that should be tested by the same person or group. In this case, within-subject design is designed to measure enjoyment in those three games that will be tested, because every study participant has to give his or her opinion about each single game. The implementation of the within-subject design is conducted twice, first in Taiwan and secondly in Indonesia. In this study, about ten game players from Taiwan and Indonesia will become the subject of the study. On the other hand, between-subject design is used to measure some variables with involving many people or groups to test a kind of item (Lambdin & Shaffer, 2009).

(a) (b) (c)

Figure 2. Game tested (a) Saving Kantan, (b) Indonesian Language Learning, (c) Eternal Grace.

The design is used to measure the cross-culture players from two different countries to determine whether the model itself is valid and reliable in different cultures or not.

Each test involves about thirty people, so the sample taken will be around 180 samples that consist of 90 samples from Taiwan and another 90 from Indonesia. Participants are randomly chosen, but with several criteria. They should have an age of between 15 to 27 years old in order to let them understand the questionnaire items well, and games that will be tested are targeting their age group. The participants should play games for longer than 10 hours per week in their daily life. This criterion is needed because they often play many games, so they can determine which game that can be enjoyable for them. Another criterion that should be fulfilled is they have never played the game tested before, in order to keep the freshness and objectivity. Each participant should answer 58 questions in order to fill the player enjoyment measurement scale. The result of each country will be compared to find the cross-culture impact for the game enjoyment by using between-subject design.

To conduct this kind of method, some questions are arranged into a questionnaire and several games are to be tested. The author made the questionnaire by using the scale development process that has been tested theoretically (Fu, Su & Yu, 2008). Each criterion will have about 5 to 10 questions.

### 3.2 *Game tested*

Three different games are developed. They will be tested as the testing product. They are shown in Figure 2. The first game is a clone of a Nintendo *Battle City*. The clone game talks about nature preservation called *Saving Kantan* (Figure 2a). The second one is an educational game to learn Indonesian covered in an adventure game entitled *Indonesian Language Learning* (Figure 2b). The third one is an action role playing game entitled *Eternal Grace* (Figure 2c).

Those games and questions will be analyzed to get the comparison value of enjoyment and to find what criteria are highly rated in each kind of game. The result is also analyzed to study whether there is a need for a model improvement in order to determine the rate of enjoyment among the games with different genres.

## 4 RESULT & ANALYSIS

A questionnaire that has a rate from 1 to 5 is used to evaluate people's enjoyment when they play a game. Three completely different games become the objects of the study, while the participants are people from Taiwan and Indonesia. The questions number 1 to 58 that can be seen in Table 1 are tested within theory analysis. Eight criteria have been tested in the previous study (Fu, Su & Yu, 2008) and several improved questions are clsely based on the questions from the previous study.

Table 1. Factor analysis of new criterion, persuasion.

| Question items | |
|---|---|
| Persuasion | Factor loadings |
| Does the game increases your knowledge and tell you about something you didn't know before? | 0.813 |
| Do you try to apply the knowledge in the game? | 0.735 |
| Do you want to know more about something in the game? | 0.801 |
| Does the game motivate you to do something? | 0.679 |
| Does the game give enough reasons to change your mind? | 0.791 |
| Can the game or some of the contents affect some aspects of your life? | 0.760 |

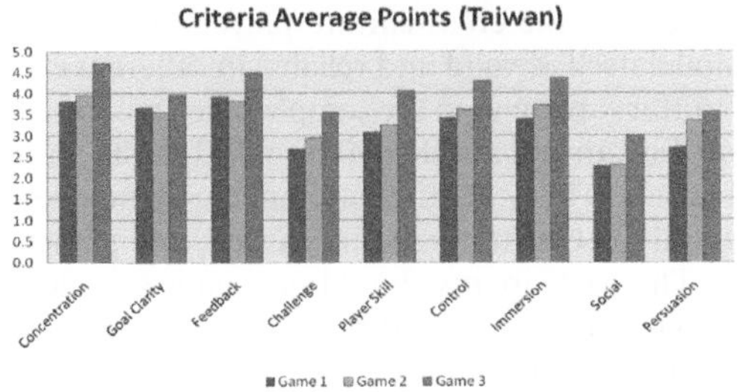


Figure 3. Criteria average points from participants in Taiwan.

Some modification of some questions are arranged to get better validity and reliability through the questionnaire. The result that will be analyzed is focused on the new criterion, persuasion. The questions in the persuasion criterion will be tested for their validity and reliability. Figure 3 represents the results in a bar chart for participants in Taiwan.

From Figure 3, some analyses can be made. Game 1 is good in goal clarity and feedback criteria because of its game design, and because of the lack of visual art and graphics, this game is weak in immersion and persuasion. Game 2 focuses on e-learning has much better score in player skill and persuasion, because it can give new knowledge, new skill and new influence to the player especially for the players from Taiwan, because the game itself is about *Indonesian Language Learning* that attracts much attention among Taiwanese players. Game 3 that concerns the *XGameFlow* criteria has the highest scores in every aspect if compared to other games but it has a large difference among other games which are identified as concentration, feedback, challenge, control and immersion, while the persuasion and goal clarity only experience a small difference.

Figure 4 represents the result in bar chart for participants in Indonesia.

From Figure 4, surprisingly Game 1 has a good rate on one of the criteria, that is concentration, while goal clarity and feedback experience have the same result as in Figure 4. If it is compared with Game 2, Indonesian players give a relatively similar rate for persuasion, but the other criteria that have significantly been rated better than Game 1 are identified as challenge, control and immersion. It is all because of the visual art, audio and level design of the game. Game 3 still leads the ratings but has a smaller difference than that in Figure 3. Taiwanese players wrote that the best criteria for Game 3 is concentration, because they need more concentration while playing Game 3. Indonesian players wrote that the best rate for Game 3 is the feedback because Indonesian players need to know what should they do next and how the game gives them feedback more than merely concentrating on finishing a mission itself.

From Figure 3 and Figure 4, it can be analyzed that Game 1 has better goal clarity and feedback from both countries' participants. The only difference is in the concentration criterion, where Indonesian participants need more concentration while playing Game 1 than those from Taiwan. Taiwanese players wrote that *Indonesian Languange Learning* (Game 2) need more attention because they need to know how to continue the game more than Indonesian players who are native speakers of Indonesian. This is supported by the persuasion criterion that has been rated higher in Game 2 than Game 1 for Taiwanese participants. All of the games have a low rate in the social interaction criterion because they are offline games, so the player focused on how to finish the games instead of interacting with other people when playing this game.

After having analyzed the impact for both countries' participants and for the three kinds of game that had been tested by the participants, the points will be computed to find whether the questionnaire results are good enough to form the criterion proposed, persuasion. Confirmatory factor analysis will be used to analyze the validity of each question.

Confirmatory factor analysis is based on principal component analysis, which is more popular and easier to calculate than common factor analysis (Gorsuch, 1997), and was used

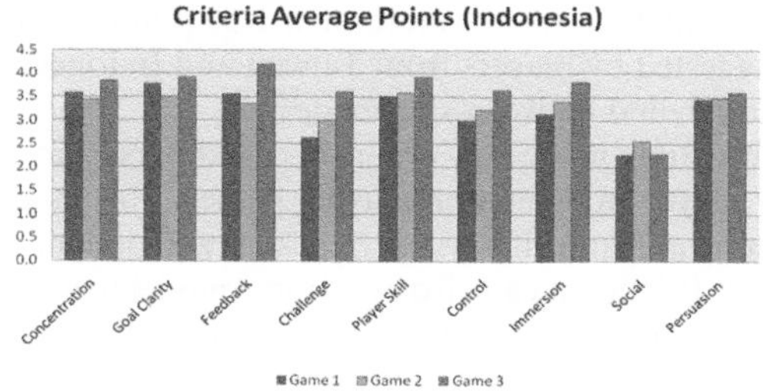


Figure 4. Criteria average points from participants in indonesia.

to determine construct validity and understand whether the questionnaire items were highly relevant to the predicted elements of this study.

The factors contained 6 items; their Eigenvalue was greater than 1.

Table 1 shows factor loading for the component matrix that has been calculated with confirmatory factor analysis. The factor loadings, also called component loadings in PCA, are the correlation coefficients between the variables (rows) and factors (columns). Analogous to Pearson's r, the squared factor loading is the percentage of variance in that indicator variable explained by the factor. To get the percentage of variance in all the variables accounted for by each factor, add the sum of the squared factor loadings for that factor (column) and divide by the number of variables. (Note the number of variables equals the sum of their variances as the variance of a standardized variable is 1.) This is the same as dividing the factor's Eigenvalue by the number of variables.

By one rule of thumb in confirmatory factor analysis, loadings should be 0.7 or higher to confirm that independent variables identified a priori are represented by a particular factor, on the rationale that the 0.7 level corresponds to about half of the variance in the indicator being explained by the factor (Fornell & Larcker, 1981; Bagozzi & Yi, 1988).

One item of the factors was lower than 0.7, the fourth question which says, "Does the game motivate you to do something?" only has value 0.679, so this item was deleted. After modifying the questionnaire, all factors loadings were greater than 0.7. Based on these measures, the construct validity of the questionnaire was judged as acceptable (Fornell & Larcker, 1981; Bagozzi & Yi, 1988). According to Table 4.5, five out of six questions are valid to be the factors of persuasion criterion because they have value more than 0.7.

## 5 CONCLUSION

According to the discussion above, some conclusions can be taken.

First, the player enjoyment measurement can be extended with one new criterion that is studied, persuasion, because it has a relationship with some other criteria, namely player skill/knowledge improvement and immersion. However, it is different from any other criterion, because it made the player think to do something instead of only memorizing it. Another difference is the player will learn something new and be persuaded to search for the relevance with the real world and then will implement it in the real world. That's why this criterion is required.

Second, the persuasion criterion is valid. It's proved by using confirmatory factor analysis where five of six factors proposed are valid after the study that involved the participants from Taiwan and Indonesia using the questionnaire. All factors loadings were greater than 0.7. Based on these measures, the construct validity of the questionnaire was judged as acceptable.

Third, *XGameFlow* model can be used as a measurement tool of player enjoyment in playing games. The conclusion can be made by analyzing the rate results of participants' answers in the questionnaire form. The results proved that the model is not very affected by the difference in a player's culture or country. It can be taken as proved because the games tested had an average rate where the difference is not more than 1.50 points or not more than 30% for each criteria when it was tested by players from Taiwan and Indonesia by using a combination of within-subject design and between-subject design.

The significance of this study is the new criterion called persuasion which can enrich the measurement model for player enjoyment in playing games. This model is important for all game designers in the game industry that want to make a well designed game without lowering the player's enjoyment. Besides, it can also be a frame model to students who are studying how to develop games that can fulfill the player enjoyment when playing their games and to teachers who are teaching the development of enjoyable games. For the players, the model can make them know how much the game can bring them enjoyment. The results of the study about player enjoyment measurement in playing are available for the game designer.

## REFERENCES


Bagozzi, R.P. & Yi, Y. (1988), On the evaluation for structural equation models. *Journal of the Academy of Marketing Science*, Vol. 16, No. 1, pp. 74–94.

Berkovsky, Shlomo, Bandari, Dipak & Kimani, Stephen. (2004). Designing Games to Motivate Physical Activity. *ACM Computer and Entertainment.*

Bogost, Ian. (2007). Persuasive Games on Mobile Devices. *Mobile Persuasion: 20 Perspectives on the Future of Behavior Change.*

Crawford, Chris. (2003). *Chris Crawford on Game Design*. U.S.A: New Riders Publishing.

Fornell, C. & Larcker, D.F. (1981), Evaluating structural equation models with unobservable and measurement errors. *Journal of Marketing Study*, Vol. 18, No. 1, pp. 39–50.

Fu, Fong-Ling, Su, Rong-Chang & Yu, Seng-Chin. (2008). *E GameFlow*: A scale to measure learners' enjoyment of e-learning games. *Science Direct*. Computers & Education, Vol. 52 (2009), pp. 101–112.

Gorsuch, R.L. (1997), Exploratory factor analysis: Its role in item analysis. *Journal of PersonalityAssessment*, Vol. 68, No. 3, pp. 532–560.

Korhonnen, Hannu, Montola, Markus & Arrasvuori, Juha. (2009). Understanding Playful User Experience Through Games. *International Conference On Designing Pleasurable Products and Interfaces, (DPPI 09)*. 13–16 October 2009, Compiegne University Of Technology, Compiegne, France, pp. 274–285.

Lambdin, C. & Shaffer, V.A. (2009). Are within-subjects designs transparent? *Judgment and Decision Making*, Vol. 4, No. 7, December 2009, pp. 554–566.

Lovlie, A.S. (2007). *The Rhetoric of Persuasive Games: Freedom and Discipline in America's Army*. http://www.duo.uio.no/sok/work.html?WORKID=54268&fid=26043. Accessed at September 13th, 2010.

Prasida, T. & Arie Setiawan. (2009). *Pengembangan Game Edukasi Online Indonesia (GEOI) dengan Menggunakan Teknologi Ajax*. Thesis. Yogyakarta: Universitas Gadjah Mada.

Scheible, J., Tuulos, V.H. & Ojala, T. (2007). Story Mashup: Design and Evaluation Of Novel Interactive Storytelling Game For Mobile and Web Users. *MUM'07*, Oulu, Finland, December, Vol. 12–14, p. 2007.

Sweetser, Penelope & Wyeth, Petra. (2005). *GameFlow*: A Model for Evaluating Player Enjoyment in Games. *ACM Computers in Entertainment*, Vol. 3, No. 3, July 2005. Article 3A.

Yannakakis, G.N. & Hallam, J. (2008). *Capturing Player Enjoyment in Computer Games.* Danish Study Agency, Ministry of Science, Technology and Innovation (project no: 274-05-0511).

Zyda, Mike. (2005). From Visual Simulation To Virtual Reality To Games. *IEEE Journal*. Vol. 38, Issue 9, pp. 25–32.

*Ergonomics in Asia – Shih & Liang (eds)*
*© 2012 Taylor & Francis Group, London, ISBN 978-0-415-68414-9*

# User research for e-reader design in higher education

Ya-Wen Chang
*Institute of Information Systems and Applications, National Tsing Hua University, Taiwan*

Yu-Chen Hsu
*Institute of Learning Sciences, National Tsing Hua University, Taiwan*

ABSTRACT: With the rapid development of handheld devices and the digitalization of images and audio, many things can be accomplished anywhere with ubiquitous computing technology. Recently, e-readers received much attention from not only the general public but also the human-computer interface (HCI) community. Within the extensive literature on usability of e-readers, relatively little effort has been done on the actual needs and behavior of users. Users were asked to provide feedback on their preference for an existing e-reader over the traditional printed books, but little about the users' needs was addressed. Therefore, the purpose of this study is to discover and discuss the users' needs for an e-reader in higher education and to investigate the factors influencing users' acceptance. The primary research questions to be addressed in this paper are user research and technology adoption. Online survey, contextual inquiry technique and in-depth interview are applied in this research. After quantitative and qualitative data analysis, we then provide a conceptual design of e-readers for higher educational uses.

*Keywords*: contextual inquiry, e-reader, HCI, higher education, user research

## 1 INTRODUCTION

With the rapid development of handheld devices and the digitalization of images and audio, many things can be accomplished anywhere with ubiquitous computing technology. In 2007, Amazon released a new e-book reader named Kindle ("Amazon Kindle,") which sold well worldwide, and Kindle soon became a trend in the information technology market. An e-book reader, also called an e-reader, is a portable device which can display text and image-based publications in digital form, called e-books (Suarez & Woudhuysen, 2010). Compared to tablet PCs, e-readers have the advantages of having a more comfortable display without flash light and have a longer battery life. Realizing the advantage of e-readers and the trend of handheld devices, more and more companies such as Barnes & Noble, Kobo and Reader PRS-500 have started developing different e-readers ("The Digital Reader from Sony; Kobo eReader Sales; Nook,").

Likewise, e-book readers received much attention from not only the general public but also the human-computer interface (HCI) community in recent years. Some studies argue that users might suffer from multiple fatigue due to the size, weight, and display of e-book readers and may have reduced reading performance (Hernon, et al., 2007; Kang, et al., 2009; Morineau, et al., 2005). There are also studies that discuss the differences between user experiences of e-book readers and conventional printed books to provide the design principles of interface (Chen, et al., 2008; Hernon, et al., 2007; Oakley & Park, 2009; Wilson, et al., 2002). In addition, some researchers also suggest possible educational applications of e-book readers (Annand, 2008; Kotsopoulos, 2005).

Education can be carried out by various means such as face-to-face discussion, distance learning via TV, e-learning via computers, and mobile learning via handheld devices.

Due to the advantage in readability, battery-saving and the convenience, e-readers possess great potential in educational applications, especially with higher education. Pilot experiments on e-readers in educational uses were carried out in Princeton University in 2009 and in Fairleigh Dickinson University in 2010. They all used the existing e-readers (Kindle DX, Sony Reader Touch etc.) to examine whether this technology is able to provide equal or improved experience than traditional printed books ("E-reader Pilot at Princeton," 2009; "E Book Reader Pilot Program At Fdu," 2010). Other researches also pay their attention to either usability tests on specific e-readers (Clark, et al., 2008; Gil-Rodríguez & Planella-Ribera, 2008; Leung, 2009; Stone, 2008; Wilson & Landoni, 2003) or the preference between e-books and conventional books (Gil-Rodríguez & Planella-Ribera, 2008; Guy & Harry, 2009).

Within the extensive literature on usability of e-readers, relatively little effort has been done on the actual needs and behavior of users. Users were asked to provide feedback on their preference for an existing e-reader over the traditional printed books, but little about the users' need was addressed. Therefore, the purpose of this study is to discover and discuss the users' need for an e-reader in higher education and to investigate the factors influencing users' acceptance. The primary research questions to be addressed in this paper are user research and technology adoption. Online survey, contextual inquiry technique and in-depth interview are applied in this research. After quantitative and qualitative data analysis, we then provide a conceptual design of e-readers for higher educational uses.

The rest of the paper is organized as follows. Section 2 describes the previous studies about e-reader in education, methodology of user research and the technology adoption model. Section 3 explains the research questions in detail. Section 4 shows the methods and participants applied in this research. Section 5 reports the experimental results of quantitative and qualitative data analysis, and Section 6 concludes the paper and lists some of the future directions.

## 2 LITERATURE REVIEW

In this section, we present related works on e-readers and user analysis in different fields to see how to conduct a user research. Then we review the technology adoption model to know about users' attitude toward technology-assisted learning and help to refine the results from user research. Finally, we refer to the interaction design method of HCI to examine the process of our research and then provide a conceptual design for e-readers used in higher education in the future.

### 2.1 *E-readers in education*

In the 1970s, there was an early e-book prototype for computers and in 1998, U.S. libraries began to provide e-books to students and the public by means of web site services (Helfer, 2000). People can download and read books as they like. In 2007, Amazon released Kindle with e-ink technology and started a new business model for e-books, and so there was an increase in studies about e-readers.

Hernon's research on undergraduate students was to investigate the use of e-books by students. They asked the participants to do some search tasks with a think aloud protocol and conducted a follow-up interview. The results showed that students access e-books for the following reasons (Hernon, et al., 2007):

- An alternative copy
- Convenience
- Cost savings
- Currency
- Efficiency
- Portability.

Janssens & Martin have also provide e-readers to students and asked them to use it for three weeks. They noted the important advantages of an e-reader the following items (Janssens & Martin, 2009):

- Convenient compact size
- Easier to take along
- Easy access to study material
- Studying any place
- Ease of use
- Accessibility files
- Ease of adding supporting study material
- Ease of use of text index.

Some studies have found that e-readers are particularly well suited to learning because of the interactive components of electronic text, and people use e-books in order to facilitate consultation, and they read then write in learning and work-related scenarios (Adler, et al., 1998; Bell, McCoy & Peters, 2002; Landoni, 2008).

### 2.2 *Methodology of user research*

The first challenge of design is to understand the users: their needs, their desires, and their approach to the work (Kim, et al., 2007). User research is a method which aims to explore users (or potential users), scenarios, use patterns and requirements of the product. User research is a reality check. It tells you what really happens in users' work practice (Nielsen, 2004). However, users sometimes don't truly understand themselves, so the best way of gathering user insights is to go to their work place and observe how they work (Kuniavsky, 2003). There are some techniques of conducting field research to gather data on users and contextual inquiry (CI) is the most frequently used method (Bondarenko & Janssen, 2005; Cui & Roto, 2008; Harmala, et al., 2010; Thursky & Mahemoff, 2007).

According to the definitive book on CI, *Contextual Design: Defining Customer-Centered Systems*, "CI is a process of field data gathering that studies a few carefully selected individuals in depth to arrive at a fuller understanding of the work practice across all users" (p. 160). Then the collected data will be transformed into five models to show the different facets of users' behavior in the interpretation section. Through inquiry and interpretation, it can reveal commonalities of the users (Beyer & Holtzblatt, 1998).

In another research into digital library learners, Edelson & Gordin also provide six facets of user research: motivation, goals, activities, background knowledge, expectation and experience needed (Edelson & Gordin, 1996). Therefore, with reference to these studies, we designed our research in the seven aspects: activity, task, scenario, motivation, goal, use of tool and user experience.

### 2.3 *Technology adoption models*

In a mobile phone adoption research, Biljon & Kotze have noted that technology adoption models specify a pathway of technology acceptance from external variables to beliefs, intentions, adoption and actual usage. They take social influence, facilitating conditions, perceived usefulness, perceived ease of use, attitude, and behavioral intention as the determining factors and three other factors (personal, demographic and socio-economic) as the mediating factors (Biljon & Kotze, 2007).

Lee et al. conducted an experiment on the mobile Internet to learn about user acceptance. They found that social influence and the self-efficacy variables have a great impact on perceived usefulness and perceived ease of use, respectively, while perceived ease of use and perceived usefulness influence actual usage frequency (Lee, et al., 2002).

Viswanath Venkatesh carried out much research about technology adoption. He found gender and age differences on the usage of technology and the perceived ease of use also play an important role in making technology adoption decisions (Morris & Venkatesh, 2000; Venkatesh, 2000; Venkatesh & Davis, 2000).

## 3 RESEARCH QUESTIONS

### 3.1 *User research*

What do college instructors and students do in teaching/learning and thus what do they need from an e-reader? In order to come up with a conceptual design of an e-reader, we have to first understand the users. Existing literature (Gunter, et al., 2003; Moore, 1993) analyzes the general learning activities in higher education contexts, but little insight of users' behavior is given. Therefore, referring to previous studies, we design our user research framework as the following Figure 1 shows.

Hence, we decided to go to the work place and observe users directly. We would like to know:

a. In what academic contexts do users do their teaching/learning activities?
b. What activities do instructors and students do in and outside the class?
c. What are their tasks in the teaching/learning activities?
d. What is their motivation and goals?
e. What kind of technology do they use to support their teaching/learning?
f. How do they use technology to support their work?

### 3.2 *Issues about e-reader adoption in higher education*

Information technology provides potential advantages to both instructors and students. However, even though we know about what users do and can provide a conceptual model that meets their requirements, this does not guarantee their acceptance of the design. There are some factors which can enhance or degrade the e-reader adoption in education. Therefore, referring to these studies and the characteristics of e-readers, we use the factors from technology adoption models in our research to explore the issues about the users' accessibility as the following Figure 2 shows.

Hence, in order to know what issues should be considered when e-readers come to higher education, we have to explore:

a. What personal factors such as socioeconomic status, self-efficacy and technology literacy can influence users' adoption of e-readers?

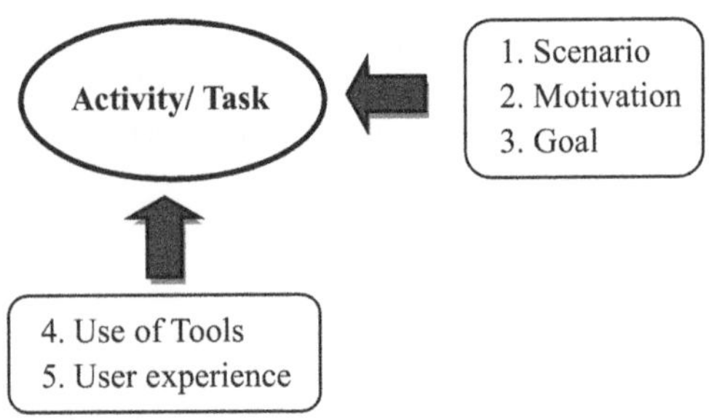


Figure 1. Framework of user research.

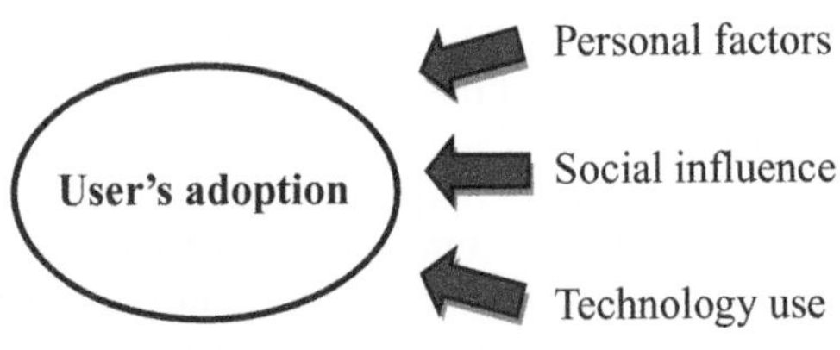


Figure 2. Three issues affecting user's adoption.

b. What social influence can facilitate/weaken technology adoption in higher education?
c. What are their past experiences in using teaching/learning tools and how do these experiences affect their acceptability to e-readers?

## 4 METHODS AND PARTICIPANTS

The study is based on the contextual inquiry technique and a semi-structured interview is used to explore the personal factors, social influence and use of technology experience about acceptability of e-readers to users.

In the early stage of this research, we conduct an online survey to find out about users' general behavior and attitude to learning and also to provide the direction for the following field study of contextual inquiry. 128 college students (51 males and 77 females) who lived in Taiwan participated in this survey. We also recruited participants for contextual inquiry through an online survey.

In the contextual inquiry, there are 10 participants including 4 college instructors and 6 students from different academic fields. Researchers first get the participant's consent then start recording and get an overview of the participant's work through an interview. Participants are observed and interviewed in the context, when doing their teaching or learning tasks. At the end of the inquiry session, researchers summarize what they have learned during interviews, and ask participants if the summary is correct.

After the contextual inquiry, participants are asked to share their experience about using tools in teaching/learning. Researchers then give a brief introduction about e-readers to participants and give them a ten-minute trial. This shows interactions between users and tools and also reveals the user's preference for e-readers.

At the end of this research, we use an affinity diagram technique and build work models (flow model, sequence model, culture model and artifact model) to interpret the e-qualitative data (Beyer & Holtzblatt, 1998). According to the affinity diagram and work model, we will generate a conceptual design of an e-reader in the form of a paper prototype and then conduct a user testing for refining the design concept. The following Figure 3 shows the process applied in this study:

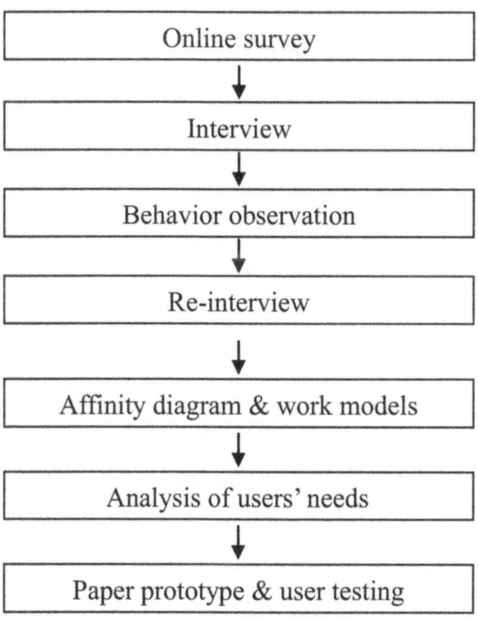


Figure 3. Research process applied in this study.

## 5 RESULTS

Data from parts of our quantitative research (questionnaire) reveals that students participate in course-related learning activities most frequently in classroom/laboratory (81%), at home/dormitory (86%) and in the library (75%). Even though Internet technology can facilitate people to participate in discussions without getting together, students tend to discuss academic issues and share their experiences in the classroom. If they have to concentrate on their individual school work, they then turn to work at home. Furthermore, the most common activity in the class is that professors deliver a lecture by PowerPoint (86%) then students take notes on their textbook (76%) or give a group presentation (70%). In tool use experience, computers (93%), electronic dictionaries (48%) and calculators (44%) are the most frequently used technology for supporting learning. This research is currently underway and further data on users will be collected later. We plan to complete it in late June.

There are several expected features for e-readers mentioned by the subjects in pilot research. The first one is that users would be able to take notes efficiently using an e-reader. Secondly, e-readers should be light and easy to be carried around. Thirdly, there shouldn't be too many entertainment functions built into e-readers; otherwise, students in class might fail to concentrate on learning. Lastly, the use of e-readers should be able to truly reduce paper consumption. These are the preliminary findings and need to be supported with more data from contextual inquiry and in-depth interviews.

## 6 CONCLUSIONS

This paper proposes a conceptual design for e-readers in higher education by user research method. We first review previous work to get a rudimentary understanding about e-readers and then design a process of contextual research method that helps us to know users' behavior and requirements in higher education. Preliminary experimental results from the questionnaire give directions for contextual inquiry. Further research including user analysis and conceptual design of e-readers is currently underway, and we plan to complete it in late June. In addition, usability problems might be other issues after using e-readers, and this may be the future work of e-reader research in HCI.

## REFERENCES

Adler, A., Gujar, A., Harrison, B.L., O'Hara, K. & Sellen, A. (1998). A diary study of work-related reading: design implications for digital reading devices, *Paper presented at the Proceedings of the SIGCHI conference on Human factors in computing systems.*

Amazon Kindle. Retrieved April 25, 2011, from http://en.wikipedia.org/wiki/Amazon_Kindle

Annand, D. (2008). Learning Efficacy and Cost-effectiveness of Print Versus e-Book Instructional Material in an Introductory Financial Accounting Course. *Interactive Online Learning*, Vol. 7, No. 2.

Bell, L., McCoy, V. & Peters, T. (2002). E-Books Go to College. *Library Journal,* Vol. 127, No. 8, 44–46.

Beyer, H. & Holtzblatt, K. (1998). *Contextual design: defining customer-centered systems.* Morgan Kaufmann Pub.

Biljon, J.v. & Kotze, P. (2007). Modelling the factors that influence mobile phone adoption, *Proceedings of the 2007 annual research conference of the South African institute of computer scientists and information technologists on IT research in developing countries.*

Bondarenko, O. & Janssen, R. (2005). Documents at Hand: Learning from Paper to Improve Digital Technologies. *Paper presented at the Proceedings of the SIGCHI conference on Human factors in computing systems.*

Chen, N., Guimbretiere, F., Dixon, M., Lewis, C. & Agrawala, M. (2008). Navigation techniques for dual-display e-book readers. *Paper presented at the Proceeding of the twenty-sixth annual SIGCHI conference on Human factors in computing systems.*

Clark, D.T., Goodwin, S.P., Samuelson, T. & Coker, C. (2008). A qualitative assessment of the Kindle e-book reader: results from initial focus groups. *Performance Measurement and Metrics,* Vol. 9, No. 2, pp. 118–129.

Cui, Y. & Roto, V. (2008). *How people use the web on mobile devices. Paper presented at the Proceeding of the 17th international conference on World Wide Web.*

E-reader Pilot at Princeton. (2009, October 27, 2010). Retrieved April 25, 2010, from http://www.princeton.edu/ereaderpilot/

E Book Reader Pilot Program At Fdu. (2010). Retrieved April 25, 2011, from http://www.slideshare.net/denoshea/e-book-reader-pilot-program-at-fdu

Edelson, D.C. & Gordin, D.N. (1996). *Adapting digital libraries for learners: accessibility vs. availability.*

Gil-Rodríguez, E. & Planella-Ribera, J. (2008). Educational Uses of the e-Book: An Experience in a Virtual University Context. In A. Holzinger (Ed.), *HCI and Usability for Education and Work* (Vol. 5298, pp. 55–62): Springer Berlin / Heidelberg.

Gunter, M.A., Estes, T.H. & Schwab, J.H. (2003). *Instruction: A models approach*: Allyn and Bacon.

Guy, J. & Harry, M. (2009). The Feasibility of E-Ink Readers in Distance Learning: A Field Study.

Harmala, M., Ko, T., Jonsson, A., Garg, G. & Chia, Y.-w. (2010). *Night Beacon: A system to empower people to walk with confidence at night.* Paper presented at the CHI 2010.

Helfer, D. (2000). E-books in libraries: Some early experiences and reactions. *Searcher,* Vol. 8, No. 9, pp. 63–65.

Hernon, P., Hopper, R., Leach, M.R., Saunders, L.L. & Zhang, J. (2007). E-book Use by Students: Undergraduates in Economics, Literature, and Nursing. [doi: DOI: 10.1016/j.acalib.2006.08.005]. *The Journal of Academic Librarianship,* Vol. 33, No. 1, pp. 3–13.

Janssens, G. & Martin, H. (2009). The Feasibility of E-Ink Readers in Distance Learning: A Field Study. *International Journal of Interactive Mobile Technologies (iJIM),* Vol.3, No.3, pp. 38–46.

Kang, Y.-Y., Wang, M.-J.J. & Lin, R. (2009). Usability evaluation of E-books. *Displays,* Vol. 30, No. 2, pp. 49–52.

Kim, H., Heo, J., Shim, J., Kim, M., Park, S. & Park, S. (2007). Contextual Research on Elderly Users' Needs for Developing Universal Design Mobile Phone. In C. Stephanidis (Ed.), *Universal Acess in Human Computer Interaction. Coping with Diversity* (Vol. 4554, pp. 950–959): Springer Berlin/Heidelberg.

Kobo eReader Sales. Retrieved April 25, 2010, from http://koboereader.com/

Kotsopoulos, D. (2005). E-Learning with visual math: An E-book review. *Canadian Journal of Science, Mathematics and Technology Education,* Vol. 5, No. 4, pp. 517–519.

Kuniavsky, M. (2003). *Observing the User Experience: A Practitioner's Guide to User Research.* San Francisco, CA Morgan Kaufmann Publishers.

Landoni, M.A. (2008). The active reading task: e-books and their readers, *Paper presented at the Proceeding of the 2008 ACM workshop on Research advances in large digital book repositories.*

Lee, W.J., Kim, T.U. & Chung, J.Y. (2002). User acceptance of the mobile Internet. *In M-Business 2002.*

Leung, L.A.M.S. (2009). Usability and usefulness of eBooks on PPCs: How students¡¦ opinions vary over time. *Australasian Journal of Educational Technology,* Vol. 25, No. 1, pp. 30–44.

Moore, M.G. (1993). Three types of interaction. *Distance education: New perspectives*, 19.

Morineau, T., Blanche, C., Tobin, L. & Guéguen, N. (2005). The emergence of the contextual role of the e-book in cognitive processes through an ecological and functional analysis. *International Journal of Human-Computer Studies,* Vol. 62, No. 3, pp. 329–348.

Morris, M.G. & Venkatesh, V. (2000). Age differences in technology adoption decisions: Implications for a changing work force. *Personnel psychology,* Vol. 53, No. 2, pp. 375–403.

Nielsen, J. (2004). Acting on User Research. *Alertbox* Retrieved April 26, 2011, from http://www.useit.com/alertbox/20041108.html

Nook. Retrieved April 25, 2010, from http://www.barnesandnoble.com/

Oakley, I. & Park, J. (2009). Motion marking menus: An eyes-free approach to motion input for handheld devices. [doi: DOI: 10.1016/j.ijhcs.2009.02.002]. *International Journal of Human-Computer Studies,* Vol. 67, No. 6, pp. 515–532.

Stone, N. (2008). The e-reader industry: Replacing the book or enhancing the reader experience? *Scroll,* Vol. 1, No. 1.

Suarez, M.F. & Woudhuysen, H.R. (2010). *The Oxford Companion to the Book.* Oxford; New York: Oxford University Press.

The Digital Reader from Sony. Retrieved April 25, 2010, from http://ebookstore.sony.com/reader/

Thursky, K.A. & Mahemoff, M. (2007). User-centered design techniques for a computerised antibiotic decision support system in an intensive care unit. *International Journal of Medical Informatics,* Vol. 76, No. 10, pp. 760–768.

Venkatesh, V. (2000). Determinants of Perceived Ease of Use: Integrating Control, Intrinsic Motivation, and Emotion into the Technology Acceptance Model. *Info. Sys. Research,* Vol. 11, No. 4, pp. 342–365.

Venkatesh, V. & Davis, F.D. (2000). A Theoretical Extension of the Technology Acceptance Model: Four Longitudinal Field Studies. *Manage. Sci.,* Vol. 46, No. 2, pp. 186–204.

Wilson, R. & Landoni, M. (2003). Evaluating the usability of portable electronic books. *Paper presented at the Proceedings of the 2003 ACM symposium on Applied computing.*

Wilson, R., Landoni, M. & Gibb, F. (2002). A user-centred approach to e-book design. [Case study]. *The Electronic Library,* Vol. 20, No. 4, p. 9.

*Ergonomics in Asia – Shih & Liang (eds)*
*© 2012 Taylor & Francis Group, London, ISBN 978-0-415-68414-9*

# The effects of WEB 2.0 interface on performance—a case study of user intuitive clearness and transparency in WEB interface

Chih-hsiung Pan
*Department of Industrial Engineering, Chung-Yuan Christian University, Taiwan*

ABSTRACT: Interactive webpage interface design emphasizes the importance of being user-oriented and being user-centered, in order to elevate the performance of interaction between user and the computer. The design standard that webpage design adopts is an evaluation method of wide availability. Under general acknowledgment, usability is the primary factor that affects the performance of a website. The obstacle that has been concerned chronically is to eliminate the tremendous gap between webpage designer and user. The current study introduces the NASA task load index to evaluate whether differences would occur when a different subject-target interface is applied to users with diverse mental models or groups. Through multi-gradient approach, participants are able to conduct a self-evaluation objectively and record their task load, calculating its mean value, standard variation, relevancy and significance. To produce a well-designed user interface, designers should have comprehensive understanding of user features and take all functions or application items, usability and overall design framework of the website into carefully consideration.

*Keywords*: webpage, interface, user-oriented, intuitive (user interface), usability, performance

## 1 GENERAL INSTRUCTIONS

Generally speaking, intuitive user-computer interaction standards or intuitive user mental models are widely adopted in daily life and academic theses.

There is no consistent definition, even for intuitive user mental models found in previous research studies to evaluate user's performance assessment toward the intuitive webpage.

The goal of the current study is to dispose the obstacles to intuitive application or interaction, in order to make them easier to use, and for enhancing definition and transparency on new webpage interface design. NASA-TLX (Task Load Index) and questionnaire are introduced to evaluate and study the intuitive user mental model. Also, a minimum list is built by users, in order to let designers take the performance evaluation and benchmark into consideration, and transform them into an intuitive user interface website.

A website interface designer will rely on actual website users for website design and experiences. A designing group often takes account of experience, cultural background, corporate principles and requirements of customers to make assumptions about the user mental model before releasing the design. This process would provoke user feedbacks: for instance, color palette dislike, privacy issues.

When users purchase products online, designers must evaluate the performance appropriately, in order to fully supply the potential market of the website and make adjustments to serve the need of users.

We are living in a globalized world but local or cultural identities strongly influence our patterns of behavior and our interpretation of behavior in others by establishing norms and values. This opening section provides an overview of website design elements and related issues. We have identified over 60 categories of websites such as art, banking, games, and politics.

These can be further subdivided into online shopping, identity (company, institution), information, education, entertainment, and community websites.

In some case, websites cover multiple categories.

The usability.gov site summarizes website design considerations as content organization, titles/headings, page length, page layout, font/text size, reading and scanning, links, graphics, search, navigation. (Nielsen 1993), and Chan et al. (2005) consider website design as the overall structure of the site, page design and content design (how the text and graphic content on each page is designed).

UI is important because it affects the feelings, the emotions, and the mood of your users. If the UI is wrong and the user feels like they can't control your software, they literally won't be happy and they'll blame it on your software. If the UI is smart and things work the way the user expected them to work, they will be cheerful as they manage to accomplish small goals.

To make people happy, you have to let them feel like they are in control of their environment. To do this, you need to correctly interpret their actions. The interface needs to behave in the way they are expecting it to behave.

## 2 LTERATURE REVIEW

The study is based on past experiences and basic intuitive assumptions. Previous research studies have shown that intuitiveness is a type of unconscious cognitive processing. Users may make use of their experiences and directly interact with the interface or product to acquire knowledge or similar experiences. When developing a new interface or product, we also take advantage of these experiences and an intuitive classification to turn them into a rule (Blackler, Popovic, and Mahar, 2003a, b, 2004a, b, 2005).

Intuitiveness depends on user's experience to become familiar with rules and dynamic status of the interface; it derives from the implicit memory of the brain and long-term experiences. Intuitions tend to come from the user's experience, which could not be described specifically; but it seems not that mysterious. Most people solve problems primarily by their everyday experience and intuitions. If an experience that relates to an event exists, one may sense it promptly. Intuition seems to appear in a field that one is familiar with. We call this phenomenon a "metaphor" (Blackler, Popovic & Mahar).

An important factor of user interface is that it affects user's perception, mood and emotions. A defective user interface may bring negative feedbacks, because it makes users feel unable to control the software. On the contrary, a brilliant user interface enables users to do everything smoothly; they even feel satisfied just doing a trifle. Users may contribute their behaviors as a "metaphor," because they feel like they are controlling the environment. To achieve this, designers must correctly interpret user's behavior, and the user interface must function in a way the users expect.

Consequently, to effectively handle complex issues derived from web user interface design, the rules of design must be revised. Compared with art and design, subjective preference is not qualified for a judgment standard of interface design. An impressive user interface may not be user-friendly, while a simple interface may be brilliant.

Also the general public often uses their "intuition" to judge the quality of a design. It is applicable to most cases.

Intuition determines whether users can predict the usage and result of a certain UI component (e.g. a link or a button) before taking any steps. A well-designed user interface should be predictable, and let users feel everything is under their control. For instance, the metaphor of the UI component from the form below, of course, is the link.

A square checkbox means a multi selection, of course. Maybe it's not the most splendid design, but at least users would not have to learn and understand the relationship between components again. Beautiful design is never a must, but usability is inevitable. To make a user interface predictable, users should at least sense the existence of UI components; if not, they wouldn't be used. For example, the form of a hyperlink should be intuitive enough to let users identify them (and clickable) and never mixed up with the text; this is so-called consistency, which prevents users from making wrong assumptions and inappropriate operations.

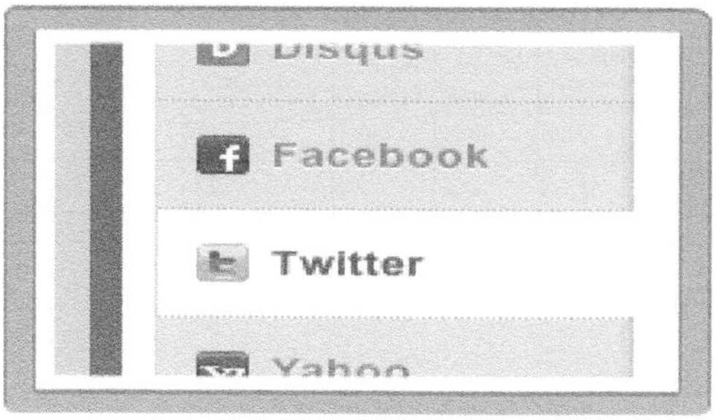


Figure 1. Show word.

Generally, the design of a website is based on elements such as visual, information and navigation. To compete with millions of website counterparts, the mission of a webpage designer is to design a website that attracts users and keeps them on it. A user may use a web service for its functions or content, or even instant messages that enables interaction with other users. In consideration of these issues, designers should apply human-machine interface and standard applications to their websites. A comprehensive website should focus on their target market more, such as keeping their users and influencing their consumer behavior. In view of the more rapid development of websites, designers are facing a refurbishment or redesign of their website.

At the later period of testing (verification period), the website should be evaluated and a standard of usability set up as well as a criterion (for example, the process time of a task). During this period, evaluation could include the joint operation of all components, in order not to release defective products into the market and to raise issues such as recall or adjustment. Therefore, an effective and instant approach should be adopted to evaluate promptly and provide specific data so that designers could redesign or make adjustments.

## 3 RESEARCH METHOD

The data of the current study are collected by giving out questionnaires, and then analyzing the relevancy and significance by SPSS. The detail is provided below.

Methodology

A certain webpage is given to participants to browse and operate. Each participant is designated to accomplish three tasks. The webpage and tasks given to each participant are the same.

After participants have finished browsing and completed the tasks, a questionnaire is assigned.

Subjects

Subjects of the current study are Internet users that surf the Internet over eight hours a day.

The number of total questionnaires given out is 30 copies.

Questionnaire Design

The study adopts questionnaires of National Aeronautics and Space Administration Task Load Index (NASA-TLX) to collect data needed, which includes three parts:

I. Eight questions about webpage interface operating, including an inverse question at question 7.
II. Load Factor weight ratio participants would compare each two of the six mental load factors including Mental Demands, Physical Demands, Temporal Demands, Effort, Performance and Frustration according to their perceptions throughout the test, and choose a more important factor. Then, participants calculate the weight of each load factor according to the NASATLX ML formula.

Relevancy Analysis of Intuitiveness Average

| | Average of Intuitiveness | Load of Mental Demand | Load of Physical Demand | Load of Performance | Load of Frustration | Load of Effort | Time |
|---|---|---|---|---|---|---|---|
| Pearson Correlation | 0.083 | 0.083 | -503* | -354 | -566* | -577* | -384 |
| Significance (Two-tailed) | 0.792 | 0.729 | .024 | .126 | .009 | .008 | .095 |
| Counts | 30 | 30 | 30 | 30 | 30 | 30 | 30 |

Figure 2. Relevancy analysis.

| Coefficient | | | | | | |
|---|---|---|---|---|---|---|
| Mode | | Non-standardized Coefficient Estimate of B | Standard Error | Standardized Coefficient Beta Distribution | t | Significance |
| 1 | Constant | .646 | .189 | | 3.427 | .006 |
| | VAR00001 | 0.14 | .025 | .094 | .556 | .589 |
| | VAR00002 | -.034 | .019 | -.339 | -1.808 | .098 |
| | VAR00003 | -.059 | .019 | -.619 | -3.020 | .012 |
| | VAR00004 | -.010 | .024 | -.083 | -.427 | .678 |
| | VAR00005 | .010 | .018 | .117 | 570 | .580 |
| | VAR00006 | -.060 | .023 | -.463 | -2.693 | .023 |
| | VAR00007 | .044 | .027 | .337 | 1.631 | .131 |
| | VAR00008 | .038 | .022 | .376 | 1.692 | .119 |

a. Dependent Variable : NASATLX

Figure 3. Significance analysis.

NASATLX ML = WMD (Weight of Mental Demand) × LMD (Load of Mental Demand) + W D (Weight of Physical Demand) × LPD (Load of Physical Demand) + WTD (Weight of Temporal Demand) × LTD (Load of Temporal Demand) + W P (Weight of Performance) × LP (Load of Performance) + WE (Weight of Effort) × LE (Load of Effort) + WF (Weight of Frustration) × LF (Load of Frustration).

Load Level: Participants grade each of the six user mental load factors according to their perceptions during the test in 10 gradations from 0 to 100.

## 4 RESULTS

The analysis result of Figure 2 shows negative correlation between degree of intuition and Physical Demand, Temporal Demand, Performance, Frustration and Effort. Positive correlation is observed between degree of intuition and Mental Demand. Furthermore, a significant difference is observed between degree of intuition and Physical Demand, Frustration as well as Effort. Result of Figure 3 reveals that "Process of webpage operating" and "Fonts and images of webpage" have a significant difference with mental load factors.

Based on the results, the easier the operating process of a webpage, the lower the mental load is observed. Webpages that presented in plain texts would bring a lower mental load than those in images.

To summarize the results of analysis above, webpage design should focus on the process of operation and font display and seek improvements.

For all teams, the results are exhibited by formally appointed leaders, or by non-appointed emerging leaders (Klimoski & Jones 1995). Another similar helping behavior between team members is supporting what could be defined as monitoring the activities of other team members, taking action to correct their errors, giving and receiving feedback in a non-defensive manner, and providing and seeking assistance or backup when needed (Smith-Jentsch, 1995; Johnston et al., 1997; Weil et al., 2004). In brief, both leadership and support would involve assisting the performance of other team members, so this study would integrate them into one aspect.

## 5 CONCLUSIONS

The aim of the current study is to provide discussions concerning the performance evaluation of intuitive webpage interface design. At present, there is still no standard evaluation method or suggestions about interface design in all dimensions. To pursue the maximum performance under limited technology, resources and budget, webpage designers must rely on experience. Hence, the result of the current study provides a strong support.

If the cognitive behavior of an application is completely in accordance with user's expectations, it is considered to be a well-designed user interface. When an application is being used for the first time, its intuitive result is especially important. If users had not had the chance to be familiar with its functions, they could not be used correctly. Based on the previous and current research study results, suggestions for future development are made below:

1. Messages should be represented with "minimum user list" if possible, in order to save time for reference. Also, the information provided in the website should be categorized so as to be presented in a simple and clear manner.
2. In this way, users could browse by clicking and cut down their time to search manually. User's mental load may also be lower.

3. Minimum user list design:

| Overall remark | Provide adequate information | Texts | Images |
|---|---|---|---|
| Consistent interface | Good instructions that cohere with user experiences | Clear fonts | Are images used coherently |
| Separate different elements clearly | Fluent and easy control | Font size is appropriate for reading | Are the images helping memory |
| The overall structure of the interface is comfortable | No burden of memorizing | Unified fonts | Easy to distinguish |
| | Auto recovery on errors | | Helping the operation |

4. The design of the interface should be improved in consideration of the theme of the webpage and the using process. These would boost the performance and lower user's mental load, as well as increasing more intuitive operation for webpage users.
5. Usability is introduced to build up standards and principles of webpage design.

## REFERENCES


Abla, G., Kim, E.N., Schissel, D.P. & Flanagan, S.M. (2010), Customizable scientific web portal for fusion research. *Fusion Engineering and Design,* Vol. 85, pp. 603–607.

Blackler, Popovic & Mahar MMI-Interaktiv, Nr. 13, Aug 2007, ISSN 1439–7854,

Carvalho, P.V.R., Dos Santos, I.L., Gomes, J.O., Borges, M.R.S. & Guerlain, S. (2008), Human factors approach for evaluation and redesign of human–system interfaces of a nuclear power plant simulator. *Displays,* Vol. 29, pp. 273–284.

Chen, S.Y. & Macredie, R.D. (2002), Cognitive Styles and Hyper media navigation: Development of a learning model. *Journal of The American Society for Information Science and Technology*, Vol. 53, pp. 3–15.

Chan, A.H.S. & Lee, P.S.K. (2005), Effect of display factors on Chinese reading times, comprehension scores and preferences. *Behaviour & Information Technology,* Vol. 24, pp. 81–91.

Cheung, B., Hofer, K., Heskin, R. & Smith, A. (2004), Physiological and Behavioral Responses to an Exposure of Pitch Illusionin the Simulator, *Aviation, Space, and Environmental Medicine*, Vol. 75, pp. 657–665.

Ducrot, S. & Pynte, J. (2002), What determines the eyes' landing position in words? Perception and Psychophysics. 64, pp. 1130–1144.

Dyson, M.C. & Haselgrove, M. (2000), The effects of reading speed and reading patterns on the understanding of text read from screen. *Journal of Research in Reading,* Vol. 23, pp. 210–223.

Dyson, M.C. & Haselgrove, M. (2001), The influence of reading speed and line length on the effectiveness of readingfrom screen. *Human-Computer Studies,* Vol. 54, pp. 585–612.
Ha, J.S., Seong, P.H., Lee, M.S. & Hong, J.H. (2007), Development of human performance measures for human factors validation in the advanced MCR of APR–1400. *IEEE Transactions on Nuclear Science,* Vol. 54, No. 6, pp. 2687–2700.
Isaac, A., Shorrock, S.T. & Kirwan, B. (2002), Human error in European air traffic management (ATM): the HERA project. *Reliability Engineering and System Safety* Vol. 75, pp. 257–272.
Jiao, K., Li, Z.Y., Chen, M. & Wang, C.T. (2004), Effect of different tvibration frequencies on heart rate variability and driving fatigue in normal drivers. *Int Arch Occup Environ Health,* Vol. 77, pp. 205–212.
Kacmar, Z. & Carey, J. (1991), Assessing the Usability of icons in UserInterface. *Behavior and Information Technology,* Vol. 10, pp. 443–457.
Kong, Y.K. & Lowe, B.D. (2005), Optimal cylindrical handle diameter for grip force tasks. *International Journal of Industrial Ergonomics,* Vol. 35, pp. 495–507.
Kuijt-Evers, L.F.M., Groenesteijn, L., de Looze, M.P. & Vink, P. (2004), Identifying factors of comfort in using hand tools. *Applied Ergonomics,* Vol. 35, pp. 453–458.
Lin, C.J., Jou, Y.T., Yenn, T.C., Hsieh, T.L. & Yang, C.W. (2009), A Study of Control Room Staffing and Workload from the Human Information Processing Perspective, *2009 IEEE International Conference on Networking, Sensing and Control,* Okayama City, Japan.
Min, B.C., Chung, S.C., Min, Y.K. & Sakamoto, K. (2004), Psycho physiological evaluation of simulator sickness evoked by a graphic simulator. *Applied Ergonomics,* Vol. 35, pp. 549–556
Neilson, J. (1993), Usabilitv Engineering, California: AP PROFESSIONAL Press., pp. 97–155.
Sanders, M.S. & McCormick, E.J. (1993), *Human Factors in Engineering and Design.* McGraw Hill, New York.
Scholtes, P.R. (1988), *The team training handbook.* Madison. WI: Joiner Associates.
Sebok, A. (2000), Team performance in process control: influences of interface design and staffing levels. *Ergonomics,* Vol. 43, No. 8, pp. 1210–1236.
Uang, S.T. & Hwang, S.L. (2002), Effects on driving behavior of congestion information and of scale of in-vehicle navigation systems. *Transportation Research Part C,* Vol. 11, pp. 423–428.
Vergara, M. & Page, A. (2002), Relationship between comfort and back posture and mobility in sitting-posture. *Applied Ergonomics,* Vol. 33, pp. 1–8.
Vitu, F., McConkie, G.W., Kerr, P. & O'Regan, J.K. (2001), Fixation Location effects on fixation durations reading: an inverted optimal viewing position effect. *Vision Research.* Vol. 41, pp. 3513–3533.
Wang, M.J., Wang, E.M. Y., Lin, Y.C. & Lin, Y.Z. (2002), Computer Aided VDT Workstation Design Based on the Anthropometric Data of Taiwanese Workers. *Ergonomic Study,* Vol. 1, pp. 47–56.
Yaginuma, Y., Yamada, H. & Nagai, H. (1990), Study of the relationship between lacrimation and blink in VDT work. *Ergonomics,* Vol. 33, pp. 799–809.
Yang, G.Z., Laura, D.M., Hu, X.P. & Rowe A. (2002). Visual search: Psychophysical models and pracital applicatios. *Image and Vision computer,* Vol. 20, pp. 273–287.

*Ergonomics in Asia – Shih & Liang (eds)*
*© 2012 Taylor & Francis Group, London, ISBN 978-0-415-68414-9*

# A proposal for three useful methods of GUI design usability evaluation

T. Yamaoka
*Wakayama University, Wakayama, Japan*

I. Hirata
*Hyogo Prefectural Institute of Technology, Hyogo, Japan*

ABSTRACT: Three GUI design usability evaluation methods are introduced in this paper. They are GUI design checklist, SUM (Simple Usability Evaluation Method) and usability task analysis. After their characteristics are explained, the three GUI designs were evaluated using them as examples. The results suggest the following. The GUI checklist is a very convenient method and can also be used as the minimum requirements to construct the GUI design. The SUM can grasp the characteristics of the GUI design and the structure of problems easily. On the other hand the usability task analysis can evaluate GUI designs quantitatively and acquire new user requirements with which to modify the bad points of usability.

## 1 INTRODUCTION

The purpose of this paper is to describe the characteristics of three GUI design usability evaluation methods. They are (1) a checklist, (2) SUM (simple usability evaluation method) and (3) usability task analysis. They were designed for designers and engineers. Generally speaking, a lot of GUI design usability evaluation methods were designed for usability professionals, not designers and engineers.

## 2 GUI DESIGN CHECKLIST

The GUI design checklist consists of 4 categories and 16 evaluation items. They are as follows.

1. The evaluation items based on the SUM (explained later).
    1. Operation time is appropriate
    2. Presentation of various information
    3. Easy to see.
2. The evaluation items based on the visualization three principles.
3. The evaluation items based on the GUI design six principles.
    4. Important information is emphasized
    5. Layout and information is simple
    6. Consistency is examined
    7. The visual clues make acquisition of information and navigation easy
       This item is also extracted from the SUM
    8. Understood terms are used
    9. The mapping among information is appropriate
10. Visual feedback.
11. The user interface is constructed based on the mental model.

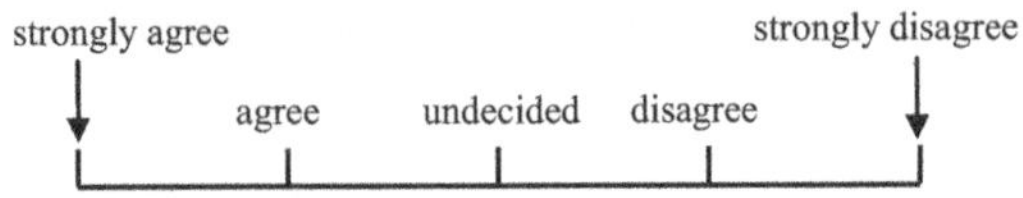


Figure 1. Five-point scale of the GUI design checklist.

4. The evaluation items regarding the general user interface.
   12. The layer of systems is understandable
   13. The operation time is understandable
   14. To grasp the system wholly is easy
   15. The system enables errors to be easily corrected
   16. The user interface is flexible and customized.

The evaluation is done using a five-point scale: strongly agree (5), agree (4), undecided (3), disagree (2), strongly disagree (1) (Figure 1).

## 3 SUM

SUM (simple usability evaluation method) (Yamaoka, 2010) is a usability evaluation method for GUI mainly based on 3P (point) task analysis (Yamaoka, 2001) and ASQ (Lewis, 1995). The SUM has three usability items namely "navigation", "terms and redundancy of information" and "operation time and others" (Table 1).

Most problems are mainly caused by navigation (relationship among operation parts), terms and mental model (functional model and structural model). Functional model means a kind of procedure, namely "How to use it". Structural model means "How it works" and shows the structure of system. Operation time and others are an evaluation item in order to extract other problems. The operation time can be understood as a clue in order to extract usability problems.

Tasks are evaluated mainly from the viewpoint of navigation, terms and redundancy of information.

Criteria of the evaluation are as follows.

1. When users can go to next task in spite of having one problem to operate, the level of the problem is defined as bad-1. If a task has one problem (bad-1) based on three criteria (navigation, term and redundancy, operation time and others), the evaluation score is 0.
2. If a task has two problems or more (bad-1) based on three criteria (navigation, terms and redundancy, operation time and others), the evaluation score is −1.
3. When users cannot go to next task by one problem, the level of the problem is defined as bad-2.
   If a task has one problem (bad-2) based on three criteria (navigation, terms and redundancy, operation time and others), the evaluation score is −1.
4. If a task has no problem based on three criteria (navigation, terms and redundancy, operation time and others), the evaluation score is +1.
5. If users take a lot of time to operate in spite of bad-1, the bad-1 becomes bad-2. Although the length of the time depends on tasks, permitted time length is usually within 60 or 120 seconds.

The procedure is as follows.

1. The scene and the tasks to be evaluated are identified.
2. The task is analyzed based on "navigation", "terms and redundancy of information" and "operation time and others". Problems are extracted based on these items.
3. The bad-1 or bad-2 problems are decided by judgment as to whether the user can go to the next task or not.
4. The evaluation score is decided based on the number of bad-1 and bad-2 problems.

Table 1. The SUM.

| Task (subtask) | Navigation<br>Functional and structural model | Terms and redundancy of information | Operation time and others | Evaluation (+1, 0–1) |
|---|---|---|---|---|
| Task-1 | Bad-1 (can go to next) | Good | Good | 0 |
| Task-2 | Bad-2 (cannot go to next) | Good | Good | –1 |
| Task-3 | Good | Bad-1 | Bad-1 | –1 |
| Task-4 | Good | Good | Good | +1 |
| Task-5 | Bad-2 | Bad-2 | Bad-1 | –1 |
| Task-n | Bad-1 | Bad-2 | Good | –1 |

5. The synthetic evaluation score is calculated by the total of task scores. As the numbers of bad-1or bad-2 regarding the three evaluation items are counted, we can understand the bad evaluation items.
6. The problems as to bad-2 are examined. Hence the important requirements are extracted.

## 4 USABILITY TASK ANALYSIS

Originally usability task analysis was developed for products. The usability task analysis described in this paper is focused on GUI design. As the usability task analysis is also a kind of task analysis, each task are evaluated from view point of the good and bad points of the task (Table 2). The participants are asked for the good and bad points regarding each task. Next they evaluate the good and bad points using strongly agree (3), agree (2) and neutral (1).

The good points become user requirements while the bad points are changed into good user requirements. The user requirements are structured. A design concept is constructed based on the user requirements.

The synthetic evaluation score is calculated on the basis of good (3), barely acceptable (2) and poor (1). Finally the synthetic evaluation of all tasks is also evaluated in the same way as the evaluation of each task. These data can be analyzed using the quantification 1 and FCA (formal concept analysis) and so on.

The evaluation score in the synthetic evaluation as the dependent variable and evaluation score in tasks as the independent variable are analyzed using the quantification 1. The important tasks which influence the evaluation score in the synthetic evaluation are chosen.

As the very important keywords are chosen by the quantification 1, the GUI design is evaluated using binary data based on the keywords. The evaluation data are analyzed using FCA. Therefore the relationship between GUI design and keywords becomes clear.

## 5 APPLYING THE GUI DESIGN CHECKLIST TO AN EVALUATION OF A GUI DESIGN

A GUI design to get a movie ticket (Figure 2) was proposed and evaluated using the GUI design checklist.

1. Method

Twelve participants were evaluated using the checklist and the protocol analysis. The participants were Wakayama University students. After they operated the GUI of the movie ticket, they evaluated the GUI using the GUI design checklist. The experimenter is a Wakayama University student who majored in design and ergonomics, and designed the GUI design of a movie ticket.

Table 2. The usability task analysis.

| Task (subtask) | GUI design (A) | | GUI design (n) | |
|---|---|---|---|---|
| Task-n | Good points | Comments *********** | Good points | Comments *********** |
| | Bad points | Comments ******** | Bad points | Comments ******** |
| | Synthetic evaluation score | score 3 | Synthetic evaluation score | score 2 |

Figure 2. A GUI design of a movie ticket (Decide the date to use the ticket).

2. Results
The experimenter' comments are as follows.

1. It was efficient and evaluated in a short time compared with the protocol analysis.
2. The cause of problems could not be identified by the checklist.
3. The checklist could find problems which could not be extracted by the protocol analysis.

## 6 APPLYING THE SUM TO AN EVALUATION OF A GUI DESIGN

A GUI design of a digital newspaper (Fig. 3) was proposed and evaluated using the SUM.

1. Method
Five participants evaluated every GUI design screen according to the SUM procedure. They filled in the blanks of the SUM format (Tables 1 and 3). The participants were Wakayama University students. The experimenter was a Wakayama University student who majored in design and ergonomics, and designed the GUI design of a digital newspaper.
2. Results
The experimenter' comments are as follows.

1. It was an easy method.
2. It was difficult for the participants to understand the meanings of the evaluation terms.
3. It is very convenient to analyze data from various viewpoints (Table 3, Figure 4).

The Table 3 shows the task-1 is bad and the task-6, 7, 8 are good. The Figure 4 shows the navigation result is not good and should be redesigned.

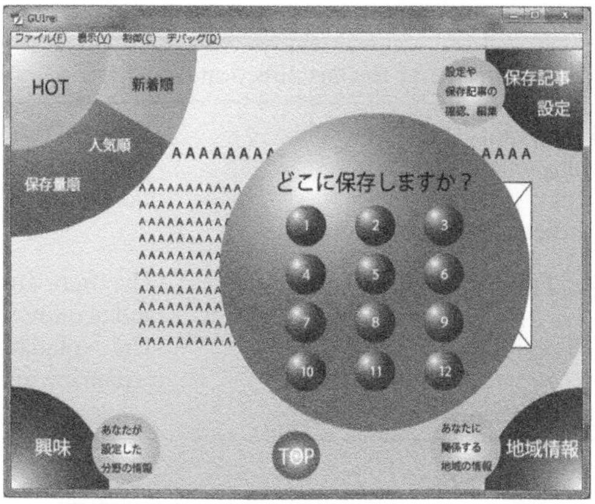


Figure 3. A GUI design of a digital newspaper.

Table 3. The results of SUM.

| | Participants | | | | |
|---|---|---|---|---|---|
| Tasks | A | B | C | D | E |
| Task-1 | –1 | 1 | –1 | –1 | –1 |
| Task-2 | 1 | 1 | 1 | –1 | 1 |
| Task-3 | 0 | 0 | 1 | 1 | 1 |
| Task-4 | 0 | 1 | 1 | 0 | 0 |
| Task-5 | 0 | 0 | 0 | 0 | 0 |
| Task-6 | 1 | 1 | 1 | 1 | 1 |
| Task-7 | 1 | 1 | 0 | 1 | 1 |
| Task-8 | 1 | 1 | 1 | 1 | 1 |

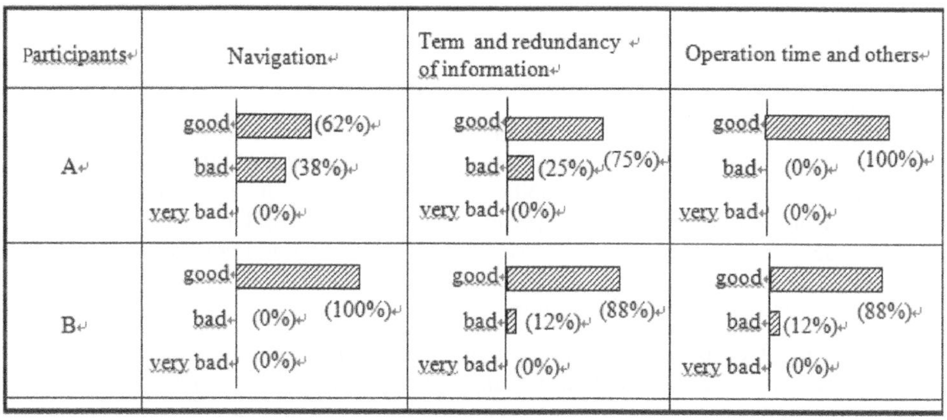


Figure 4. The ratio of good, bad and very bad results for three criteria for the digital newspaper.

## 7 APPLYING THE USABILITY TASK ANALYSIS TO AN EVALUATION OF A GUI DESIGN

A GUI design of a bus direction board (Fig. 5) was proposed and evaluated using the usability task analysis.

Table 4. The results of usability task analysis.

| Tasks | The GUI design of the bus direction board | |
|---|---|---|
| Task-1 | Good points | The icon is understandable |
| | Bad points | The "return" button should be located on the right side |
| | Synthetic evaluation score: 3 | |
| Task-2 | Good points | The users can operate accurately and acquire the information |
| | Bad points | The explanation and functional expansion are a problem |
| | Synthetic evaluation score: 4 | |

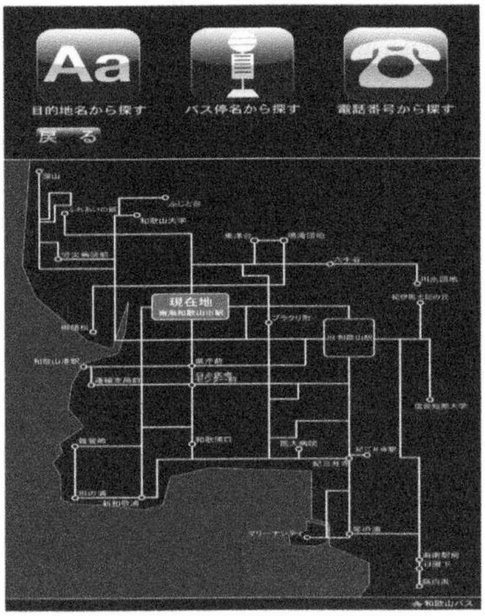


Figure 5. A GUI design of a bus direction board.

1. Method

Five participants evaluated the GUI design using the usability task analysis. After they answered the good points and bad points, and evaluated every task (GUI screen), an experimenter filled in the blanks of the usability task analysis format. The participants were Wakayama University students. The experimenter is a Wakayama University student who majored in design and ergonomics and designed the GUI design of a bus direction board.

2. Results

The experimenter' comments are as follows.

1. It is a very effective method as the good and bad points are extracted easily.
2. The data can be analyzed using the quantification 1 and FCA and so on quantitatively.
3. The usability and design of system are grasped roughly.

## 8 DISCUSSION

The findings extracted from the results of three usability method tests are as follows.

1. The three methods can be analyzed quantitatively at small cost compared with the protocol analysis. The protocol analysis is a typical usability evaluation method which has

been employed by designers and engineers. However it takes a lot of time to analyze data because of there being no rule and no format to analyze. As the three methods have a rule and format to analyze data, it is easy and efficient for them to acquire and analyze data. These findings are supported based on the experimenter's opinion.
2. The GUI design checklist is a very convenient method and can also be used as the minimum requirements to construct the GUI design. The SUM can grasp the characteristics and the structure of problems of the GUI design easily. On the other hand the usability task analysis can acquire new user requirements with which the bad points are modified.
3. The criteria enable us to select three methods:

    1. If designers or engineers want to evaluate a GUI design easily and quickly, the GUI design checklist is recommended.
    2. If designers or engineers want to grasp the characteristics of a GUI design easily and identify problematical tasks, the SUM is recommended.
    3. If designers or engineers want to evaluate GUI designs quantitatively and acquire new user requirements, the usability task analysis is recommended.

## 9 CONCLUSION

The GUI design checklist, SUM and usability task analysis are very easy and efficient methods at small cost.

1. The GUI design checklist is recommended for easy evaluation in a short time.
2. The SUM is recommended for grasping the problems of usability and characteristics of a GUI design easily.
3. The usability task analysis is recommended for evaluating GUI designs quantitatively and acquiring new user requirements.
4. When designers and engineers evaluate GUI designs quickly, the GUI design checklist and the SUM are recommended.
5. When designers and engineers evaluate GUI designs deeply and acquire new user requirements, the usability task analysis is recommended.

## REFERENCES

Lewis, J.R. (1995). IBM computer usability satisfaction questionnaires: psychometric evaluation and instructions for use. *International Journal of Human–Computer Interaction,* Vol. 7, No. 1, pp. 57–78.

Yamaoka, T. (2001). Human Design Technology as a new product design method, *First International Conference on Planning and Design,* Taipei: CD JP003-F 01–10.

Yamaoka, T. & Tukuda, S. (2010). A proposal of simple usability evaluation method and its application, *Proceedings of the 9th Pan-Pacific Conference on Ergonomics,* pp. 63–66, ISBN 978-0-415-58608-5, Kaohsiung, Taiwan, November 7–10, 2010.

# *Part IV: Biomechanics and anthropometry*

*Ergonomics in Asia – Shih & Liang (eds)*
*© 2012 Taylor & Francis Group, London, ISBN 978-0-415-68414-9*

# Taiwan high-tech industries hiring staff with physical and mental disabilities to work in clean-room: Experience sharing by CMI Corp

Jyh-Chau Wang, Haw-Kuen Liu, Yu-Hua Chang, Hua-Chang Huang, Harry Liao, Yung-Fen Lin, Wen-Chuan Chen & Cho-Fan Hsu
*Chimei-Innolux Corp., Tainan Site Health Management Department, Taiwan*

Chen-Yang Shih, Wan-Ting Wei & Ya-jen Chang
*Bureau of Employment and Vocational Training, Council of Labor Affairs - Employment Services Center, Yunlin-Chiayi-Tainan region, Taiwan*

ABSTRACT: Taiwan high-tech industries carry social responsibilities and practice the concept of sharing: hiring staffs with physical and mental difficulties (PMD), while pursuing happiness and economic well-being. Take CMI for example, since 1998 CMI has been positively hiring staff with PMD. The year 2010, CMI overcame the barriers that the staff with PMD can only be given clerical jobs, and the issue concerning working safety in the clean-room. For that, CMI cooperated with government and through its support and supervision, CMI worked on a large cross-sector cooperation project on hiring employees with PMD. CMI have worked on PDCA procedure in reaching the goal. At CMI, staff with disabilities working in clean-room account for, up to date, 65% of all workers with disabilities PMD. It shows that "If the corporations show willingness and determination, there are no obstacles". Taiwan high-tech industries will be the pioneer and take good care of the disadvantaged groups, staffs with PMD, secondary employment, aboriginal people, and so on. Taiwan high-tech industries will keep up the compassion, faith and determination to reach the goal for integration, responsibility, improving, mutual benefit and sharing, and at the same time reach the state of "Three happiness goals for both company and employees: wealth, stability and development."

*Keywords*: staff with physical and mental difficulties (staff with PMD), AGV (Automated Guided Vehicle)

## 1 INSTRUCTION

It is important in order to meet a corporation's social responsibility goals to practice the concept of sharing while at the same time the high-tech industries in Taiwan are pursuing happiness and economic well-being.

Chimei-Innolux Corp. (CMI) is one of the major display manufacturing companies of the world. As a vital link in the global optoelectronics supply chain, CMI is deeply aware of its corporate duties and responsibilities. The company also firmly believes that, aside from pursuing profits, it is important to positively bear the corporation social responsibilities (CSR CMI, 2009). To achieve this goal, CMI enhance the working rights of staff with PMD (Physical and Mental Difficulties), and the company have been hiring staff with PMD since 1998.

The American Disabilities Act (ADA) in the U.S.A in 1990 and the People with Disabilities Rights Protection Act in Taiwan of 2007 are regulations to protect staff with PMD, by giving them equal employment opportunities so that they can fully develop their potential and achieve self-realization of their goals in life (NTCU, 2007).

Based on the 38th Article of "People with Disabilities Rights Protection Act": any given private school, association, or private business agency/organization/institution whose total number of employees is no less than 67, shall employ staff with PMD who have the capability of working, and the number of employee with PMD shall be no less than 1% of the total number of the employees, and no less than 1 person (Ministry of the Interior, 2011).

Compared to other corporations which were penalized for inadequately hiring staffs with PMD (Shu-Fen Tseng, 2001), CMI went much more positively. From 1999 to 2011, CMI has become the pioneer of the private sector, employing the maximum number of staff with PMD in Taiwan high-tech industries.

## 2 OBJECTIVE

In the year 2010, CMI overcame the barriers that staff with PMD can only be given clerical jobs (Ministry of the Interior Department of Statistics, 2001), and the issue concerning working safety in the clean-room. For that, CMI cooperated with government. Through its support and supervision, CMI worked on a large cross-sector cooperation project on hiring employees with PMD.

## 3 RESEARCH METHOD

The project was to cooperate with the Manufacturing department and the Safety, Health and Environment Group division, and staffing and ER division in CMI. The team used PDCA procedure to define problems and establish improvement plans. The practice was targeted at staff with PMD who worked in the clean room (IOSH, 2000; Yang etc., 2002; Chi, C.F. etc., 2004).

Since 1998 CMI have practiced progressively hiring staff with PMD in their new plant. The company considered providing possibly a barrier-free working environment for the staff with PMD. CMI officially began to hire the staff with PMD to work in the clean room in 2010.

### 3.1 *Plan period (Plan)*

In CMI, the Manufacturing department and Safety, Health and Environment Group division planned to cooperate before practicing this project and the steps were as follows.

#### 3.1.1 *Establish consensus of all staff*

This project of hiring staff with PMD is different from the past, this time mainly for them to participate in the clean room operation. It needs not only the support of the supervisors, but also requires recognition of all employees within the workplace.

In order to avoid alienation of employees leading to rejection or negative reaction, CMI applied communication and advocacy to all employees before the project started, which included: informing the employees about the situation of staff with PMD and working in a production line; treating them with respect, rather than sympathy; understanding the characteristics of staff with PMD through assisting their training process, and through proper training so that they would be competent to work in the clean room.

Establishing full consensus not only promotes the concept of social care as everyone's responsibility, but also brings realization that having the chance to help staff with PMD promotes workplace happiness.

#### 3.1.2 *Put-up safety and health signs and labels*

From the point of view of the staff with PMD, hazard awareness and response is different from common people, so CMI put up more safety labels before they participated in the clean room. In addition, it added more messages reminding staff with hearing difficulties, as follows.

1. Emergency alarm system: to prevent staff with hearing difficulties from delaying escaping time, (1) setting central alarm system in the guardroom and the vibration monitor connected with the central alarm system (Figure 1). (2) emergency evacuating alarm signal

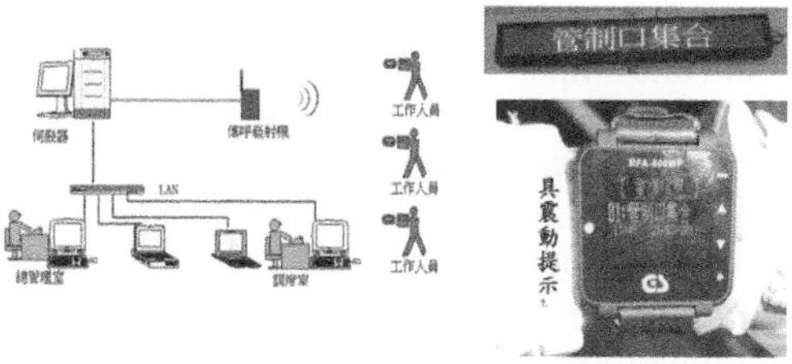


Figure 1. Vibration paging system and the vibration monitor.

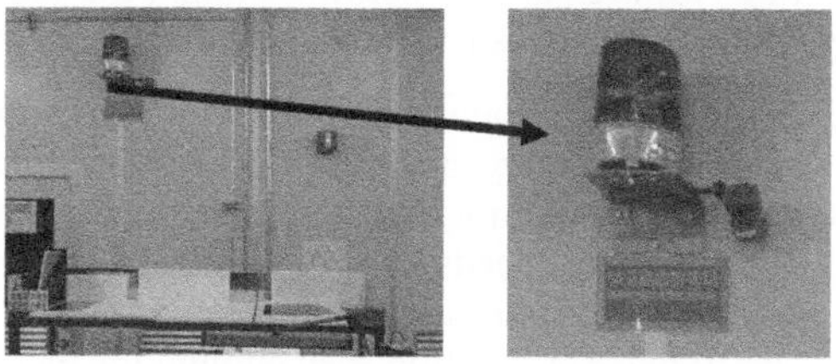

Figure 2. Emergency evacuation alarm signal.

(Figure 2). CMI reinforced emergency escape notices for staff with hearing difficulties. LED scrolling text marquee and vibration monitor are available for workers with PMD to receive the alarm messages at their workplace.

2. AGV (Automated Guided Vehicle) mapping signs: considering the alarm volume (the low volume or the AGV frequency that the hearing impaired staffs cannot receive), in order to remind staff to move through the AGV line carefully with added LED scrolling text marquee and safety labeling along the moving line; these methods are used to protect staff with hearing difficulties and for them to be fully aware and notice the signs.

3.1.3 *Job qualification and selection of suitable candidates*

Staff with PMD have a certain level of limitations; therefore, they require a different standard. Before they participate in clean room work, the plant must first understand which staff are qualified for the position. ("Physical and Mental Difficulty Rights Protection Act" Ministry of the Interior, 2011).

After final analysis and consideration of job content and emergency evacuation in the clean room workplace (IOSH, 2000), the staff with PMD who can be recruited to the clean room are: certain level of physical disabilities, hearing/language impairment, visual impairment and organ impairment staff.

Although these PMD staff could meet the qualifications of the clean room work, the safety, quality, productivity and other issues still need to be considered before participating in the manufacturing (IOSH, 2000; Gary Yang Zhong, etc. 2002; Ji Jiafen, 2001).

CMI has considered the following aspects:

1. Safety: improving the workplace safety and increasing emergency response equipment.
2. Quality: reinforcing pre-job training; ensuring quality to meet the production standards.
3. Productivity: through job design, selecting appropriate working stations and job content to maximize the staff's function.

Depending on the disability types and degrees of staff, the plant managers considered the job content and workplace landscape, such as moving line distance/width, height of operating, tools placement, the sight of drinking water, volume, powerful magnets, weight of loading, emergency evacuation and other related factors were analyzed for the site to assess the suitable job content and workplace. Then, the plant managers selected the best job content and location of work for the staff in order to bring the maximum benefits for both employees and the company.

1. For example, the staffs with auditory disability or vocal malfunction: considering staffs with the severe or moderate impairments are having difficulty to communicate with others, so that only the staffs with mild impairments are allowed to operate machinery. On the other hand, various degrees of impairments will not affect the work of the panel inspection station; the staff could participate without physical challenges on the job.
2. For the staff with physical difficulties: considering the risk of machine operation, only the staff with mild upper and lower limb disabilities can operate machinery, but they will still need to stand for a long time at work. It is necessary to assess staff with lower limb difficulties before participating in the workplace.
3. On the checking station: there is no operational problems for any kind caused by staff with any degrees of difficulties. However, at the checking station, the staff need to use tools for work. Staffs with upper limbs difficulties will be evaluated before being assigned those jobs.

Concerning the issues of emergency evacuation, staff with PMD are on the first priority. They are stationed as close as to the emergency exits as necessary. In addition, reinforce (1) for the hearing impaired staff who could not clearly receive the warning broadcast, whether the evacuation messages can be received needs to be considered, (2) due to the mobility problem of staff with lower limb disabilities, the escaping distance, map, or auxiliary placement needs to be assessed in order to make evacuation easil enough for them.

3.1.4 *Redesign and modify job operation*

Before the staff with PMD enter the workplace, employers should establish obstacle-free environments to help the staff work stably and to reduce the worries for the employer. (Birchall & Wild, 1973; Council of Labor Affairs, 2011). The improved items are as follows:

1. Establish a safe working environment and identification: staff with PMD are required to wear a pink hat to be identified and easy to be assisted during emergency evacuation (Fig. 3).
2. Manufacturing Department improved working equipment or tools by using flash cards. For example, plant managers created product defect coding notes and flash cards for operating key words (Fig. 4), instead of talking, staffs are now using the cards of work order or figures to communicate in clean-room.

Adjusting working procedure or content.

I. Reorganize job content or procedure improvement.
II. Establishing one-on-one instructor system and agent system to help staff with PMD resolving problems.
III. Prohibiting staff with physical difficulty from climbing or carrying heavy objects. When such needs occur, they should request support.

3. Import Structured On-the-Job Training (SOJT).

In considering the difference between staff with PMD and other staff, plant managers revised pre-job training courses and materials to meet those needs. For example, wearing gloves; if workers with PMD cannot wear them properly it may affect their safety and productivity (Fig. 5). Moreover, the plant established mentoring programs and an agent

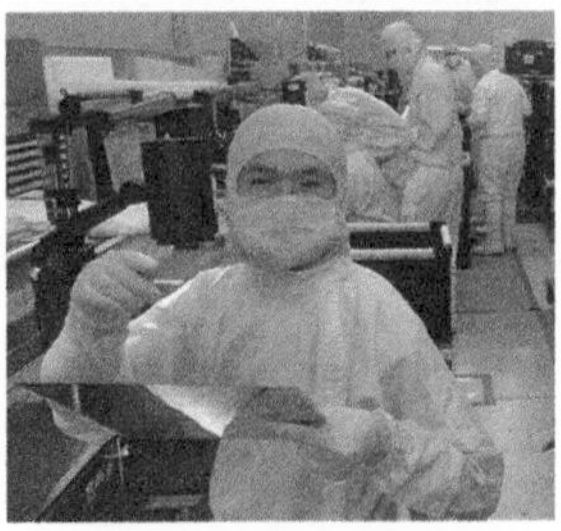

Figure 3. Staff hate to wear pink to enhance safety identification.

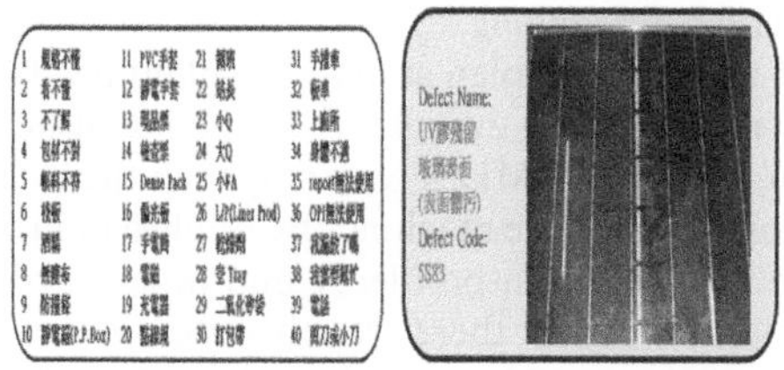


Figure 4. The product defect coding notes and flash cards for operating key words.

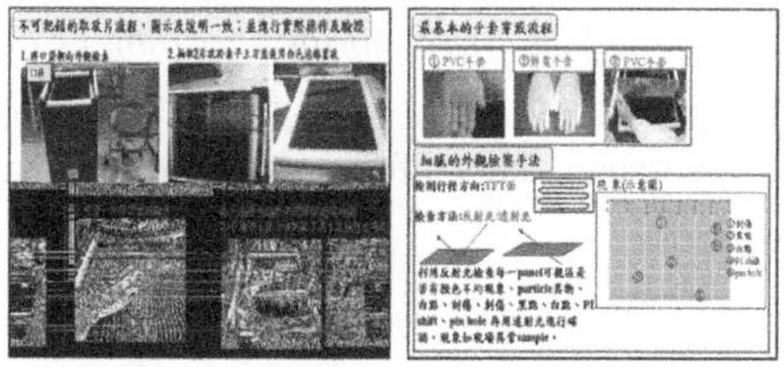

Figure 5. Re-edited SOJT teaching materials.

system by selecting and training one-on-one instructors to assist on-the-job training for staffs with PMD and for the following issues that may occur after work.

Furthermore, efforts are made in the training program for:

I. Changing the training center from the clean room to the office area to reduce the anxiety of the staff with PMD, and to make the training course easier for the hearing impaired staffs. The plant also extended the schedule of pre-job training and certification.

II. The project team hired a sign language teacher, to translate the training content to hearing impaired staff, their work-partners, instructors, supervisors and other related members.

III. The plant changed the teaching material from paper reading to hands-on real material practice training.

IV. Staff have to pass all exams of pre-job training before they could actually work.

#### 3.1.5 *Making safety and health rules and standards*

The plant made safety and health rules and standards to avoid accidents in the workplace and to guarantee operational security, especially for staff with PMD. For example, staff with physical disabilities are not allowed to climb high and staff with severe upper limb difficulties are not allowed to move heavy objects. When the needs occur, they should request support.

For emergency evacuation, before operating in the clean room, staff with PMD must pass the requirements of the emergency evacuation standards test.

### 3.2 *Conducting period (Do)*

CMI HR & ER Division and Employment Services Center of Yunlin-Chiayi-Tainan region teamed up to accomplished job qualification and job matching.

#### 3.2.1 *Continue recruiting staffs with PMD*

The Employment Services Center of Yunlin-Chiayi-Tainan, region assisted CMI with interviewing employees with PMD.

#### 3.2.2 *Implementing the operating training with high-tech keywords in sign language*

Teaching sign language and training communication skills with high-tech keywords for clean room operation: the plant invited sign language teacher to translate the training content for staff with PMD, their work-partners, instructors, supervisors and other related members.

### 3.2.3 *Job qualification and job distributing*

For staff with PMD, CMI followed the plan above. The plant needs to consider the impacts of restrictions, emergency response and evacuation, job content and the workplace layout, and daily life situations for staff with PMD. For example, in the clean room, could the staff with vocal or hearing impairment operate machines by English interface? Can they keep their focus on jobs? Are there communication problems, and other barriers? How do they commute from their home to plant? Whether there are other factors or other issues, the company needs to give special care to other problems in their daily lives; the plant should carefully select suitable candidates for the jobs.

## 3.3 *Risk assessment, prevention and improvement for operating (Check and Action)*

When the staff with PMD entered the work place, the plant would keep taking care of them on the work adaptation issues, daily life, on-job training, and physical and mental health care.

### 3.3.1 *Training and examination of safety and health issues*

For protecting the operating safety issues, after staff with PMD enter the workplace, the company helps increase their hazard prevention knowledge and skills through monthly examination, operating practice, qualifications examination in safety guidance system.

### 3.3.2 *Psychological counseling, emergency response and evacuation issues*

1. Health and psychological care: staff with PMD when assigned on jobs, will be arranged to a first counseling with nurses of the Health Center within the first month. The counseling includes personal health, the need for auxiliary equipment or work adapting. If the staff have need of psychological counseling, they will be referred to the psychological counselor for further counseling.
2. Emergency evacuation standards examination: staffs with PMD need to complete training courses before entering the clean room. They are required to complete annual fire prevention training, and the semi-annual evacuation drills to increase their emergency response skills and concepts.
3. Management support: In the plant, colleagues receive not only support in daily work, but also regular informal meetings with their supervisors. For example, (1) the director shows his constant warm caring through discussing with staff with PMD to understand their needs, suggestions, and other matters. (2) The informal meetings with the manager focus on finding out the professional development and proposed improvement for staff with PMD. (3) Group leaders of the production line aimed at checking their attendance to provide quality assistance to enhance their capability, and give them consulting when they meet problems in the production line to help them solve the problems.

### 3.3.3 *Plant completed initial improvement*

After staffs with PMD entered the clean room to work, CMI still kept safety care and safety interviews regularly to understand their working conditions and potential problems they might face. First, the ergonomic self-improving team of CMI entered the clean room to do working conditions analysis and safety observation. Then, the ergonomic self-improvement team and manufacturing department proposed and completed an initial improvement plan.

### 3.3.4 *Guidance by Taiwan experts in occupational service*

In March 2011, experts of Government, (1) Bureau of Employment and Vocational Training, Council of Labor Affairs, Executive, (2) Southern Taiwan Science Park Administration, (3) Institute of Occupational Safety & Health, Council of Labor Affairs, (4) Employment Services Center of Yunlin-Chiayi-Tainan region, and CMI group led by the Minister of Council of Labor Affairs and the president of CMI entered the clean room to ensure the safety of the on-site working conditions and assess issues for staff with PMD.

Furthermore, CMI invited ergonomic experts to continue giving guidance on the safety operating and health care to prevent potential hazards and protect the work safety and health of staffs with PMD.

### 3.3.5 *Guidance by ergonomic experts*

Furthermore, CMI invited the ergonomic experts of Institute of Occupational Safety & Health, and Council of Labor Affairs, to enter the clean room to do risk assessment and human engineering improvement for the workplace. Within the second safety observation and assessment, CMI improved the workplace and operating procedures more.

## 4 RESULTS

In CMI, all staff with PMD have passed the time-limited evacuation training, and have completed the semi-annual evacuation drill. The safety prevention signal system and protection equipment have also been set up and the situation is continuing to improve.

During the process of training staff with PMD before entering the clean room, CMI found the upside of the program, that the staff with PMD held a positive attitude toward the occupational field. They are highly stable on jobs, and have no major problems of learning. The company needs to put up more complete resources into PMD staffs' emergency evacuation issues.

The workers with hearing difficulties were strongly focused on working, highly adaptable and stable. However, there are communication problems, learning difficulties and problems with receiving sounds of alarm broadcasts. A complementary system of evacuation and additional setups for escaping are needed.

Although the staff with PMD have the disadvantages as mentioned, the obstacles can be eliminated or improved through interviews, adjusting job content, work-partner system, job qualification assessment, psychological counseling, and installing more emergency response equipment and other related facilities to improve the working conditions for them.

With multilateral cooperation with all divisions, CMI has completed the initial integration. Staffs with PMD could participate in the clean room. Since May 2010, CMI has continued to employ staff with PMD with the help of the employment service station of Yongkang. In CMI, staff with PMD who work in the clean room account for the up-to-date percentage of all staff with PMD at 65%. It shows that "If the corporations are willing and determined to do good deeds, there are no obstacles."

## 5 CONCLUSIONS

To complete the goal, staff with PMD could work in the clean-room without difficulties. CMI has been putting in a lot of additional education, training resources, and supporting counseling system to benefit people with difficulties who work in the clean room. CMI not only provide staff with PMD with more professional jobs, but also understand that a complete solution needs a long-term effort and plan. Another challenge is to change the attitude of the companies not employing staff with PMD to no longer resist such employees working in the clean room. Engaged in high-tech jobs, more staff with PMD could create greater value if the corporation and the Government could provide more complete supporting system for them.

Taiwan high-tech industries will be the pioneer to take good care of the disadvantaged groups, for example mental and physical difficulties, secondary employment, aboriginal people, and so on. Taiwan high-tech industries will keep up the compassion, faith and determination to reach the goals of integration, responsibility, improving, mutual benefit and sharing. And the same time, reach the state of "Three happiness goals for both company and employees: wealth, stability and development."

## ACKNOWLEDGEMENT

Especial thanks for

1. Sign language teacher: Ma-li Liang, Wei-Hua Tsai and Ling-Hua Zhuang. Employment Services Center, Yunlin-Chiayi-Tainan region and Bureau of Employment and Vocational Training Chen-Yang Shih, Council of Labor Affairs, Executive of Yong-Kan: Wan-yin Chi.
2. Ergonomic experts of Institute of Occupational Safety & Health: Chi-Yuang Yu, sub-professor of Department of Industrial Engineering and Engineering Management National Tsing Hua University, and Jer-Hao Chang, Doctor of Occupational therapy department of National Cheng-Kung University, and Institute of Occupational Safety & Health, Council of Labor Affairs: Chiu-Jung Chen and Yi-Tsung Pan.
3. Southern Taiwan Science Park Administration: section chief, lobor affairs & environment division, Chia-Ming Kuan Chun-Wei Chen, Yung-Shou Lin, Pai-Hung Chen and Chun-Te Tung.
4. Zoe Hsu gave us invaluable comments and suggestions to revise our English grammar.
5. Council of Labor Affairs, Executive Ju-Hsuan Wang and San-Kuei Lin.

In CMI cooperation

1. Manufacturing department: Manager Shi-wen Chen, sub-manager Chia-xi Young, etc.
2. Executive department: Manager Yu-yu Chen, Tsui-min Hu, etc.
3. Health & Safety Environment Group division: Manager Ming-Piao Wu, Manager Chin-Lien Tsai, Pei-Hsun Lin, Wan-Jun Yang, Hung-Kai Huang, Yi-Kai Wang, Hsin-Jan Huang, etc.
4. Occupational nurses of Health Management department.
5. Safety engineer of Manufacturing department: Chi-long Lin.

## REFERENCES

Birchall, D. & Wild, R. (1973). Job restructuring amongst blue collar workers. *Personnel Review,* Vol. 2, pp. 40–56.

Chia-Fen Chi, Jang Y, Shiu-Lin Liu, Jun-Teng Chen, Wen-Yu Yeh & Yen-Hui Lin. (2001). Evaluation and Improvement of Disabilities' workplace. *Journal of Occupational Safety and Health,* Vol. 10, pp. 185–197.

Chi, C.F., Jung-Shung Pan, Tzu-Hsin Liu & Yuh Jang. (2004). The Development Of A Hierarchical Coding Scheme And Database Of Job Accommodation For Disabled Workers. *International Journal of Industrial Ergonomics,* Vol. 33, pp. 429–447.

Chimei Innolux. (2009). Corporate Social Responsibility Report.

Council of Labor Affairs, Bureau of Employment and Vocational Training, (2001). job re-designed.

Institute of Occupational Safety & Health, IOSH, (2000). Guidelines of Workplace safety for persons with disabilities.

Ministry of the Interior, (2001). People with Disabilities Rights Protection Act.

Ministry of the Interior Department of Statistics, (2001). Disabilities Life needs survey in Taiwan, volume of Employment Services and Job Training.

National Taichung University of Education Special Education Center, (2007). Theory and Practice of Re-designed Jobs' Content for Physical and Mental difficult staffs.

NTCU (National Taichung University of Education) Special Education Center), (2007).

Robert Heron. 2005. JOB AND WORK ANALYSIS Guidelines on Identifying Jobs for Persons with Disabilities, ILO Skills and Employability Department:http://www.ilo.org/wcmsp5/groups/public/—ed_emp/—ifp_skills/documents/publication/wcms_111484.pdf

Shi-Chian Wang. (2006). Sharpen employees' profession at work.—Structured On-the-Job Training. Training & Development Fashion, Vol. 49, pp. 1–12.

Shu-Fen & Tseng. 2001. Limbs disabilities workplace safety and health needs survey, 16_7.pdf http://www.iosh.gov.tw/userfiles/File/workshop/data/w900516_7.pdf)

Sweo-Chung Yang & Zhou Zaiqi. (2002). workplace safety and health promotion for Disabilities. Brief Journal of Occupational Safety and Health, p. 54.

*Ergonomics in Asia – Shih & Liang (eds)*
 *ISBN 978-0-415-68414-9*

# The effect of load intensity on back muscle oxygenation in static muscle work

M. Movahed, H. Izumi & M. Kumashiro
*Department of Ergonomics, Institute of Industrial Ecological Sciences, University of Occupational and Environmental Health, Kitakyushu, Japan*

J. Ohashi
*Department of Management and Business, Kinki University, Iizuka, Japan*

N. Kurustien
*Faculty of Physical Therapy, Mahidol University, Bangkok, Thailand*

ABSTRACT: Any muscle fatigue during static contraction is important; previous studies suggest that fatigue is associated with musculoskeletal injury and low back pain. Among local factors that affect muscle fatigue are blood perfusion of the muscle and level of its oxygenation. To investigate the effect of load intensity on the low back muscle during repeated static work, 11 young male subjects aged between 23–27 years old and no history of low back pain during the last 12 months participated. Subjects held a load in their hands whilst sustaining trunk flexion at 30° which was equal to 10 and 40% of their maximal voluntary force till they felt tired in the low back region. The task was repeated for 12 times with resting time between each task. Oxyhemoglobin (Oxy-hb) and deoxy-hemoglobin (Deoxy-hb) of the erector spinae muscle (ESM) at the third level of the vertebral column was evaluated using near infrared spectroscopy. As a result Oxy-hb showed significantly larger decreases by repetition in 40% MVC than in 10% MVC. On the contrary Deoxy-hb showed significantly larger increases by repetition in 40% MVC than in 10% MVC.

*Keywords*: near infrared spectroscopy, muscle oxygenation, low back muscle, trunk flexion

## 1 INTRODUCTION

Localized muscle fatigue can be a limiting factor for prolonged static work at the workplace. It is important to study the etiology of fatigue; previous literatures suggest that fatigue is associated with musculoskeletal injury and low back pain (Gorelick et al., 2003; Kankaanpää et al., 1998; Kumar et al., 2001; Kumar & Narayan 1998; Payne et al., 2000; Tsuboi et al., 1994).

A definition of muscle fatigue is a decrease in the maximal muscle force or power and it appears gradually after the onset of sustained physical activity (Enoka & Duchateau 2008), or sensation of fatigue which is believed to be a conscious awareness of changes in the homeostatic system (St Clair Gibson 2003). The low back muscles are initiating and controlling of major movement and also stabilizing of the spine. They maintain posture by an extension moment during forward bending and lifting (Gilchrist et al., 2003). However, fatigued muscles are less effective at generating the required extensor moment: forcing the intervertebral discs and ligaments to face a relatively greater bending stress, therefore making them more susceptible to injury (Mannion 1999).

Reduced blood flow or decreased muscle oxygenation is one of the most important factors underlying muscle fatigue (Grandjean 1988; Murthy et al., 2001). With near-infrared spectroscopy (NIRS), muscle oxygenation measurements can now be made non-invasively. Earlier studies have shown changes in low back muscle oxygenation during static muscle contractions (Albert et al., 2004; Jensen et al., 1999; Kell et al., 2004; McGill et al., 2000; Yoshitake et al., 2001). In our previous paper (Movahed et al., 2011) we evaluated the changes of low back muscle oxygenation during repeated static muscle work under different load intensity. Task duration was based on the subjective fatigue sensation. The results showed that in both levels of 10% and 40% of maximum voluntary contraction (MVC), oxy-hemoglobin (Oxy-hb) decreased significantly in each static task. On the contrary deoxy-hemoglobin (Deoxy-hb) increased significantly in each contraction. We concluded that decreases of Oxy-hb and increases of Deoxy-hb are important factors in inducing muscle fatigue in both level of contractions. The aim of the present paper is to compare the effect of different load intensity on the changes of low back muscle oxygenation after 12 times repetition of static muscle work.

## 2 METHOD

### 2.1 *Participants*

Eleven healthy males with no history of low back pain during the last 12 months, participated in this experiment. The average physical characteristics were age 23.9 (SD: standard deviation 1.2) yr; height 1.72 (SD 0.07) m; body mass 64.3 (SD 8.5) kg, and BMI 21.5 (SD 2.5) $kg/m^2$. The Medical Ethics Committee at the University of Occupational and Environmental Health, Japan, approved the study. All participants provided written informed consent before testing.

### 2.2 *Procedure*

The subject stood in front of an iron chain anchored to an aluminum base with an intervening load cell (TU-BR5 KN, TEAC Corporation, Tokyo, Japan) at one end and an industrial size handle at the other. In 30° bending posture, the subject was required to pull upward on the handle to obtain his isometric maximum pulling force (MVC). After that the subject repeated a static contraction task of holding a box in 30° trunk flexion posture for 12 repetitions interrupted by rests whose duration was equal to the duration of each preceding contraction (Figure 1). Each task was stopped at strong fatigue sensation level, which corresponded to Borg's CR-10 5 (Borg 1998). The repeated task was performed at two contraction levels of 10 and 40% MVC on separate days (fully described in our previous paper; Movahed et al., 2011).

#### 2.2.1 *Near-infrared spectroscopy*

A near-infrared spectroscope (BOM-L1 W, Omegawave, Inc., Tokyo, Japan) was used to measure muscle oxygenation of the erector spinae muscle (ESM). NIRS is based on the differential absorption properties of hemoglobin and myoglobin in the near-infrared range of the absorption spectrum. At 760 nm, these light-absorbing compounds are in the deoxygenated state (Deoxy-hb), whereas at 830 nm they occur in the oxygenated form (Oxy-hb). The difference in tissue absorbency between these two wavelengths reflects the

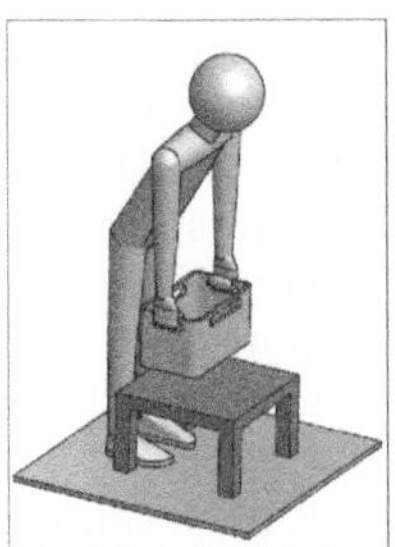

Figure 1. Illustration of the experimental setup.

relative change in oxygen saturation at the small blood vessels. The NIRS probe was attached to the skin on the right side of the ESM at the third level of lumbar vertebra (L3), 3 cm from the vertebral column using a sticky gel sheet.

#### 2.2.2 *Subjective measurement*

Borg scale (CR-10): The subjects rated low back muscle fatigue on the Borg category-ratio scale. The rating was made once at the start of contraction and every 30 s during the contraction until the onset of the target point (Borg's CR-10 5). The subjects were given the same standard instruction before the experiment (Borg 1998).

### 2.3 *Data analysis and statistics*

The sampled values of muscle Oxy-hb and Deoxy-hb were averaged. The variables were calculated every 250 ms and summarized (averaged) to be adjusted to the time units used in the statistics and graphs. The signals in the first 4 s and last 2 s of each contraction were discarded to remove fluctuant parts. The average of the first 60 s in 10% MVC and 30 s in 40% MVC of the first repetition (C1) was used as a baseline to reduce fluctuation in levels of the measurements among experiments. The variables were calculated as the difference from the baseline.

Wilcoxon pair test was used to see the effect of load intensity by repetition; the sum of the durations for a total of 12 repetitions was calculated. Then, for this amount of duration the changes of each variable to the baseline were obtained. The significant level was set at $p < 0.05$. All statistics were calculated using SPSS version 11.5 (Chicago, USA).

## 3 RESULTS

Oxy-hb showed significantly larger decreases by repetition in 40% MVC than in 10% MVC (*p* value < 0.05). On the contrary, Deoxy-hb showed significantly larger increases by repetition in 40% MVC than in 10% MVC (*p* value < 0.05) (Table 1 and Figure 2).

Table 1. Wilcoxon pair test: Comparison of variables in different load intensity.

| | Oxy-hb | Deoxy-hb |
|---|---|---|
| 40% MVC-10% MVC | – | + |

– and + are written only in significant conditions ($p < 0.05$). +: larger increase; –: larger decrease.

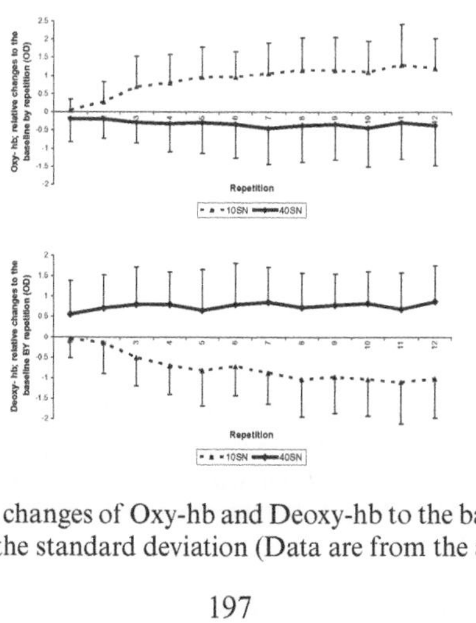


Figure 2. The group averaged changes of Oxy-hb and Deoxy-hb to the baseline in 10 and 40% MVC by repetition. Error bars indicate the standard deviation (Data are from the average of each contraction).

## 4 DISCUSSION

The aim of this paper is to compare the changes of muscle oxygenation after 12 times repetition of static muscle work between two different levels of load intensity. The load levels were determined as the conditions of incomplete occlusion of blood flow in low back muscles and near complete occlusion, and being repeatable more than ten times. The fatigue level was determined as a reasonable condition in actual work. Although NIRS cannot measure blood flow directly, it can measure the changes in tissue oxygenation in real-time. An increase in Oxy-hb and/or a decrease in Deoxy-hb mean either the increase of oxygen supply to the monitoring site increased or there was a decrease of oxygen demand or consumption at the site, which may be interpreted that sufficient new blood has come to the muscle. A decrease in Oxy-hb and/or an increase in Deoxy-hb indicate opposite conditions, which may be seen in fatiguing phase.

Previous studies showed that during forward trunk flexion, the intramuscular pressure in the ESM increases (Konno et al., 1994; Miyamoto et al., 1999; Mueller et al., 1998), which has been reported to compromise the muscle blood flow during static contraction (Jensen et al., 1999; Kramer et al., 2005; Styf & Lysell., 1987). Earlier studies showed decreases of oxygenation in the ESM at forces as low as 2% of MVC and also between 10 and 30% of MVC. It was suggested that this would lead to fatigue during sustained activity (Kagaya & Homma 1997; McGill et al., (2000); Wesche 1986). Two other studies on ESM activity reported decreases of muscle Oxy-hb in approximately 50% MVC contractions (Kell et al., 2004; Yoshitake et al., 2001). In our previous paper (Movahed et al., 2011), we showed that Oxy-hb was decreased and Deoxy-hb was increased significantly in each contraction in both 10 and 40% MVC but changes appeared larger as a function of contraction intensity. This seems reasonable, since intramuscular pressure is increasing more by increasing load intensity (Mueller et al., 1998). Also Masuda et al., 2006 studied maximal forward bending in volunteers holding up to 20 kg. They found muscle oxygenation in the ESM was the most decreased during the heaviest load.

However when the effect of different load intensity on the changes of muscle oxygenation after the end of 12 times repetition was compared, Oxy-hb showed significantly larger decreases by repetition in 40% MVC than in 10% MVC ($p$ value < 0.05). On the contrary Deoxy-hb showed significantly larger increases by repetition in 40% MVC than in 10% MVC ($p$ value < 0.05). It showed that blood flow to the muscle during rest periods increased by repetitions in 10% MVC to supply oxygen for the greater energy consumption during muscle contractions. Also since the resting duration in 10% was more than 40% MVC, it seems that the muscle had more time for supplying energy. Our result showed that rest time is important, especially in the lower load intensity, for the muscle recovery and to prevent accumulated muscle fatigue which may lead to muscle injury.

## 5 CONCLUSION

A finding of this study provides the evidence that muscle has different response to the task (from the viewpoint of oxygenation changes) based on the load intensity. It seems that in low level of intensity, muscle has the ability to supply oxygen during for the rest time to compensate for the energy demand which increases by muscle contraction.

## REFERENCES

Albert, W.J., Sleivert, G.G. & Neary, J.P. (2004). Monitoring individual erector spinae fatigue responses using electromyography and near infrared spectroscopy. *Can J Appl Physiol,* Vol. 29, pp. 363–379.

Borg, G. (1998). *Borg's perceived exertion and pain scales*. Champaign (IL): Human kinetics.

Enoka, R.M. & Duchateau, J. (2008). Muscle fatigue: what, why and how it influences muscle function. *J Physiol,* Vol. 586, pp. 11–23.

Gilchrist, R.V., Frey, M.E. & Nadler, S.F. (2003). Muscular Control of the Lumbar Spine. *Pain Physician,* Vol. 6, pp. 361–368.
Gorelick, M., Brown, J.M. & Groeller, H. (2003). Short-duration fatigue alters neuromuscular coordination of trunk musculature: implications for injury. *Appl Ergon,* Vol. 34, pp. 317–325.
Grandjean, E. (1988). *Fitting the Task to the Man.* 4th ed. London: Taylor & Francis.
Jensen, B.R., Jørgensen, K. & Hargens, A.R. (1999). Physiological response to submaximal isometric contractions of the paravertebral muscles. *Spine,* Vol. 24, pp. 2332–2338.
Kagaya, A. & Homma, S. (1997). Brachial arterial blood flow during static handgrip exercise of short duration at varying intensities studied by a Doppler ultrasound method. *Acta Physiol Scand,* Vol. 160, pp. 257–265.
Kankaanpää, M., Taimela, S. & Laaksonen, D. (1998). Back and hip extensor fatigability in chronic low back pain patients and controls. *Arch Phys Med Rehabil,* Vol. 79, pp. 412–417.
Kell, R.T., Farag, M. & Bhambhani, Y. (2004). Reliability of erector spinae oxygenation and blood volume responses using near-infrared spectroscopy in healthy males. *Eur J Appl Physiol,* Vol. 91, pp. 499–507.
Konno, S., Kikuchi, S. & Nagaosa, Y. (1994). The relationship between intramuscular pressure of the paraspinal muscles and low back pain. *Spine,* Vol. 19, pp. 2186–2189.
Kramer, M., Dehner, C., Hartwig, E., Völker, H.U., Sterk, J. & Elbel, M. (2005). Intramuscular pressure, tissue oxygenation and EMG fatigue measured during isometric fatigue-inducing contraction of the multifidus muscle. *Eur Spine J,* Vol. 14, pp. 578–585.
Kumar, S. (2001). Theories of musculoskeletal injury causation. *Ergonomics,* Vol. 44, pp. 17–47.
Kumar, S. & Narayan, Y. (1998). Spectral parameters of trunk muscles during fatiguing isometric axial rotation in neutral posture. *J Electromyogr Kinesiol,* Vol. 8, pp. 257–267.
Kumar, S., Narayan, Y., Stein, R.B. & Snyders, C. (2001). Muscle fatigue in axial rotation of the trunk. *Int J Ind Ergon,* Vol. 28, pp. 113–125.
Mannion, A.F. (1999). Fiber type characteristics and function of the human paraspinal muscles: normal values and changes in association with low back pain. *J Electromyogr Kinesiol,* Vol. 9, pp. 363–377.
Masuda, T., Miyamoto, K. & Shimizu, K. (2006). Intramuscular hemodynamics in bilateral erector spinae muscles in symmetrical and asymmetrical postures with and without loading. *Clin Biomech,* Vol. 21, pp. 245–253.
McGill, S.M., Hughson, R.L. & Parks, K. (2000). Lumbar erector spinae oxygenation during prolonged contractions: implications for prolonged work. *Ergonomics,* Vol. 43, pp. 486–493.
Miyamoto, K., Iinuma, N., Maeda, M., Wada, E. & Shimizu, K. (1999). Effects of abdominal belts on intra-abdominal pressure, intramuscular pressure in the erector spinae muscles and myoelectrical activities of trunk muscles. *Clin Biomech,* Vol. 14, pp. 79–87.
Movahed, M., Ohashi, J., Kurustien, N., Izumi, H. & Kumashiro, M. (2011). Fatigue sensation, electromyographical and hemodynamic changes of low back muscles during repeated static contraction. *Eur J Appl Physiol,* Vol. 111, pp. 459–467.
Mueller, G., Morlock, M.M., Vollmer, M., Honl, M., Hille, E. & Schneider, E. (1998). Intramuscular pressure in the erector spinae and intra-abdominal pressure related to posture and load. *Spine,* Vol. 23, pp. 2580–2590.
Murthy, G., Hargens, A.R., Lehman, S. & Rempel, D.M. (2001). Ischemia causes muscle fatigue. *J Orthop Res,* Vol. 19, pp. 436–440.
Payne, N., Gledhill, N., Katzmarzyk, P.T. & Jamnik, V. (2000). Health-related fitness, physical activity, and history of back pain. *Can J Appl Physiol,* Vol. 25, pp. 236–249.
St Clair Gibson, A., Baden D.A., Lambert, M.I., Lambert, V. & Harley, Y.X.R., et al. (2003) The conscious perception of the sensation of fatigue. *Sports Med,* Vol. 33, pp. 167–176.
Styf, J. & Lysell, E. (1987). Chronic compartment syndrome in the erector spinae muscle. *Spine.* Vol. 12, pp. 680–682.
Tsuboi, T., Satou, T., Egawa, K., Izumi, Y. & Miyazakim, M. (1994). Spectral analysis of electromyogram in lumbar muscles: fatigue induced endurance contraction. *Eur J Appl Physiol Occup Physiol,* Vol. 69, pp. 361–366.
Wesche, J. (1986). The time course and magnitude of blood flow changes in the human quadriceps muscles following isometric contraction. *J Physiol,* Vol. 377, pp. 445–462.
Yoshitake, Y., Ue, H. & Miyazaki, M. (2001). Assessment of lower-back muscle fatigue using electromyography, mechanomyography, and near-infrared spectroscopy. *Eur J Appl Physiol,* Vol. 84, pp. 174–179.

*Ergonomics in Asia – Shih & Liang (eds)*
*© 2012 Taylor & Francis Group, London, ISBN 978-0-415-68414-9*

# Differences in lifting strengths and postures between workers and novices among Taiwanese females

Yi-Lang Chen
*Department of Industrial Engineering and Management, Mingchi University of Technology, Taipei, Taiwan*

Yu-Chi Lee
*Department of Industrial Engineering and Engineering Management, National Tsing Hua University, Hsinchu, Taiwan*

ABSTRACT: The effect of working experience on lifting strengths for males had been verified by Chen et al. (2011). This paper extends that study to compare the lifting strengths and the postures at various exertion heights between experienced workers and novices among Taiwanese females. Forty-six experienced workers and novices (23 of each) were recruited and were required to determine their static-lifting strengths under various height levels (10~150 cm in increments of 10 cm) using two exertion methods (vertically upward lifting, VUL, and toward body lifting, TBL). Results showed that the VUL forces were much higher than the TBL at 15 height positions ($p < 0.001$). At lower heights (≦50 cm), workers' VUL forces were 5.67–7.40 kg higher than novices' and no differences were found in TBL at these heights (less than 2 kg). A reverse trend was found when lifting ranges were equal to or higher than 90 cm. Forces in all 30 task combinations showed no difference between the two groups at heights of 60–80 cm. Workers' strengths under all 15 lifting heights showed a relatively small fluctuation compared to novices' and they tended to adopt a fully squat and a more erect posture at lower and higher heights, respectively. The results suggested that lifting strength data generally collected on students should be carefully used in the task design.

*Keywords*: lifting postures, experienced worker, novice, exertion heights

## 1 INTRODUCTION

Logistics is one of the largest industries in Taiwan with over 500,000 workers employed in various logistic capacities (CEPD 2004). Most of the activities in this industry require muscular strength under various height settings; and these activities, especially the lifting, are routine in warehouses, supermarkets, shopping malls, and distribution centers, etc. Lifting may be conducted at sub-maximal weight and as non-gender-specific tasks, and the lifting height varies as the companies attempt to economize on storage spaces. A recent investigation into the musculoskeletal disorders of Taiwanese warehouse workers reported that, instead of the lower back, the shoulders and arms were the most uncomfortable body regions (Chen 2008). This result is undoubtedly connected to the logistic tasks these workers were asked to perform.

Many studies have established human strength data under various task variables, which were usually derived from student participants (Yates et al., 1980, Lee 2004). These studies conducted on student participants are on the basis of: ease in recruitment, greater reliability, lower cost, and availability for a long-term study (Mital 1987). However, it is essential to identify the suitability of an individual's capabilities before directly applying the results, conducted on students, to the entire industrial population.

Much of the literature has reported that the lifting techniques of highly skilled workers substantially differ from those of novices (Authier et al., 1996, Gagnon et al., 1996). Field studies have also shown that workers are using methods other than the recommended ones (Kuorinka et al., 1994, Baril-Gingras and Lortie 1995), and there is a lack a consensus on the best methods (Authier et al., 1996). For a given task, in general, the experienced workers employ lower biomechanical spinal loads (Marras et al., 2006), less back-muscle activities (Keir and MacDonell 2004), more psychophysically accepted lifting-weights (Mital 1987), and higher subjective discomfort thresholds (Parakkat et al., 2007) than the novices. On the contrary, some investigations suggest that the response patterns of industrial and non-industrial workers to task variables in manual lifting activities are similar (Mital and Manivasagan 1983). The controversies may be attributed to the diverse experimental settings among these studies.

One of the factors affecting lifting strength is exertion height (Yates et al., 1980, Mital et al., 1993, Lee 2004). This is extremely realistic in the service warehouse stores as the storage spaces have to be economized and maximized. Investigations on lifting strength are usually confined to some specific height levels (e.g., floor, knuckle, and shoulder). Height of lifting a workload, in addition to the posture adopted during a lift, also influenced human strengths (Yates et al., 1980, Lee 2004), but also posture strategy (Chen 2000). Vertically upward lifting (VUL) has been accepted as the standard method in previous strength measurement studies (Ayoub et al., 1978, Chaffin 1975, Kumar 1991, Lee 2004). However, another human strength measurement that usually adopted is toward body lifting (TBL). In a study conducted by Garg et al. (1983), their participants pulled the load toward the body while lifting psychophysically determined maximum weights. Lifting at an angle resulted in a decrease in moment at the upper arms and spinal L5/S1 joints and an increase in moment at the knees and ankles. Lee and Chen (1996) requested the participant to pull on the handle toward the body and found that the lifting strength can be a better predictor for the MAWL than other isometric strength. Thus, identification of strength differences generated by different exertion directions deserves further study. More exhaustive strength data are also needed for ergonomic considerations at the workplace.

Recently, Chen et al. (2011) systemically assessed the simulated lifting strengths of male workers and novices under full-range heights and found that inherent differences in lifting strengths did exist between the two groups. This paper therefore extends the study of Chen et al. to compare the lifting strengths and the postures at various exertion heights between experienced workers and novices of Taiwanese females.

## 2 METHODS

### 2.1 *Participants*

Forty-six experienced female workers and novice participants (23 of each), with no prior history of any musculoskeletal disorder, volunteered for this study and received an hourly wage for finishing all test conditions. Twenty-three workers, with at least two-year experience levels, were chosen as experienced participants. They were selected from a large hypermarket, and their duties consisted mainly in varied replenishing tasks onto the shelves. Another 23 novice participants (with no manual material handling experience) were recruited from university students. The mean (SD) age, stature, and body weight was 28.6 (6.4) years, 158.9 (5.3) cm, and 60.4 (12.5) kg for the experienced group, and 22.3 (3.2) years, 160.2 (4.1) cm, and 51.3 (6.5) kg for the novice group, respectively.

### 2.2 *Experimental apparatus*

The static lifting strengths were measured using a Static-Lifting Strength Tester (SLST), as illustrated in Figure 1. The SLST, with an incremental height setting for measuring strengths, consisted of a standing platform, a steel frame with 20 positioning holes (ranging from 10 cm to 200 cm in increments of 5 cm), a sliding height stopper along the frame, and a 55 cm handle bar (diameter: 3.5 cm) attached to the stopper. The force applied onto the handle bar was measured by a load cell that connected with the bar and was rigidly placed on the stopper, and then the strength signal (60 Hz) was transferred into an A-D converter and a digital readout unit (JSES Model 32628, U.S.A.). The A-D converter was calibrated eight times prior to the testing against known static loads. The accuracy of the measurement was within ±1% of the lifting strength.

### 2.3 *Body posture recording*

The testing postures of randomly sampled participants (one third of participants) were videotaped in this study. Six adhesive reflective markers were attached to the right side of the participant's wrist, elbow, shoulder, hip, knee, and ankle joints, as is also shown in Figure 1. In the study, the joint angles were all defined as the inter-joint angles between the two adjacent segments (Figure 2). The digital video camera, which was set up at a distance of 5 m from participants, was used to record the positions of each joint marker when participants performed the strength test. The camera heights were always adjusted to orthogonally align the participant's hip positions to eliminate as much as possible the distortion errors resulting from the varied testing heights. Image processing software was then used to calculate the relevant joint angles.

### 2.4 *Experimental design*

This study examined the differences in static-lifting strengths between female workers and novices under 15 lifting heights as well as under different exertion methods. Fifteen lifting heights, ranging from 10 to 150 cm and spaced 10 cm apart, were set by the SLST. There were two exertion methods performed by each participant at all 15 height levels. One exertion method was vertically upward lifting (VUL) and the other was toward body lifting (TBL). The VUL has been generally accepted as the standard method in previous strength measurement studies, whereas the TBL refers to the finding of a study by Garg et al. (1983). Garg et al., observed that the subjects pulled the load toward the body when lifting psychophysically determined maximum weights. Lifting at an angle resulted in a decrease in moment at the upper arms and spinal L5/S1 joints and an increase in moment at the knees and ankles. In this study, we would like to identify the strength differences caused by different exertion directions. All participants were requested to perform the maximum lifting strengths for each task combination for at least three repetitions. Each strength measurement was

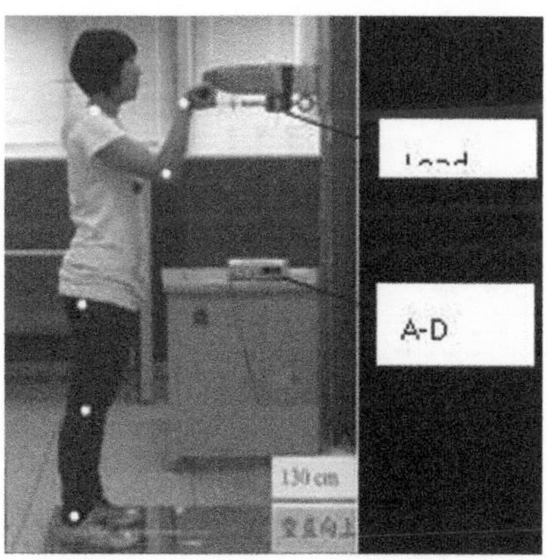


Figure 1. The device for lifting strength and schematic testing posture in the study (h = 130 cm).

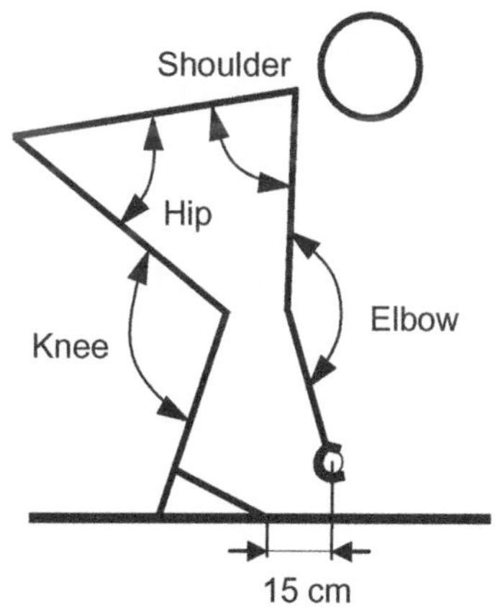


Figure 2. The definition of the body joint angles while testing in this study.

repeated until three readings were obtained, which were consistent within a range of 10%. As a result, a total of 4,140 strength data (46 participants × 2 exertion methods × 15 heights × 3 repetitions) was collected.

### 2.5 *Experimental procedure*

All participants were familiarized with the experimental procedures, and stretched themselves at least 10 min before the data were collected. During the experiment, the participants wore light clothing and rubber gym shoes. They were randomly requested to perform all maximum strengths at a specific exertion height and direction with free style but symmetrical lifting postures. The strength testing procedure was performed according to the methodology used by Chaffin (1975) and Ayoub et al., (1978). A period of 5 s was chosen as sufficient time to build up and maintain a constant force, while being brief enough to be endured without perceiving any muscular fatigue. A minimum rest period of 2 min was required between successive trials. The horizontal projection distance between the tip of the shoes of the participants and the handle bar was set at 15 cm. As was observed in a recent field study, this distance took the most advantageous testing posture to apply force by participants for full-range heights.

While determining the lifting strength, a nested design was used for strength data analysis. Each participant was considered as a block. We analyzed the strength data using a two-way ANOVA and used Duncan's multiple-range test for post-hoc comparisons. Differences in participants' strengths of VUL and TBL were checked by an independent t-test. An alpha level of 0.05 was selected as the minimum level of significance.

## 3 RESULTS

### 3.1 *Difference in strengths between the groups*

The lifting height influenced the force exerted during two lifting methods among workers and novices ($p < 0.05$), as shown in Figure 3. Table 1 shows the summary t-test result of lifting strengths with different lifting methods and heights between two female participant groups. As shown in the table, the VUL forces were much higher than the TBL at 15 height positions ($p < 0.001$). At lower heights (≦50 cm), workers' VUL forces were higher than novices' and no differences were found in TBL at these heights (less than 2 kg). A reverse trend was found when lifting ranges were equal to or higher than 90 cm. Forces in all 30 task combinations showed no difference between the two groups at heights of 60–80 cm.

Different strength profiles under various exertion heights can also be seen in Figure 3. The VUL forces were always higher than the TBL, regardless of working experience. As shown in Table 1, the novices' strength values were significantly higher at 10–50 cm than workers' (differences ranged from 5.67 to 7.40 kg) and oppositely, the workers' strength values were higher than novices' (differences ranged from 2.14 to 3.32 kg) at 90–150 cm.

### 3.2 *Posture analyses*

Comparisons of lifting postures while exerting maximum strength indicated that there were higher knee, shoulder, and elbow angles at lower height levels (i.e., 10–50 cm) in the novices'

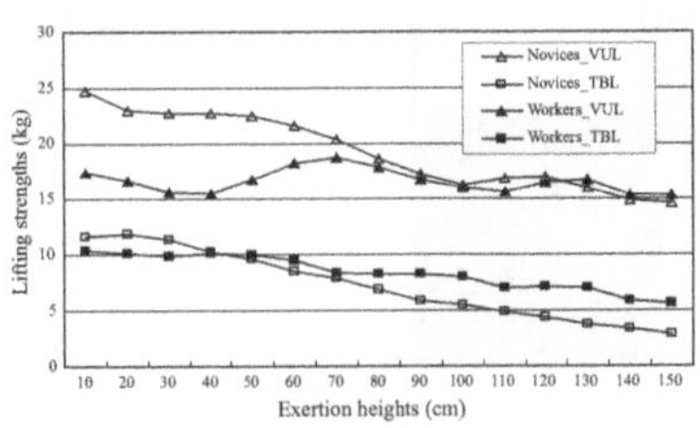


Figure 3. Lifting strength profiles under various exertion heights and methods.

Table 1. Differences in lifting strengths between female workers and novices.

| Height (cm) | ΔVUL (kg) | t | p | ΔTBL (kg) | t | p |
|---|---|---|---|---|---|---|
| 10 | 7.40 | –3.95 | <0.001 | 1.27 | –1.09 | ns |
| 20 | 6.40 | –3.51 | <0.001 | 1.72 | –1.46 | ns |
| 30 | 7.14 | –3.79 | <0.001 | 1.42 | –1.19 | ns |
| 40 | 7.33 | –3.40 | <0.001 | 0.19 | –0.42 | ns |
| 50 | 5.67 | –3.18 | <0.005 | –0.45 | 0.42 | ns |
| 60 | 3.36 | –1.93 | ns | –1.05 | 1.04 | ns |
| 70 | 1.63 | –0.95 | ns | –0.56 | 0.58 | ns |
| 80 | 0.68 | –0.39 | ns | –1.34 | 1.51 | ns |
| 90 | 0.39 | –0.25 | ns | –2.39 | 3.13 | <0.005 |
| 100 | 0.25 | –0.16 | ns | –2.48 | 3.13 | <0.005 |
| 110 | 1.26 | –0.83 | ns | –2.22 | 2.99 | <0.001 |
| 120 | 0.50 | –0.31 | ns | –2.79 | 3.80 | <0.001 |
| 130 | –0.78 | 0.49 | ns | –3.32 | 4.12 | <0.001 |
| 140 | –0.57 | 0.41 | ns | –2.46 | 2.90 | <0.001 |
| 150 | –0.76 | 0.50 | ns | –2.75 | 3.29 | <0.005 |

ΔVUL = Novices' VUL—Workers' VUL; ΔTBL is the same. ns, non-significant.

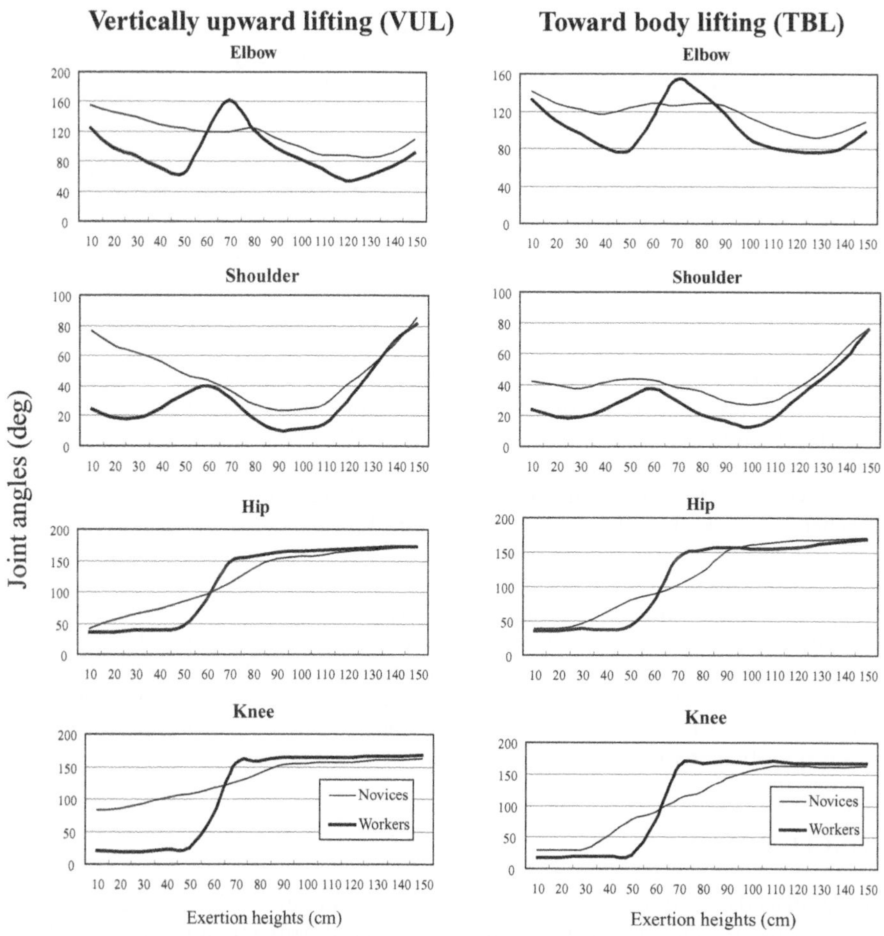


Figure 4. Comparison of the lifting postures while participants performed the VUL and TBL.

VUL tests, as shown in Figure 4. This means that a more stooped technique was chosen by novices while lifting upwardly from near-floor levels, as opposed to a more flexed knee strategy by workers. From the figure, it can also be seen that workers employed relatively consistent patterns of joint angles than did the novices, for whichever of VUL or TBL that was performed.

## 4 DISCUSSION

This study was the first to investigate the static lifting strength associated with a full range of lifting heights between female workers and novices. The results show that there exists a discrepancy in strength during VUL and TBL between the two female participant groups at the lower (≦50 cm) and higher (≧90 cm) positions, respectively. Similar trends in strength with increased heights were exhibited by novices' VUL (range 10.14 kg) and TBL (range 9.03 kg), whereas workers' strengths revealed a relatively small fluctuation (VUL: range 3.39 kg; TBL: range 5.45 kg). The findings were very different from the result of previous strength data which were collected on Taiwanese males (Chen et al., 2011). Chen et al., found that differences in strengths between male workers and novices only existed in VUL task at the heights from 100 to 120 cm. However, the testing protocol adopted by the present study was the same as that in Chen et al. Gender differences in lifting strategies at full lifting range between workers and novices need to be further clarified.

Some investigators have suggested that the response patterns of MAWL determination of workers and novices to task variables in lifting activities are similar (Garg and Saxena 1979, Mital and Manivasagan 1983). Their results may not be completely true as compared to our study, especially for some specific task conditions (e.g., lifting heights). In the present study, it is difficult to understand why the dramatic decrease obtained in VUL forces among novices (range 10.14 kg) was not observed among the experienced workers (range 3.39 kg). One possible reason can be attributed to differences in their adopted postures. As shown in Figure 4, workers' knee and hip joints always maintained fully flexed (at 10–50 cm) or fully extended (at 70–150 cm). For the VUL test, workers chose a nearly knee-flexion strategy, whereas the novices tended to choose a more stooped one. It is generally accepted that a stooped lifting posture, which can provide higher lifting strengths, would impose more compressive forces on the spinal L5/S1 disc. This would lead the workers to be more safety conscious than would the novices, even resulting in lower strength values of workers.

During the TBL at higher positions, female workers' strengths were significantly greater than female novices'. This result did not match that of Chen et al's study (2011) for male participants. As shown in Figure 4, the postures adopted by workers and novices were similar when the positions are higher than 100 cm. That is, with this nearly erect body posture, the strengths were reasonably determined by their upper extremities. In our study, the greater forces generated by workers during higher TBL positions may be due to the worker's lifting experiences for satisfying the job demands.

It is worthwhile to point out that lifting strengths toward the body were much lower than during VUL. One possible reason is that in this experiment, the participants were required to voluntarily exert their maximum exertions without losing their balance. Unlike the VUL method that completely counteracted with the ground during exertion, the TBL method generates resistance between the ground, and the participant's shoes. This mechanism results in a lower strength measure because more force for body stabilization is needed during lifting, and this was particularly true as the lifting height increased from 10 to 150 cm.

## 5 CONCLUSION

This paper extends the study of Chen et al. (2011) to compare the lifting strengths and the postures at various exertion heights between experienced workers and novices of Taiwanese females. The findings of this study were not in accordance with those of Chen et al.

The results show that there exists a discrepancy in strength during VUL and TBL between the two female participant groups at the lower (≦50 cm) and higher (≧90 cm) positions, respectively. Workers' strengths under all 15 lifting heights showed a relatively small fluctuation compared with novices' and tended to adopt fully squat and fully erect postures at lower and higher heights, respectively. The results of this study provide an inference that experienced workers may have learned a protective strategy and a more effective lifting technique to satisfy their daily job demands.

## REFERENCES


Authier, M., Lortie, M. & Gagnon, M. (1996). Manual handling techniques: Comparing novices and experts. *International Journal of Industrial Ergonomics*, Vol. 17, pp. 419–429.

Ayoub, M.M., Bethea, N.J., Deivanayagam, S., Asfour, S.S., Bakken, G.M., Liles, D., Mital, A. & Sherif, M. (1978). Determination and modeling of lifting capacity. *Final Report, HEW [NIOSH]*, Cincinnati, OH.

Baril-Gingras, G. & Lortie, M. (1995). The handling of objects other than boxes: univariate analysis of handling techniques in a large transport company. *Ergonomics*, Vol. 38, pp. 905–925.

Chen, Y.L., Lee, Y.C. & Chen, C.J. (2011). Differences in lifting strength profiles between experienced workers and novices at various exertion heights. *International Journal of Industrial Ergonomics*, Vol. 41, pp. 53–58.

Council for Economic Planning and Development (CEPD) (2004). *Guidelines and Action Plans for Service Industry Development*. Policy Report, Taipei, Taiwan, R.O.C.

Chaffin, D.B. (1975). Ergonomics guide for the assessment of human static strength. *American Industrial Hygiene Association Journal*, Vol. 36, pp. 505–511.

Chen, C.J. (2008). *Investigation of musculoskeletal disorders for the warehouse workers in Taiwan*, Master Dissertation, Mingchi University of Technology, Taipei, Taiwan.

Chen, Y.L. (2000). Optimal lifting techniques adopted by Chinese men when determining their maximum acceptable weight of lifting. *American Industrial Hygiene Association Journal*, Vol. 61, pp. 642–648.

Gagnon, M., Plamondon, A., Gravel, D. & Lortie, M. (1996). Knee movement strategies differentiate expert from novice workers in asymmetrical manual materials handling. *Journal of Biomechanics*, Vol. 29, pp. 1445–1453.

Garg, A., Sharma, D., Chaffin, D.B. & Schmidler, J.M. (1983). Biomechanical stresses as related to motion trajectory of lifting. *Human Factors*, Vol. 25, pp. 527–539.

Garg, A. & Saxena, U. (1979). Effects of lifting frequency and technique on physical fatigue with special reference to psychophysical methodology and metabolic rate. *American Industrial Hygiene Association Journal*, Vol. 49, pp. 894–903.

Kuorinka, I., Lortie, M. & Gautreau, M. (1994). Manual handling in warehouses: the illusion of correct working postures. *Ergonomics*, Vol. 37, pp. 655–661.

Lee, T.H. (2004) Static lifting strengths at different exertion heights. *International Journal of Industrial Ergonomics*, Vol. 34, pp. 263–269.

Lee, Y.H. & Chen, Y.L. (1996). An isometric predictor for maximal acceptable weight of lift for Chinese men. *Human Factors*, Vol. 38, pp. 646–653.

Marras, W.S., Parakkat, J., Chany, A.M., Yang, G., Burr, D. & Lavender, S.A. (2006). Spine loading as a function of lifting frequency, exposure duration, and work experience. *Clinical Biomechanics*, Vol. 21, pp. 345–352.

Keir, P.J. & MacDonell, C.W. (2004). Muscle activity during patient transfers: a preliminary study on the influence of lift assists and experience. *Ergonomics*, Vol. 47, pp. 296–306.

Kumar, S. (1991). Arm lift strength in work space. *Applied Ergonomics*, Vol. 22, pp. 317–328.

Mital, A. (1987). Patterns of differences between the maximum weights of lift acceptable to experienced and inexperienced materials handlers. *Ergonomics*, Vol. 30, pp. 1137–1147.

Mital, A., Garg, A., Karwowski, W., Kumar, S., Smith, J.L. & Ayoub, M.M. (1993). Status in human strength research and application, *IIE Transactions*, Vol. 25, pp. 57–69.

Mital, A. & Manivasagan, I. (1983). Maximum acceptable weight of lift as a function of material density, center of gravity location, hand preference, and frequency. *Human Factors*, Vol. 25, pp. 33–42.

Parakkat, J., Yang, G., Chany, A.M., Burr, D. & Marras, W.S. (2007). The influence of lift frequency, lift duration and work experience on discomfort reporting. *Ergonomics*, Vol. 50, pp. 396–409.

Yates, J.W., Kamon, E., Rodgers, S.H. & Champney, P.C. (1980). Static lifting strength and maximal isometric voluntary contractions of back, arm and shoulder muscles. *Ergonomics*, Vol. 23, pp. 37–47.

*Ergonomics in Asia – Shih & Liang (eds)*
*© 2012 Taylor & Francis Group, London, ISBN 978-0-415-68414-9*

# The effect of wearing spandex underwear on the activity of low-back and hip muscles

Jung-Yong Kim, Seung -Nam Min, Min-Ho Lee & Ju-Hyen Jung
*Industrial and Management Engineering, Hanyang University, Ansan, Korea*

Mi-Sook Kim
*Clothing & Textiles, Kyung Hee University, Seoul, Korea*

ABSTRACT: The purpose of this study was to analyze the effect of wearing spandex underwear on the symmetric activity of low-back and hip muscles. Ten males and ten females participated in the experiment and electromyography (EMG) was measured from their trunk and hip flexor/extensor muscles. Data were collected during dynamic and static flexion/extension motions. The degree of asymmetry measured by EMG was used as a dependent variable and the spandex underwear and gender were used as independent variables. In the results, a significant improvement was found in symmetric muscle activity when wearing spandex underwear. That is, the spandex underwear with a specially designed pattern can help subjects to equally use their right and left muscles. This result can be further investigated in order to find the possibility of using the spandex underwear as an assistive garment for rehabilitating LBP patients with anatomical or functional asymmetry of the trunk and hip.

*Keywords*: EMG, back and hip muscle, asymmetry, rehabilitation

## 1 INTRODUCTION

Excessive work, psychological tension, stress, and lack of exercise increase the incidence of spine-related disorders (Kim, 2007). A healthy person often maintains a correct posture, but a person who suffers from low-back disorders may have certain symptoms due to asymmetry of the body. In fact, a medical corset is used to help patients rehabilitate in spondylopathy treatment (Na, et al., 1999). The functional underwear (hereinafter, 'spandex underwear') was tailored and sold to correct the posture, and the effects of posture correction and rehabilitation were reported informally. Therefore, a scientific verification is necessary if people indeed need to use the spandex underwear for posture correction.

There is a medical back belt used at hospitals and workplaces for the prevention of back pain. Researches have been conducted, and it was reported that there was a considerable disagreement in the effect of back belt (Earle-Richardson, et al., 2005, Carter and Birrell, 2000). Perkins and Bloswick (1995) suggested that the use of the back belt reduced the load on the muscles when stretching one's back and eventually relieved muscle fatigue. The back belt reduced the pressure imposed on the vertebra, protecting the back against the risk of back injury.

A sit to stand (STS) protocol was used to examine the normality of muscle function. STS is basic movements like walking, which is an essential motion in our everyday lives (Nuzik, et al., 1986). In fact, STS is a prerequisite for locomotion (Jo, 2007).

Therefore, in this study, the spandex underwear was examined to see whether or not it would be really effective in rehabilitation treatment, as many users were reporting based on their personal experiences.

## 2 METHOD

### 2.1 *Participants*

Participants in the present study were 10 males and 10 females who were in good health and without a history of back pain (Table 1).

### 2.2 *Motions and posture tested*

Two dynamic motions and two static postures were selected for an experiment. These were a full squat simulating sitting to standing, dynamic flexion of the trunk, and flexion and extension statically maintained (Table 2).

### 2.3 *Muscle*

Four muscles were selected in order to measure the asymmetry of right and left muscles. For hip flexor and extensor, the quadriceps femoris (QF) and the biceps femoris (BF) were selected. For trunk flexor and extensor, the rectus abdominis (RA) and the erector spinae (ES) were selected. To measure the EMG, surface electrodes were attached (Figure 1).

### 2.4 *Equipment*

A Mega Electronics ME6000 (8-channel wireless EMD measuring instrument) (Figure 2) was used to measure participants' muscle activity. Sampling rate per EMG channel was 1000 Hz.

Table 1. Participants' demographics.

| | Age (years) | Height (cm) | Weight (kg) |
|---|---|---|---|
| Male (N = 10) | 35.5 ± 12.8 | 170.5 ± 5.4 | 69.5 ± 13.1 |
| Female (N = 10) | 33.0 ± 11.1 | 160.1 ± 3.4 | 56.8 ± 8.2 |

Table 2. Posture.

| Posture | Description |
|---|---|
| Fully squat | Dynamically sit and stand |
| Flexion | Dynamically bend 90° degree of the trunk and straighten up |
| Fixed flexion | Statically maintain 45° degree flexion |
| Fixed extension | Statically maintain 30° degree extension |

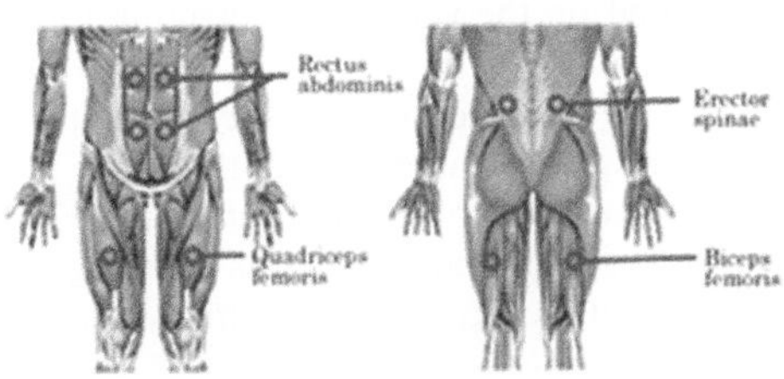


Figure 1. Selected muscles and the location of EMG attached.

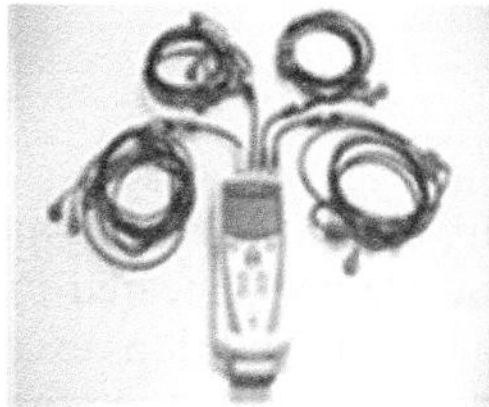

Figure 2. 8 channels electromyography.

### 2.5 *Experimental design*

A RMS (root mean square) value was used to compute the degree of asymmetry as a dependent variable to evaluate EMG. Gender (male, female) and wear of spandex underwear (wear, non-wear) were used as independent variables. Four experimental conditions were used as a repeated condition—Fully Squat, Flexion, Fixed Flexion, and Fixed Extension. A $2 \times 2 \times 4$ mixed-factors design was used. Latin squared design was used to reduce the carry-over effect (Kim, 2001).

### 2.6 *Procedure*

Before starting each experiment, the objective and procedure of the experiment were explained to the participants. Before measuring the EMG signal, surface electrode attaching points were wiped to minimize skin resistance errors. After attaching the electrodes and visually checking the stable EMG signal from each muscle, participants took a rest while they were instructed about the four motions/postures in the experiment. After the instruction, participants performed each task for five times in three minutes, and had 10 minutes of rest to avoid muscle fatigue accumulation (Dolan, 1995). All experimental performances were recorded on a video camera. After watching the recorded video, three repetitions out of five were selected for final analysis.

### 2.7 *Analysis*

Data in this experiment were analyzed by analysis of variance (ANOVA) through statistical package SPSS 17 for windows. For electrical noise reduction of EMG, 60 Hz frequency band signals were removed by using a Notch filter. During the each 5-minute measurement, the performing parts were extracted and analyzed. For analysis of EMG signal, the root mean square (RMS) value was used. Since the personal skin characteristics can change the RMS value, Equation 1 was used for normalization in order to prevent measuring error due to personal differences.

$$NEMG_{Posture} = \frac{EMG_{(W,Posture)} - EMG_{(WO,Standing)}}{EMG_{(WO,Standing)}} \qquad (1)$$

NEMG: Normalized EMG
EMG, W: EMG with spandex underwear
EMG, WO: EMG without spandex underwear

These normalized data were used to compare the difference of EMG from the left and right muscles, which was defined as the degree of asymmetry in Equation 2.

$$NEMG_{Asymmetry(i)} = |NEMG_{Left(i)} - NEMG_{Right(i)}| \qquad (2)$$

$NEMG_{Asymmetry}$: Asymmetry of Normalized EMG
$NEMG_{Left}$: Left normalized EMG
$NEMG_{Right}$: Right normalized EMG
i: Rectus abdominis, Erector spinae, Quadriceps femoris, Biceps femoris.

## 3 RESULTS

### 3.1 *Muscle activity*

A paired t-test was used to compare the muscle using pattern of 20 participants. In the results, the degree of asymmetric muscle use appeared to be significantly lower when the participants wore spandex underwear in all conditions except the biceps femoris in the fixed extension ($p < 0.05$) (Table 3). This result shows that the spandex underwear has the ability to compensate the asymmetric muscle activity of the trunk and hip although the specific causal relationship between design pattern and the effect has not been established.

### 3.2 *The degree of asymmetry*

To examine the degree of asymmetry, ANOVA was conducted. F-values were reported and the p values were expressed as the number of asterisks in Table 4.

The results suggested that RA muscle showed a significant interaction effect between motion/posture and clothing condition ($p < 0.01$, Figure 3).

Post hoc comparison was conducted by using a Scheffe test with respect to various postures in the RA muscle. As a result, the fixed extension (2.06) posture showed a significant difference from other postures ($p < 0.05$, Table 5). That is, the degree of asymmetry was higher in the fixed extension posture than in any other postures.

As post hoc results, the wearing of spandex underwear (0.50) and non-wearing of spandex underwear (0.90) showed significant differences ($p < 0.05$, Table 6); In other words, the symmetry between left and right muscles appeared to be higher when they wore spandex underwear.

Table 3. Paired difference T-test of muscle activity.

| Posture | | RA | ES | QF | BF |
|---|---|---|---|---|---|
| Fully squat | Left | W<WO** | W<WO** | W<WO** | W<WO** |
| | Right | W<WO** | W<WO** | W<WO** | W<WO** |
| Flexion | Left | W<WO** | W<WO** | W<WO** | W<WO** |
| | Right | W<WO** | W<WO** | W<WO** | W<WO** |
| Fixed flexion | Left | W<WO* | W<WO** | W<WO* | W<WO** |
| | Right | W<WO* | W<WO** | W<WO** | W<WO** |
| Fixed extension | Left | W<WO* | W<WO* | W<WO** | W=WOa |
| | Right | W<WO* | W<WO* | W<WO* | W=WOa |

*:p < 0.05, **: p < 0.01, a: NS (Non-significant).
W: With spandex underwear, WO: Without spandex underwear, RA: Rectus Abdominis, ES: Erector Spinae, QF: Quadriceps Femoris, BF: Biceps Femoris.

Table 4. ANOVA results (F-value) for the degree of asymmetry in each muscle.

| Source | RA | ES | QF | BF |
|---|---|---|---|---|
| Motion/posture | 10.988** | 17.175** | 19.672** | 15.449** |
| Gender | 2.139a | 0.853a | 1.390a | 10.468** |
| Wear or Not | 20.676** | 10.362** | 17.361** | 5.930* |
| Motion/posture × Gender | 2.347a | 1.459a | 0.327a | 5.680** |
| Wear or Not × Gender | 1.947a | 0.006a | 2.026a | 0.847a |
| Motion/posture × Wear or Not | 5.785** | 0.838a | 3.076* | 0.638a |
| Motion/posture × Wear or Not × Gender | 0.408a | 0.699a | 7.449a | 0.347a |

*:p < 0.05, **:p < 0.01, a: NS (Non-significant).

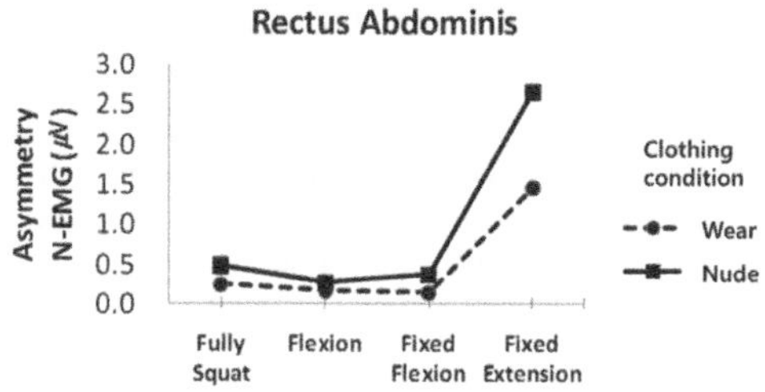


Figure 3. Interaction effects of N-EMG on RA between various posture and clothing condition (Wear or Not).

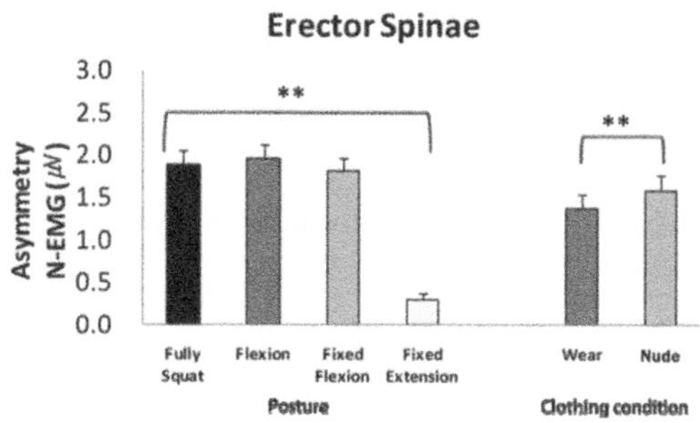


Figure 4. Main effect on the degree of asymmetry by posture and clothing condition (Wear or Not) for ES.

Table 5. Scheffe (post hoc test) of asymmetric RA activity for various motions and postures.

| | Fully squat | Flexion | Fixed flexion | Fixed extension |
|---|---|---|---|---|
| Scheffe grouping | A | A | A | B |
| N | 40 | 40 | 40 | 40 |
| Mean | 0.36 | 0.22 | 0.25 | 2.06 |

Table 6. Scheffe (post hoc test) of asymmetric RA activity for clothing condition.

| | Wear | Nude |
|---|---|---|
| Scheffe grouping | A | B |
| N | 40 | 40 |
| Mean | 0.50 | 0.90 |

In the ES muscle, the symmetry between left and right muscles appeared to be higher in the fixed extension posture than other postures ($p < 0.01$, Figure 4).

With respect to post hoc comparison of posture for the ES muscle, the fixed extension (0.30) showed significant differences ($p < 0.05$, Table 7) from other postures. In other words, the ES muscle showed the most symmetry among the postures.

A significant interaction effect between postures and clothing condition was found in the QF muscle, ($p < 0.05$, Figure 5).

In the QF muscle, the flexion and fixed flexion postures showed higher left and right muscle symmetry in post hoc comparison of postures.

In the RA muscle, the left and right muscle symmetry was shown to be greater when spandex underwear was worn ($p < 0.05$, Table 9).

Table 7. Scheffe (post hoc test) of asymmetric ES activity for various motions/postures.

| | Fully squat | Flexion | Fixed flexion | Fixed extension |
|---|---|---|---|---|
| Scheffe grouping | A | A | A | B |
| N | 40 | 40 | 40 | 40 |
| Mean | 1.90 | 2.00 | 1.80 | 0.30 |

Table 8. Scheffe (post hoc test) of asymmetric QF activity for various postures.

| | Fully squat | Flexion | Fixed flexion | Fixed extension |
|---|---|---|---|---|
| Scheffe grouping | A | B | B | C |
| N | 40 | 40 | 40 | 40 |
| Mean | 23.31 | 1.54 | 1.67 | 11.48 |

Table 9. Scheffe (post hoc test) of asymmetric RA activity for clothing condition.

| | Wear | Nude |
|---|---|---|
| Scheffe grouping | A | B |
| N | 80 | 80 |
| Mean | 8.20 | 10.80 |

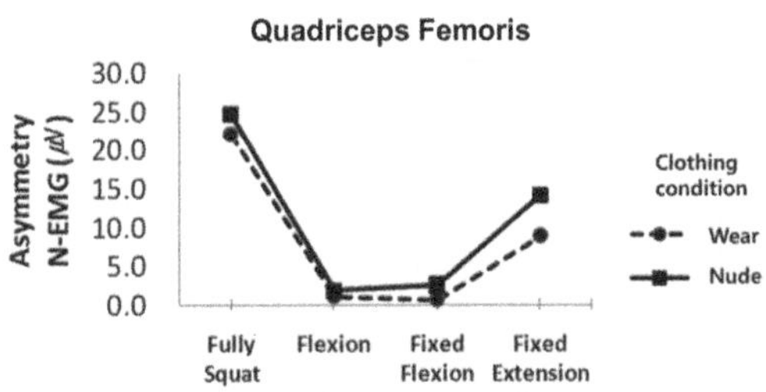


Figure 5. Interaction effects of N-EMG on QF between posture and clothing condition (Wear or Not).

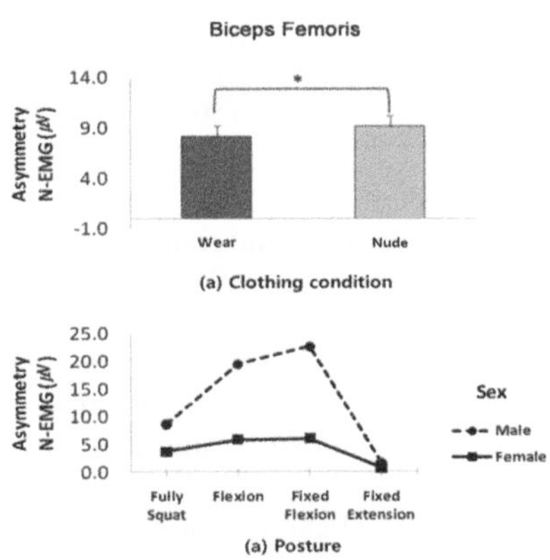


Figure 6. Main effect of clothing condition and interaction effect between posture and sex on BF muscle.

Table 10. Scheffe (post hoc test) of asymmetric BF activity for various postures.

| | Fully squat | Flexion | Fixed flexion | Fixed extension |
|---|---|---|---|---|
| Scheffe grouping | A | B | B | A |
| N | 40 | 40 | 40 | 40 |
| Mean | 6.17 | 12.72 | 14.36 | 1.26 |

Table 11. Scheffe (post hoc test) of asymmetric RA activity for gender.

| | Male | Female |
|---|---|---|
| Scheffe grouping | A | B |
| N | 80 | 80 |
| Mean | 13.1 | 4.1 |

The BF muscle showed higher left and right symmetry when spandex underwear was worn (Figure 6a), and the interaction effect between gender and posture on asymmetry was significant ($p < 0.05$, Figure 6b).

In the BF muscle, a post hoc comparison of posture, the fully squat (6.17), fixed extension (1.26) showed significant differences from the flexion (12.72), and fixed flexion (14.36) ($p < 0.05$, Table 10). In other words, the fully squat and fixed extension showed higher left and right muscle symmetry.

In BF muscle, males showed higher left and right muscles symmetry than females did (Table 11).

## 4 DISCUSSION

This study examined the effect of wearing the spandex underwear on pelvic posture correction. As a result, it was found that the spandex underwear symmetrically guided the muscle activity of left and right muscles. In particular, such effect was greater when the dynamic fully squat and flexion motion were practiced.

This indicates that spandex could be now used an as assistive garment for correcting asymmetry of posture. This study expands the usage of the spandex corset to a medical purpose and showed its potential for medical effects on rehabilitation of low-back and hip muscles. The spandex garment with a special construction pattern indeed showed a posture correcting function in this study.

## 5 CONCLUSION

Further study is necessary to find out which construction pattern or material of the spandex underwear was able to help symmetric motion. As was the case of the back belt, this hip supporting garment may have a contributing nature for LBP patients with an asymmetric functionality of muscle or asymmetric posture as the cause of back pain. A longitudinal observation of using the spandex underwear could provide information that can be compared with the cross-sectional approach in this study for further validation.

## REFERENCES

Calmels, P. & Fayolle-Minon, I. (1996). An update on orthotic devices for the lumbar spine based on a review of the literature. *Revue du rhumatisme*, Vol. 63, No. 4, pp. 285–291.

Carter, J.T. & Birrell, L.N. (2000). *Occupational health guidelines for the management of low back pain at work-principle recommendation.* London: Faculty of occupational medicine, (www.facoccmed.ac.uk).

Dolan, P., Mannion, A.F. & Adams, M.A. (1995). Fatigue of the erector spinae muscles: a quantitative assessment using "Frequency Banding" of the surface electromyography signal. *Spine*, Vol. 20, No. 2, pp. 149–159.

Earle-Richardson, G., Jenkins, P., Fulmer, S., Mason, C., Burdick, P. & May, J. (2005). An ergonomic intervention to reduce back strain among apple harvest workers in New York State. *Applied Ergonomics*, Vol. 36, No. 3, pp. 327–334.

Jo, G.G. (2007). A study on waist and lower extremities muscle activities according to foot's positions during flexion/extension of legs. *The Korean Society of Sports Science*, Vol. 16, No. 3, pp. 737–747.

Kim, H.W. (2007). The effect of a yoga correction exercise program for scoliosis. *Korea Coaching Development Center*, Vol. 9, No. 4, pp. 93–101.

Kim, J.Y. (2001). A study of trunk muscle fatigue and recovery time during isometric extension tasks. *Journal of the Ergonomics Society of Korea*, Vol. 21, No. 2, pp. 25–33.

Na, S.Y., Seo, J.H. & Kim, Y.H. (1999). Effect of abdominal corset on pulmonary function and oxygen consumption in cervical spinal cord-injured patients. *Korean Academy of Rehabilitation Medicine*, Vol. 23, No. 4, pp. 756–761.

Nuzik, S., Lamb, R., Vansant, A. & Hirt, S. (1986). Sit-to-stand movement pattern: A kinematic study, *Physical Theraphy*, Vol. 66, No. 11, pp. 1708–1713.

Perkins, M.S. & Bloswick, D.S. (1995). The use of back belts to increase intraabdominal pressure as a means of preventing low back injuries: A survey of the literature. *International Journal of Occupational and Environmental Health*, Vol. 1, No. 4, pp. 326–335.

*Ergonomics in Asia – Shih & Liang (eds)*
*© 2012 Taylor & Francis Group, London, ISBN 978-0-415-68414-9*

# Changes in finger joint angles while using scroll wheels: Comparison of scroll up and down motions

Kentaro Kotani, Takafumi Asao & Satoshi Suzuki
*Department of Mechanical Engineering, Kansai University, Japan*

Makoto Osada
*Graduate School of Science and Engineering, Kansai University, Japan*

ABSTRACT: Finger joint angles while using a scroll wheel were monitored by two types of scrolling motions (scroll up and scroll down), using a high-speed motion analysis microscope. The finger joint angles were separately analyzed with two phases (wheel rotating phase and rewinding phase) during scrolling. Based on the changes in finger joint angles for 14 participants, it was revealed that MP joint angles during scroll up motions were 4.3% more over-extended than those during scroll down motions. Such overextensions were observed when they perform scroll up motion especially in rewinding phases to prepare for the next wheel rotating phases. Subjective discomfort level in scroll up motion was significantly higher than that in scroll down motion. Observed overextension during rewinding phase in scroll up motion may be the potential factor for the subjective discomfort. It was expected that changes in the location and size of scroll wheels may reduce the overextension and discomfort levels.

*Keywords*: scroll wheel, mouse, musculoskeletal discomfort, repetitive motions, finger

## 1 INTRODUCTION

The scroll wheel has been playing a major role as an input function for the PC mouse since the mid 90s. Prior research mostly focused on its usability. Kobayashi and Igarashi (2006) developed software for operating a scroll wheel with multi-mode function to improve the usability. Lee and Lee (2010) proposed a virtually multiple mouse with capacitive touch sensors by the side of scroll wheel, where different functions were assigned by different scrolling strategies and scrolling positions of the finger around the scroll wheel. Shih, et al. (2009) replaced a new mouse driver to detect poking action on the scroll wheel for enhance usability to assist people with disabilities with minimal motor behavior. However, no significant improvement was associated with finger motions during scroll wheel operations. Wheel scrolling requires repetitive finger motions, which may induce musculoskeletal discomfort, yielding the potential risk of musculoskeletal injuries.

Results of the literature review revealed that the risk of using the mouse itself has been analyzed, (Lee, et al., 2007; de Korte, et al., 2008) however no focus was given to the finger movement for operating scroll wheels. If the appropriate moving strategy could be found, the usability as well as the risk of musculoskeletal injuries can be minimized.

Thus the objective of this study was the design of an ergonomically comfortable mouse, especially focusing on the optimal size and location of scroll wheel to the mouse. For this purpose, this current study reports the results of observations of the finger movement for the initial phase of this project.

## 2 METHODS

### 2.1 *Participants*

A total of 14 male university students participated in the study. All subjects were right-handed and with a daily use of a computer mouse. None reported previous physical disabilities in the upper extremity.

### 2.2 *Apparatus*

To record the operating motions of the finger, a high-speed motion analysis microscope (VW-6000, Keyence) was used. Frame rates of the microscope were set to 125 fps. The mouse used was a typical USB two-button optical mouse with a scroll wheel. The dimension and the shape of the mouse are shown in Figure 1. The experimental setting was illustrated in Figure 2. Distance between the subject and the display was 700 mm. A keyboard was not used in the experiment to ensure clear coverage of the finger motions.

### 2.3 *Experimental procedure*

5 mm white markings were attached at the middle point of the fingertip, DIP joint, PIP joint, MP joint, and the middle point of the metacarpal bone. The subject was seated at the desk and put the right hand on the mouse, while the left hand was positioned freely. A web browser (Windows Explorer 8, Microsoft) was used in full screen. Prior to the trials, it was confirmed that the markings were lined up in a straight line. If not, the markers were

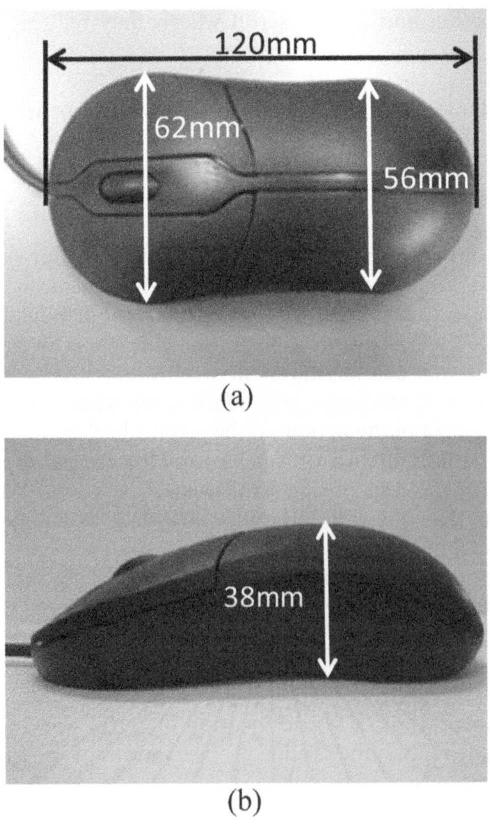


Figure 1. USB optical mouse used in this study (a) top view, (b) side view.

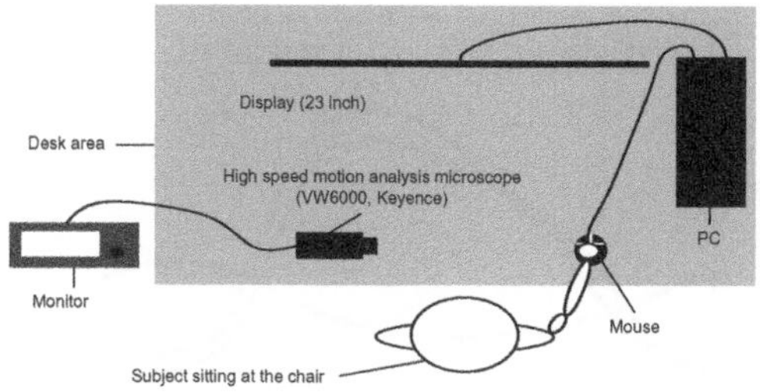


Figure 2. Schematic diagram of experimental settings.

adjusted straight while the subject's finger was at the neutral position. Then with a brief rest, the subject started scrolling web pages downward by moving the scroll wheel. The subject performed 10 trials of down-scrolling. After the trials, subjective discomfort ratings of the hand were obtained by a VAS-based form. Next, the subject performed scrolling web pages upward in the same manner as downward scrolling. Pace in rotating the scroll wheel was not specified so that the subjects performed scrolling wheels at their own paces. After the trials the hand anthropometric data were collected to investigate relationships with finger joint motions. The data included index finger length, fingertip-DIP joint length, DIP-PIP joint length, PIP-MP joint length, palm length, and the distance between the nose of the mouse and the location of the thumb on the mouse.

### 2.4 *Data analysis*

Changes in joint angles of the index finger were analyzed by two motions, i.e., scrolling upwards (SU) and downwards (SD). Each motion was separated into two phases, wheel rotating phase and rewinding phase. During the wheel rotating phase, the index finger was attached to the scroll wheel whereas the finger backed up onto the starting position of wheel rotation during rewinding phase. Each joint angle was normalized by its maximum range of motions to obtain % of maximum moveable angles (%MMA) with reference to the neutral hand position.

## 3 RESULTS AND DISCUSSION

### 3.1 *Changes in %MMA for SU and SD motions*

Figures 3 and 4 show typical trends in %MMA of each joint angle in SU and SD motions, respectively. In both figures, positive values in %MMA represent joint flexions and negative values represent joint extensions. The white area in the figures denotes the wheel rotating phase and the shaded area denotes the rewinding phase. Two figures appeared to be matched identically if one is upturned the other way since the finger movement for SU and SD motions should be theoretically the other way around; one goes clockwise and the other goes counterclockwise. All three joint angles appeared to be changing smoothly over time, confirming that the index finger was moved in cyclic motion, without having any sudden gaps to switch over between two phases in scrolling behavior. All three joints tended to extend from the neutral position for most of the cycle, only the DIP and PIP joints have the period of joint flexions approximately 10 to 20%MMA from the neutral position. DIP and PIP joints have their peak extension position at the point of changing from wheel rotating phase to rewinding phase to lift the finger upwards from the scroll wheel. On the contrary, as shown in the figure, it was revealed that MP joints tended to be overextended, approximately 10%MMA, as approaching the wheel rotating phase, which can be seen between frames 48 and 63 in the Figure 3. Such "preparing" over-extension was not found, as shown in Figure 4, when the subject performed SD motions.

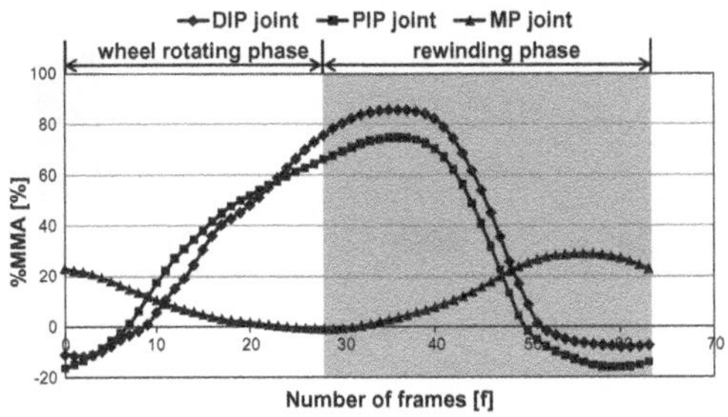


Figure 3. Changes in finger joint angles with time during SU motion.

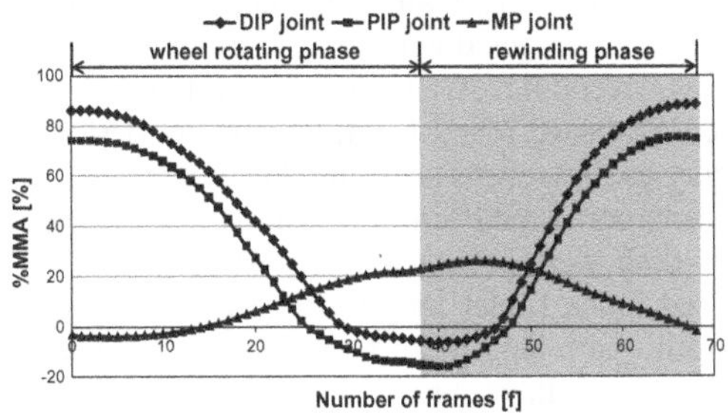


Figure 4. Changes in finger joint angles with time during SD motion.

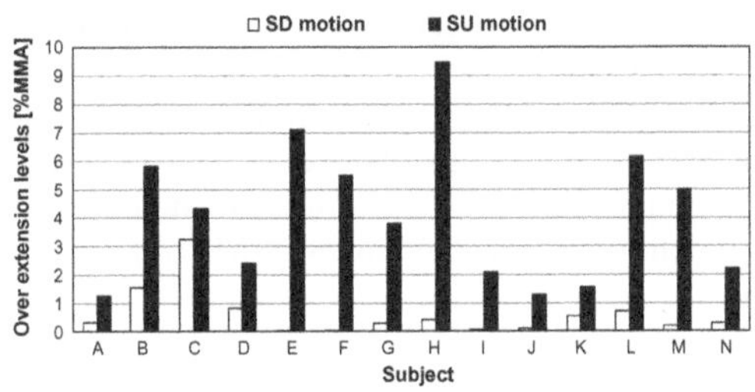


Figure 5. Comparison of overextension levels between two motions by subjects.

Figure 5 shows individual overextension levels by two motions. Apparently most subjects exhibited higher extension while preparing the next phase. Average extension level was 4.3%MMA in SU motion, which was significantly higher than that in SD motion (0.7%MMA).

### 3.2 *Correlation between hand parameters and overextension rates in SU motion*

As described in the previous section, trends in two finger motions were similar except overextension of the MP joint in SU motion. Thus, the correlation between overextension levels of MP joint and the hand anthropometric data was analyzed. Table 1 shows correlation between hand parameters and MP joint overextension levels during rewinding phase. Palm length was significantly negatively correlated with overextension levels of MP joint, followed by index finger length and fingertip-DIP joint length. No other anthropometric data were correlated with overextension levels. The results implied that, the longer the palm and the finger length were, the less overextension was required for operating the scroll wheel. In this study, only one type of mouse was used, thus the coupling effect between the mouse used and the hand may exist. Therefore, these correlation coefficients may be different if the dimensions of the mouse were be different than those used in this study.

Table 1. Correlation between hand parameters and overextension levels of the MP joint during rewinding phase.

| Hand parameters | r-values |
|---|---|
| Index finger length | –0.589* |
| Fingertip-DIP joint length | –0.586* |
| DIP-PIP joint length | –0.459 |
| PIP-MP joint length | –0.258 |
| Distance between nose of mouse and the thumb | 0.252 |
| Palm length | –0.822** |

*p < .05, **p < .01.

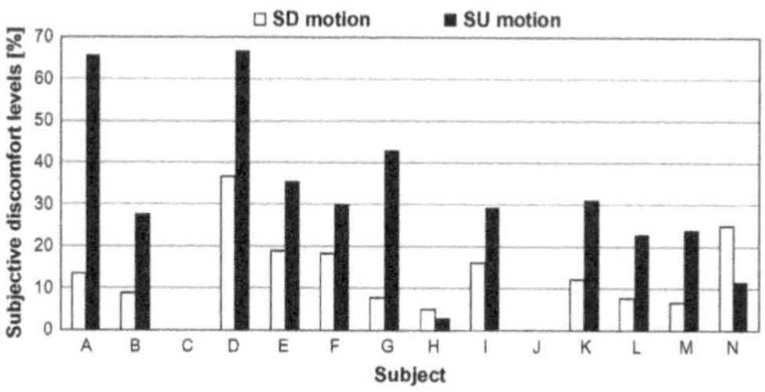


Figure 6. Comparison of subjective discomfort levels between two motions by subjects.

### 3.3 *Subjective discomfort levels for two motions*

Figure 6 shows individual subjective discomfort levels by two motions. 12 subjects reported equal or higher discomfort levels when they performed SU motions. Average discomfort level in SU motion was 27.5%, which was significantly higher than that in SD motion (13.0%).

## 4 CONCLUSION

Results for changes in joint angles in two motions showed that overextension was present during the latter part of the rewinding phase only in SU motion. Other than that, both tendencies were very similar, confirming that finger motions for rotating scroll wheel were cyclic and the two motions were symmetrical. Subjects reported relatively high discomfort levels during SU motion. Palm length, index finger length, and fingertip-DIP length were negatively correlated with overextension levels, but seemed unrelated to the location of the thumb for gripping the mouse.

Currently, we are developing a mouse prototype, where the size (diameter) and the location of the scroll wheel with reference to the mouse is changeable, leaving the shape and the size (diameter) of the mouse the same. Thus, the optimum scroll wheel location and the size would be found empirically. Also, observing finger motions in using different size and shape of the mouse would answer the importance of the hand anthropometry to the discomfort for operating scroll wheels. It is also important for the further study that measurement of electromyography in the upper extremity would confirm muscular loads during scroll wheel operations.

## ACKNOWLEDGEMENT

This study was partially supported by Ecological Interface Design Study Group, Kansai University.

## REFERENCES


de Korte, E.M., de Kraker, H., Bongers, P.M. & van Lingen, P. (2008). Effects of a feedback signal in a computer mouse on movement behavior, muscle load, productivity, comfort and user friendliness, Ergonomics, Vol. 51, No. 11, pp. 1757–1775.

Kobayashi, M. & Igarashi, T. (2006). MoreWheel: Multimode Scroll-Wheeling Depending on the Cursor Location, UIST, pp. 15–18.

Lee, D.L., Fleisher, J., Mcloone, H.E., Kotani, K. & Dennerlein, J.T. (2007). Alternative computer mouse design and testing to reduce finger extensor muscle activity during mouse use, Human Factors, Vol. 49, No. 4, pp. 573–584.

Lee, K.W. & Lee, Y.C. (2010). Design and validation of virtually multiple mouse wheels, International Journal of Industrial Ergonomics, Vol. 40, pp. 392–401.

Shih, C.H., Chang, M.L. & Shih, C.T. (2009). Assisting people with multiple disabilities and minimal motor behavior to improve computer pointing efficiency through a mouse wheel, Research in Developmental Disabilities, Vol. 31, No. 7, pp. 1378–1387.

*Ergonomics in Asia – Shih & Liang (eds)*
*© 2012 Taylor & Francis Group, London, ISBN 978-0-415-68414-9*

# Evaluating arch dimension difference by using 3-D foot scanning data

Yu-Chi Lee, Gloria Lin & Mao-Jiun J. Wang
*Department of Industrial Engineering and Engineering Management, National Tsing Hua University, Hsinchu, Taiwan*

ABSTRACT: The objective of this study was to use 3D scanning data to evaluate gender and foot size differences in arch dimensions. 150 males and 150 females were recruited in this study. Six arch dimensions including arch length, arch depth, arch height, distance of navicular (%FL), distance of metatarsal tibiale (%FL) and distance of the most medial point of medial border line (%FL) were used to determine the arch shape and position. Results showed that the gender difference was significant in arch length, arch height, distance of navicular (%FL) and distance of the most medial point of medial border line (%FL), except for arch depth and distance of metatarsal tibiale (%FL). In addition, arch length would increase with foot size for both genders, but arch depth would decrease with foot size only for males. Therefore, designing insoles based on arch dimensions for males and females separately is necessary.

*Keywords*: arch, gender difference, foot dimension, footwear

## 1 INTRODUCTION

The human foot has a complex structure. The arch as a special structure of the feet plays an important role in supporting the weight of human body in the erect posture. It balances and enables the body to stand. The arch structure confers a mechanical advantage to the leg muscles permitting the body to be lifted off the ground and propelled forward (Ker et al., 1987). It can also spread ground reaction forces as springs and thus reduce the risk of musculoskeletal injuries. In addition, the arch can absorb and dissipate the shock waves during walking. In general, the arch of the foot provides a unique foundation for stable support, mechanical leverage, shock dissipation, and balance (Saltzman and Nawoczenski 1995).

For arch dimension measurements, some studies have used different methods to capture the arch characteristics, such as optical pedobarograph (McCrory et al., 1997), caliper measurement (Hawes et al., 1992) and arch index (Cavanagh and Rodgers 1987). Also, some studies have used different arch dimensions to cluster arch types (Razeghi and Batt 2002, Xiong et al., 2010). With the advancement of optoelectronic technologies, the scanning technique was used to collect foot dimensions and arch dimensions. Lee et al. (2010) evaluated the gender differences on foot dimensions by using 3D foot scanning data and found significant gender differences on 14 foot dimensions (included arch height). Witana et al. (2009) also used 3D foot scanning to collect foot dimensions and showed that arch shape was not influenced by fore foot and hind foot dimensions.

Moreover, arch dimensions have been used to design shoes and to diagnose foot abnormalities such as flat foot. However, these studies only used the arch length and height for analysis. Very little information is available on evaluating the gender difference on arch dimensions. In this study, an experiment was conducted to collect Taiwanese young adults' arch dimensions by using a 3D foot scanner, and to evaluate the gender and foot size effect on arch dimensions.

## 2 METHOD

### 2.1 *Participant*

Three hundred students (150 males and 150 females) from the National Tsing Hua University, aged from 18 to 37 years old (mean age = 21.8 ± 2.7 years old), were recruited for this experiment. None of them had any history of visible foot abnormalities or foot illnesses. The descriptive statistics of the participants are given in Table 1.

### 2.2 *Experimental apparatus*

A 3D foot scanner (INFOOT USB scanning system, IFU-S-01, I-Ware Laboratory Co., Ltd, Japan) was used to collect arch dimensions (as shown in Figure 1). There were 8 CCD cameras and 4 laser projectors to construct a foot model in the scanning system. The accuracy of the foot scanner was within 1.0 mm (Kouchi and Mochimaru 2001)

### 2.3 *Definition of arch dimensions*

The arch length, arch depth and arch height were used to represent the arch shape. Also, the distance of navicular (%FL), distance of metatarsal tibiale (%FL) and distance of the most medial point of medial border line (%FL) were used to determine the arch position. Since these distance dimensions may be influenced by foot length, the percentage of foot length was used. The definitions of the arch dimensions involved in this study are shown in Table 2 and illustrated in Figure 2.

### 2.4 *Experimental procedure*

All the participants were asked to wash their feet before data collection. This procedure can avoid particles on foot surface influencing the scanning quality. A well trained experimenter placed six anatomical markers on each participant's foot after washing. The six marker positions are shown in Fig 3. These markers can enhance the precision of scanning data and help the scanning system to automatically measure foot dimensions. Each participant was requested to scan twice with a natural standing posture and to avoid any foot and body swing during scanning. This procedure can ensure the scanning quality for future data analysis. A successful scanning process took about 10s.

Table 1. Descriptive statistics of the participants (Mean (Standard deviation)).

| Characteristics | Males (N = 150) | Females (N = 150) |
|---|---|---|
| Age (year) | 21.8 (2.7)* | 21.8 (2.8) |
| Stature (cm) | 174.0 (5.4) | 161.1 (5.1) |
| Weight (kg) | 67.7 (9.1) | 51.9 (6.0) |
| BMI | 23.4 (3.0) | 20.0 (2.2) |

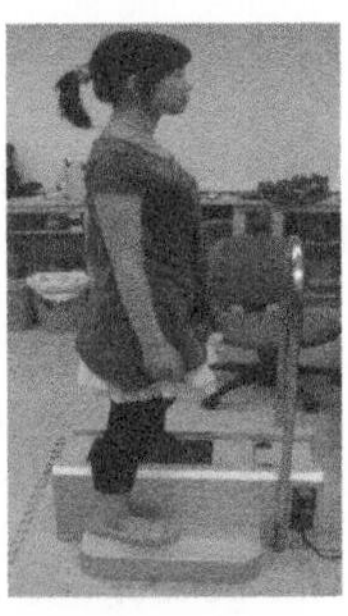

Figure 1. The 3D foot scanner.

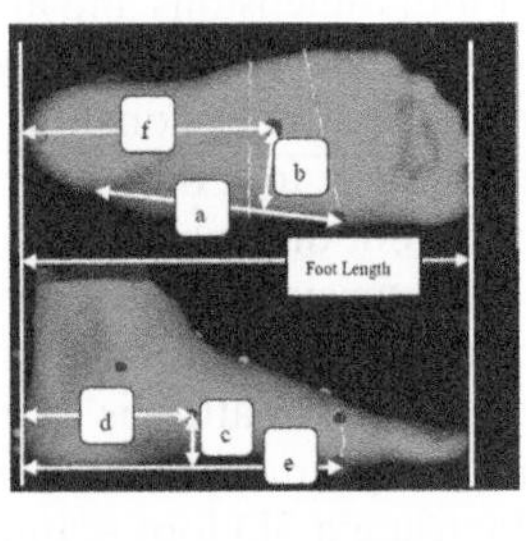


Figure 2. Illustration of the definition of the six arch dimensions.

Table 2. The definition of arch dimensions.

| Dimensions | Definition |
|---|---|
| a. Arch length | The distance from the 1st metatarsal point protrusion to the landing point of medial heel (Luximon et al., 2003). |
| b. Arch depth | The vertical distance from the most medial point of medial border line to the line of arch length. |
| c. Arch height | The height of navicular (Xion et al., 2010). |
| d. Distance of navicular (%FL) | The straight distance of navicular to heel divided by foot length. |
| e. Distance of metatarsal tibiale (%FL) | The distance from the 1st metatarsal point protrusion to the heel divided by foot length (Witana et al., 2009). |
| f. Distance of the most medial point of medial border line (%FL) | The straight distance from the most medial point of medial border line (Hawes et al., 1992) to heel divided by foot length. |

Table 3. The mean, standard deviation and ANOVA test results.

| | Arch length (mm) | Arch depth (mm) | Arch height (mm) | Distance of navicular (%FL) | Distance of metatarsal tibiale (%FL) | Distance of the most medial point of medial border line (%FL) |
|---|---|---|---|---|---|---|
| Males | 147.5 (7.4) | 35.7 (8.4) | 37.0 (5.1) | 37.3 (2.1) | 73.4 (1.6) | 50.7 (4.4) |
| Females | 134.9 (6.3) | 35.4 (7.3) | 33.1 (4.9) | 36.6 (2.5) | 73.3 (1.5) | 52.8 (3.3) |
| p-value | *** | NS | *** | ** | NS | *** |

NS: non-significant; ** $p < 0.01$; *** $p < 0.001$

### 2.5 *Statistical analysis*

Statistical analysis was conducted through the use of SPSS 13.0 software package. The ANOVA test was conducted to evaluate the gender difference on selected dimensions. The significance level was set at $\alpha = 0.05$.

## 3 RESULTS AND DISCUSSION

Table 3 shows the gender differences in the 6 arch dimensions. Results showed that a significant gender difference was found in arch length, arch height, the distance of navicular (%FL) and distance of the most medial point of medial border line (%FL), except for arch depth and distance of metatarsal tibiale (%FL). Males had greater arch length and height than females. It has been reported that males had greater foot dimensions in foot length, ball of foot length, navicular height and medial ankle height than females (Lee et al., 2010). On the other hand, it is interesting to note that males had a smaller distance in medial point of medial border line (%FL) than females. Thus, using the geometric grading method to design insoles for both genders may not be adequate. Designing the insoles based on arch dimensions for males and females separately is necessary.

Further, the 6 arch dimensions among different foot sizes were also calculated. The foot size of the subjects was represented by shoe size. The shoe size of the subjects ranged from US 7 to 11 for males and US 4 to 8 for females. The arch dimensions of males and females among different shoe sizes are shown in Table 4. As can be seen, arch length is increased with increasing foot length for both genders. For males, the arch length ranged from 138.1 mm to 162.4 mm. For females, the arch length ranged from 122.6 mm to 144.6 mm. Moreover, the male's arch depth is decreased with increasing foot length. The greatest arch depth was found in US 7 (37.7 mm) and the smallest arch depth was found in US 11 (30.3 mm) for males. Besides, there is no obvious trend between foot length and arch height, the distance of

Table 4. The arch dimensions among different shoe sizes for females (N = 150) and males (N = 150).

| Shoe sizes (US) | | Arch length (mm) | Arch depth (mm) | Arch height (mm) | Distance of navicular (%FL) | Distance of metatarsal tibiale (%FL) | Distance of the most medial point of medial border line (%FL) |
|---|---|---|---|---|---|---|---|
| Females | 4 | 122.6 (3.2) | 35.2 (5.4) | 32.5 (4.2) | 35.8 (2.2) | 52.2 (3.8) | 72.8 (1.4) |
| | 5 | 129.7 (3.8) | 32.8 (8.4) | 33.1 (4.8) | 36.6 (2.5) | 52.0 (3.8) | 73.4 (1.5) |
| | 6 | 135.0 (3.4) | 35.9 (6.8) | 32.8 (5.2) | 36.9 (2.5) | 53.2 (3.0) | 73.6 (1.6) |
| | 7 | 139.1 (3.0) | 36.7 (6.1) | 34.4 (4.1) | 35.9 (2.3) | 52.6 (3.3) | 72.7 (1.5) |
| | 8 | 144.6 (3.7) | 35.7 (10.5) | 32.4 (5.9) | 36.9 (2.8) | 53.1 (3.3) | 73.2 (1.4) |
| Males | 7 | 138.1 (4.2) | 37.7 (6.2) | 38.1 (6.0) | 37.4 (2.2) | 73.0 (1.4) | 51.7 (3.1) |
| | 8 | 145.2 (4.0) | 36.8 (7.9) | 36.3 (4.8) | 37.4 (2.2) | 73.5 (1.7) | 51.3 (4.3) |
| | 9 | 150.7 (4.1) | 35.5 (8.5) | 37.6 (5.0) | 37.0 (2.0) | 73.4 (1.5) | 51.0 (4.2) |
| | 10 | 157.0 (3.4) | 31.3 (10.5) | 36.9 (4.7) | 37.4 (2.0) | 73.5 (1.6) | 47.9 (5.4) |
| | 11 | 162.4 (4.1) | 30.3 (9.3) | 35.1 (4.9) | 36.4 (2.6) | 73.4 (1.2) | 49.3 (6.0) |

navicular (%FL), distance of metatarsal tibiale (%FL), distance of the most medial point of medial border line (%FL) for both genders.

## 4 CONCLUSION

In the study, a 3-D foot scanner was used to collect 6 arch dimensions in the study, including arch length, arch depth, arch height, distance of navicular (%FL), distance of metatarsal tibiale (%FL) and distance of the most medial point of medial border line (%FL). For gender effect, significant differences were found in arch length, arch height, distance of navicular (%FL) and distance of the most medial point of medial border line (%FL). The arch length is increased with increasing foot length from shoe size US 7 to 11 for males and from US 4 to 8 for females. The males' arch depth is decreased with increasing foot length. Thus, designing insoles based on arch dimensions for males and females separately is necessary.

## REFERENCES


Cavanagh, P.R. and Rodgers, M.M. (1987). The arch index: a useful measure from footprints. Journal of Biomechanics, 20(5), pp. 547–551.

Hawes, M.R., Nachbauer, W., Sovak. D. and Nigg, B.M. (1992). Footprint parameters as a measure of arch height. Foot Ankle, 13(1), pp. 22–6.

Ker, R.F., Bennett, M.B., Bibby, S.R., Kester, R.C. and Alexander, R.M. (1987). The spring in the arch of the human foot. Nature, 325(6100), pp. 147–149.

Kouchi, M. and Mochimaru, M. (2001). Development of a low cost foot-scanner for a custom shoe making system. In 5th ISB Footwear Biomechanics, pp. 58–59.

Lee, Y.C., Lin, G. and Wang, M.J. (2010). Evaluating Gender Differences in Foot Dimensions. The 9th Pan-Pacific Conference on Ergonomics.

Luximon, A., Goonetilleke, R.S. and Tsui, K.L. (2003). Foot land marking for footwear customization. Ergonomics, 46(4), pp. 364–383.

McCrory, J.L., Young, M.J., Boulton, A.J.M. and Cavanagh, P.R. (1997). Arch index as a predictor of arch height. The Foot, 7(2). pp. 79–81.

Razeghi, M. and Batt, M. (2002). Foot type classification: a critical review of current methods. Gait & Posture, 15(3), pp. 282–291.

Saltzman, C.L. and Nawoczenski, D. A. (1995). Complexities of foot architecture as a base of support. Journal of Orthopaedic & Sports Physical Therapy. 21(6), pp. 354–360.

Witana, C.P., Goonetilleke. R.S., Xiong. S. and Au, E.Y. (2009). Effects of surface characteristics on the plantar shape of feet and subjects' perceived sensations. Applied Ergonomics, 40(2), pp. 267–279.

Xiong, S., Goonetilleke, S.R., Witana, C.P., Weerasinghe, T.W. and Au, Y.L. (2010). Foot arch characterization, Areview, a new metric, and a comparison. Journal of the American Podiatric Medical Association, 100(1), pp. 14–24.

*Ergonomics in Asia – Shih & Liang (eds)*
*© 2012 Taylor & Francis Group, London, ISBN 978-0-415-68414-9*

# 3D anthropometry for clothes of Taiwanese 4 to 6 year old children

Yu-Cheng Lin
*Department of Industrial Engineering and Management, Overseas Chinese University, Taichung City, Taiwan*

ABSTRACT: With social development and economic improvement, the growth and proportions of body for children in Taiwan have changed gradually, especially for preschool children. The past anthropometric data and sizing system for Taiwanese preschool children does not meet most of the needs and the applications nowadays. The purpose of this study was to collect the 3D digital body data from preschool children aged from 4 to 6 years in Taiwan with a portable 3D body scanner and to measure useful dimensions from scans for children's clothes. Fourteen typical clothing dimensions were retrieved from the collected digital body scans. One hundred and eleven boys and 105 girls aged from 4 to 6 years completed the measurements. Principal component analysis was conducted to retrieve important shirt and pants dimensions. By considering the explained variation and loading factor, chest circumference and shoulder breadth were chosen as the major dimensions for shirts and blouse. The dimensions for pants and trousers are crotch height and hip circumference. Based on the results, the shirt sizing system and pants sizing system for preschool children will be obtained.

*Keywords*: preschool children, 3D anthropometry, sizing system

## 1 INTRODUCTION

During recent years, the quality of life has been enhanced by the development of society and economy in Taiwan. The change in body size and shape for preschool children is obvious. Furthermore, the body and segment proportion has varied with the improvement in nutrition and life style. Therefore, previous anthropometric data of Taiwanese children has become out of date. The demand for new children's anthropometric data for designing and manufacturing more convenient children's products is urgent. One of the urgent requests is the anthropometric data for a children's sizing system. The tailoring firms in Taiwan usually adopt other a country's children's sizing system to manufacture domestic adult and child garments because of the incompleteness and outmoded domestic sizing system. They usually adopt the anthropometric data of other countries close to Taiwan. However, the body dimensions and proportions between peoples in East Asia are different (Lin, 2004). Besides, there are various considerations in formulating the domestic children sizing systems for different countries (Chung, 2003). Without exact data, no applicable sizing system could be achieved. Proper anthropometric data should be sufficient, correct and up to date. Most of the past anthropometric surveys were done for the adults, workers, elders, and students in Taiwan but rarely for preschool children and infants. The study of establishing Taiwanese anthropometric database conducted a large-scale survey to collect about 10 thousand people's data (Wang et al., 1999 and Wang et al., 2001). The elementary student sample is also an important part of the study (Wang et al., 2002). However, only students aged above 6 were measured in that study. Thus, an up-to-date anthropometric survey and sizing system to meet the preschool children's body shape and size is needed in Taiwan.

The optical 3D body scanning system is one of the major measuring instruments now (Tsai, 1997). Adopting optical measurement methods can obtain body surface data clouds quickly and exactly within seconds and then construct the 3D digital model (Lin et al., 2006). Although the model may be fragmented because of shaded surfaces and poor reflection quality, the applications of the 3D model are more extensive than traditional anthropometric methods. In order to avoid the fragmented scanned images, the standing posture was suggested (Daanen, 1998). It is not easy to establish an optimal sizing system for children aged from 4 to 6 years since the difficulty is to measure dimensions and the high variation in body shape and proportion. The difference between regions is also important (Tsai, 2000). Therefore, the subjects in north, central and south Taiwan should be sampled and measured separately. Three indices are commonly used to evaluate a sizing system. The indices are the number of sizes, the coverage of the sizing system, and the proportion of fitness (Chung, 2003). The data provided by a robust sizing system should be the body dimensions in place of real clothes dimensions in order to avoid the problems of applications (Hung, 2001).

The purpose of this study was to collect the 3D digital body data from Taiwanese north, central and south preschool children aged 4–6 with a 3D body scanner and to measure useful dimensions from the digital scans for children's clothes. A new sizing system was developed according to these data. The results were provided to the children's garment manufacturers for reference.

## 2 METHODOLOGY

### 2.1 *Portable 3D body scanner*

The portable 3D body scanner manufactured by LT-tech Co. was employed to retrieve the children's body surface data. A set of scanners contain three scan modules (Figure 1) and the relative locations for the 3 modules is shown in Figure 2. The precision is ± 1.0 mm and the resolution is 2.5 mm. About eighty thousand 3D data points were taken within 5 seconds. The scanned images were edited and modified by the software, LTBODYCAM, and anthropometric dimensions were retrieved from the 3D digital images. The scanning process was conducted in a darkroom to avoid the influence of light.

### 2.2 *Measured dimensions*

Fourteen typical clothing dimensions, including 9 dimensions for shirts and jackets, and 5 for pants and stature, were retrieved from the collected digital body scans. These dimensions are important and useful for manufacturing children clothes and establishing a sizing system. The 14 dimensions are shown in Table 1 and Figure 3.

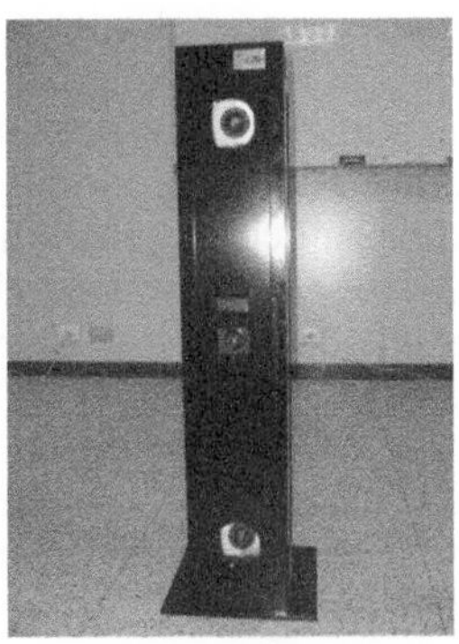

Figure 1. The scanning modules of the portable 3D body scanner.

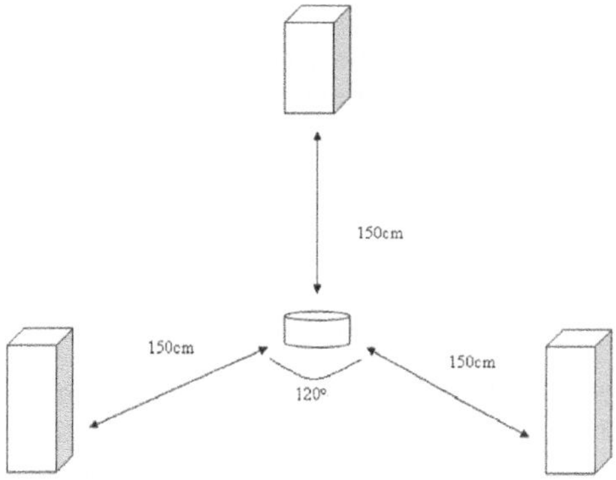


Figure 2. Relative locations of the three modules.

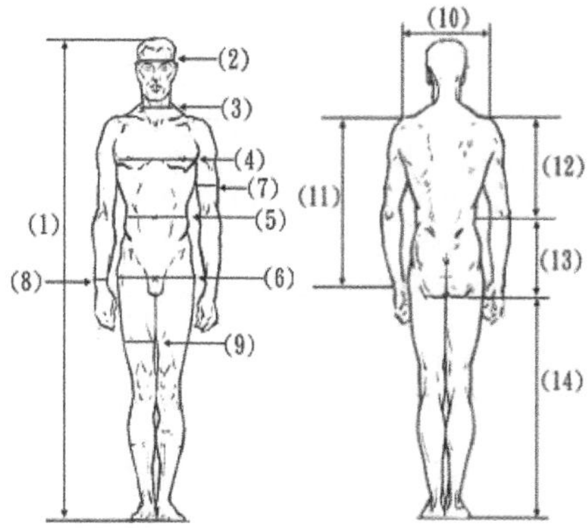


Figure 3. T The illustration of measured dimensions.

Table 1. The 14 typical dimensions for children's clothing.

| Index | Dimension |
|---|---|
| 1 | Stature |
| 2 | Head circumference |
| 3 | Neck circumference |
| 4 | Chest circumference |
| 5 | Waist circumference |
| 6 | Hip circumference |
| 7 | Axillary arm circumference |
| 8 | Wrist circumference |
| 9 | Thigh circumference |
| 10 | Shoulder breadth |
| 11 | Sleeve length |
| 12 | Shoulder to waist length |
| 13 | Waist to crotch length |
| 14 | Crotch height |

## 2.3 *Sample*

Three hundred and fifty preschool children aged from 4 to 6 years were invited randomly from 18 kindergartens in north, central and south Taiwan, and authorization letters to the children's parents were taken home by these invited children before experiment. Only those children who have parent's consent would be measured. Finally, 111 boys and 106 girls were completed the measurements. Total time to complete one subject was 5 minutes approximately. Because these preschool children are vivacious and interested, their teachers had to keep with them during the experiment.

## 3 RESULT

### 3.1 *Retrieving anthropometric data from 3D scanned models*

Each scan module of the 3D body scanner took a 3D image simultaneously and all images were combined to a 3D digital model. The 14 anthropometric data were then measured from the 3D model for every subject. The average age is 5.26. Table 2 shows the results of the anthropometric data. Only four out of 14 dimensions show significant differences. Except stature, the three significant dimensions are all related to children's trunk.

### 3.2 *Principal component analysis to retrieving major dimensions*

The 14 dimensions were separated into two groups. The first group relates to shirts and blouse and contains 9 dimensions. The second group relates to pants and trousers and contains 5 dimensions. The principal component analysis was employed to obtain the major dimensions for the two groups. The results of the principal component analysis are listed in Table 3 and Table 4. By considering the explained variation, only the first two principals were chosen (Table 3 and Table 4). According to their loading factors, chest circumference and shoulder breadth were chosen as the major dimensions for shirts and blouse. The dimensions for pants and trousers are crotch height and hip circumference (Table 5).

Based on above results of the principal component analysis, the development of a sizing system for Taiwanese preschool children is practicable. To implement the sizing classification, the following issues should be noticed. First, the sizing system should be applied to most of children aged 4–6. Second, the fitness and growth speed should be both considered. Therefore, the maximal union classification method is able to be utilized to implement the sizing system by the major dimensions obtained by this study.

Table 2. The 14 typical clothing dimensions for children.

| | | Boy | | Girl | | |
|---|---|---|---|---|---|---|
| | Dimension | Mean | Std | Mean | Std | Sig ($\alpha = 0.05$) |
| 1 | Stature | 112.35 | 5.53 | 108.56 | 4.84 | * |
| 2 | Head circumference | 50.58 | 2.01 | 51.11 | 2.51 | |
| 3 | Neck circumference | 27.03 | 2.12 | 27.65 | 2.32 | * |
| 4 | Chest circumference | 56.88 | 2.86 | 56.20 | 3.58 | |
| 5 | Waist circumference | 57.34 | 4.48 | 56.94 | 4.71 | |
| 6 | Hip circumference | 63.63 | 4.01 | 63.22 | 4.09 | |
| 7 | Axillary arm circumference | 20.89 | 2.63 | 21.01 | 2.99 | |
| 8 | Wrist circumference | 13.96 | 1.52 | 13.31 | 1.01 | * |
| 9 | Thigh circumference | 31.79 | 3.58 | 31.05 | 2.49 | |
| 10 | Shoulder breadth | 26.86 | 2.43 | 24.84 | 2.89 | * |
| 11 | Sleeve length | 28.56 | 2.45 | 27.87 | 2.32 | * |
| 12 | Shoulder to waist length | 26.45 | 2.35 | 24.53 | 3.35 | * |
| 13 | Waist to crotch length | 10.87 | 2.20 | 10.89 | 2.17 | |
| 14 | Crotch height | 51.86 | 4.15 | 51.03 | 3.43 | |

Table 3. Principal component analysis result for dimensions about shirts and blouses (gender-mixed).

| | Dimension | Loading factors Prin. 1 | Prin. 2 |
|---|---|---|---|
| 1 | Head circumference | 0.378 | 0.286 |
| 2 | Neck circumference | 0.312 | 0.066 |
| 3 | Chest circumference | **0.872** | –0.034 |
| 4 | Waist circumference | 0.804 | –0.251 |
| 5 | Axillary arm circumference | 0.721 | –0.419 |
| 6 | Wrist circumference | 0.717 | –0.218 |
| 7 | Shoulder breadth | 0.453 | **0.684** |
| 8 | Sleeve length | 0.520 | 0.463 |
| 9 | Shoulder to waist length | 0.608 | –0.273 |
| Explained variation | | 63.2% | |

Table 4. Principal component analysis result for dimensions about pants and trousers (gender-mixed).

| | Dimension | Loading factors Prin. 1 | Prin. 2 |
|---|---|---|---|
| 1 | Waist circumference | 0.786 | 0.398 |
| 2 | Hip circumference | **0.906** | 0.201 |
| 3 | Thigh circumference | 0.639 | 0.423 |
| 4 | Waist to crotch length | –0.212 | **0.767** |
| 5 | Crotch height | 0.465 | –0.719 |
| Explained variation | | 65.5% | |

Table 5. Major dimensions.

| | Major dimensions |
|---|---|
| For shirts and blouses | Chest circumference<br>Shoulder breadth |
| For pants and trousers | Crotch height<br>Hip circumference |

## 4 CONCLUSION

In order to establish an appropriate and up-to-date sizing system for preschool children aged from 4 to 6 years in Taiwan, this study conducted an anthropometric survey to collect preschool children's 3D digital models with the portable 3D body scanner. According to the anthropometric data, principal component analysis was applied to retrieve the major dimensions for shirts/blouses and pants/trousers. Finally, the major measurements for shirts and trousers are chest circumference and shoulder breadth and those for pants and trousers are crotch height and hip circumference. Therefore, a new sizing system for Taiwanese preschool children will be achieved and will be helpful to reduce the cost and enhance the incentives for product manufacture.

## ACKNOWLEDGEMENT

This work was supported by National Science Council of Taiwan (NSC 98-2221-E-240-001).


## REFERENCES


Chung, M.J. 2003. Establish sizing system for boys and girls aging 6–18 by using cluster analysis. Master thesis, Tsing Hua University, Taiwan.

Daanen, H.A.M. & Jeroen, G. 1998. Whole body scanners. Displays 19: 111–120.

Hung, K.L. 2001. Establish sizing system and standard dress form with 3D human data. Master thesis, Tsing Hua University, Taiwan.

Lin, Y.C. et al, 2004. The comparisons of anthropometric characteristics among four peoples in East Asia. Applied Ergonomics 35: 173–178.

Lin, Y.C. et al, 2006. An approach for analyzing breadth and depth measurement errors of a 3D whole body scanner, WSEAS Transactions on Systems 5 (6): 1321–1327.

Tsai, C.C. 2000. Apply 3D anthropometric data to establish sizing system for elementary and high school students. Master thesis, Tsing Hua University, Taiwan.

Tsai, C.Z. 1997. Apply anthropometric data to establish sizing system for woman's trousers. Master thesis, Tsing Hua University, Taiwan.

Wang, E.M.Y. et al, 1999. Development of anthropometric work environment for Taiwanese workers. International Journal of Industrial Ergonomics 23: 3–8.

Wang, M.J. et al, 2001. Anthropometric data book of the Chinese people in Taiwan. Ergonomics Society of Taiwan.

Wang, M.J. et al, 2002. The anthropometric database for children and young adults in Taiwan. Applied Ergonomics: 33: 583–585.

*Ergonomics in Asia – Shih & Liang (eds)*
*© 2012 Taylor & Francis Group, London, ISBN 978-0-415-68414-9*

# Evaluation of dining utensil design for children

Bor-Shong Liu
*Department of Industrial Engineering and Management, St. John's University, New Taipei City, Taiwan*

Yu-Lin Chen
*Department of System and Social Information, Nagoya University, Aichi, Japan*

ABSTRACT: The present study was divided into three stages. The first stage is the observational study. In the second stage an anthropometric survey of the hand for children aged 3–12 years olds was conducted. In the final stage, a new spoon for children was designed. The results of behavior analysis revealed that applied postures for chopstick and spoons were different between subjects. Some children held the middle of chopstick for dining. Some subjects always took the tail position of a spoon and more flexion occurred on the wrist while ladling out rice. In addition, two hundred and thirty-five participants whose who age ranged from 3 to 12 years old were separated into eight stratifications for analysis. Three dimensions of hand were measured by digital caliper. Results of ANOVA showed that all hand dimensions had a significant age effect. Further, post test revealed that mean length of hand could be divided into five subsets. Manufacturers should make products available in various sizes to accommodate different users at least five sizes for children, and various sizes based on hand length plus 30 mm. Spoon design for right-hand or left-hand users should be considered for convenience.

*Keywords*: anthropometry, chopsticks, spoon, dining behaviors

## 1 INTRODUCTION

It is well known that chopsticks are often used for dining in Asia. Our review of the literature found that little research has examined the application of chopsticks for dining behaviors of children. Shigetoshi and Shigeji (2009) and Shigetoshi, Nobuko and Shigeji (2010) reported that most Japanese schools used plastic utensils and little attention is paid to use and design of dining utensils. Thus, the purpose of this study was to determine the key factors of length for chopsticks or spoons for children. These involve anthropometrical measurement of the hand for children. It was expected that the results would lead to improvements in dining comfort and ensure greater safety.

Use of chopsticks is well known to the Chinese. Use of chopsticks for the pleasure of enjoying a meal within either a family or a social context is deeply conceived in the heart of Chinese culture. Chinese children begin using chopsticks at about the age (standard deviation) of 4.6 years (1.13) and throughout life chopsticks are the preferred eating utensils (Wong et al., 2002). However, manipulating chopsticks requires fine motor control and skills (Sandra, Donna & Jenna, 2002). Therefore, chopsticks users such as children under 4 years old, or those with cervical spinal cord injury (SCI) or other types of hand impairments frequently are forced to switch to using spoons.

From Japanese experience, the design of Japanese dining utensils needs to consider the eating of various types of foods, i.e., Japanese foods, Western foods and Chinese foods. Thus, the design idea focused on functional and convenient shape and structure while dining. Modern gastronomic culture reflects the contents of various cultures. Though East Asia is the center of chopsticks culture, western styled cutlery such as knives, forks and spoons are commonly used for daily eating life. In modern eating life, it can be said that any kind of cutlery including Western styled might be used that are suitable for the types of food served. For the new opportunity to start designing, our attempt was to realize the main themes of "Easy-to-use cutlery, simple shaped but with beauty suitable for any cuisines Japanese, Western and Chinese" as well as "Human ergonomics considering comfort". For the package of the products, our design concept was set to apply as many reusable materials and printing inks from the standpoint of environment consciousness. Our new proposal was reflected to the appearance of the product, and another theme, zero emissions, was materialized in the relation between the cutlery product and its package (Kunimoto, 2006). For universal soft spoon dessert in Nonozi shop, three designs are considered: round and shallow spoon bowl for children's mouths, easy to use long-handled grip handle design, and right/left hand side user.

Carruth and Skinner (2002) reported that at 12 months 43% of children were self-feeding with spoons, and 43% of mothers had special spoons that their children used. As early as 10 months of age the children were making hand and body movements indicating their desire to eat from the table and especially their parents' plates (Skinner et al., 1998). By 24 months, 80% of the study children were self-feeding. The children's transition from using specific types of baby spoons to adult spoons began around 16 months of age with over 54% using adult spoons at 24 months. The type of spoon chosen by mothers may reflect their children's ability to grasp a spoon using thumb and fingers to pick up and hold a spoon (palm down) compared to the ability to hold a spoon like a pencil (first two fingers and thumb with palm turned up or towards face). At a mean age of 14.37 months, the children could bring the side of a spoon to their mouth, but within the same age range they also were using their fingers to self-feed. The age of transition from special baby spoon to adult spoon may be associated with the child's receiving food from the parent's plate (using adult spoons) which would provide a differing sensory experience than a child's special spoon. Thus, eating and drinking utensils need design for users associated with motor development.

## 2 METHODS

The present study was divided into three stages. The first stage is the observational study. In the second stage we conducted the anthropometric survey of the hand for children aged 3–12 years old. In the final stage, a new spoon for children was designed.

### 2.1 *Observational study of dining behaviors*

A digital camera (HDR-HR11, Sony, Japan) mounted on a tripod was used to record the dining behavior of children. Thirty children had been observed for further analysis.

### 2.2 *Anthropometric survey*

Two hundred and thirty-five participants were randomly sampled from kindergarten and elementary school in Taipei and all participants were with no history of trauma or congenital anomalies on hands. The participants, whose age ranged from 3 to 12 years old, were separated into ten stratifications for further analysis.

### 2.3 *Methodology of measurement*

The main instruments used in the present survey were the electronic digital caliper and height measurement. An accuracy of 0.5 mm was the objective and all measurements were recorded in millimeters. Body weight was measured using a portable weighting scale (in kg). Altogether three anthropometric characteristics of the hand were measured as follows:

a. Length of hand;
b. Length between thumb and index finger;
c. Breadth of hand;

### 2.4 *Data analysis*

All data were coded and summarized using SPSS software for Windows. An analysis of variance (ANOVA) was utilized to determine the effects of gender and age on hand dimensions. Where statistically significant differences were determined, the Duncan post hoc test was performed. In addition, the present study was to compared the dimensions of various spoons in marketplace products with this anthropometric database and suggested the optimal dimensions.

## 3 RESULTS

Results of ANOVA showed that there were statistically significant differences in all dimensions among age groups.

### 3.1 *Results of observational study*

Results of behavior analysis revealed that applied postures of chopstick and spoons were different between subjects. Some children held the middle of the chopsticks for dining (Figure 1). Some subjects always took the tail position of a spoon and so more flexion occurred on the wrist while ladling out rice. Subjects raised their forearm higher when using longer spoons. Thus, the length of spoon could affect the dining behavior (Figure 2). However, the sizes of spoon were not the high priority in purchasing for children (Liu, Tseng, Wu & Peng, 2009). Therefore, manufacturers should make products available in various sizes to accommodate different users.

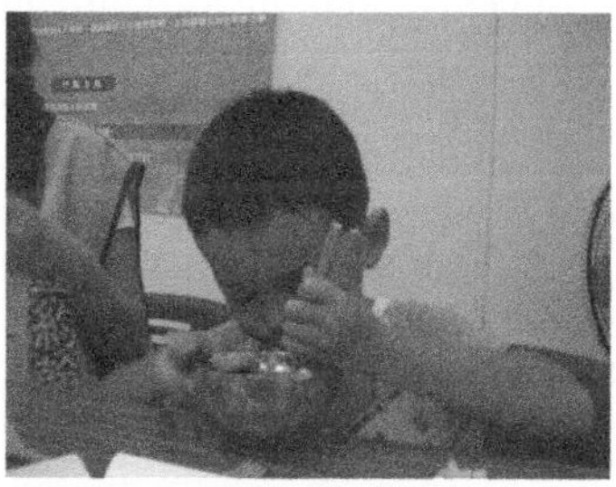

Figure 1. Dining behaviors of children by chopsticks.

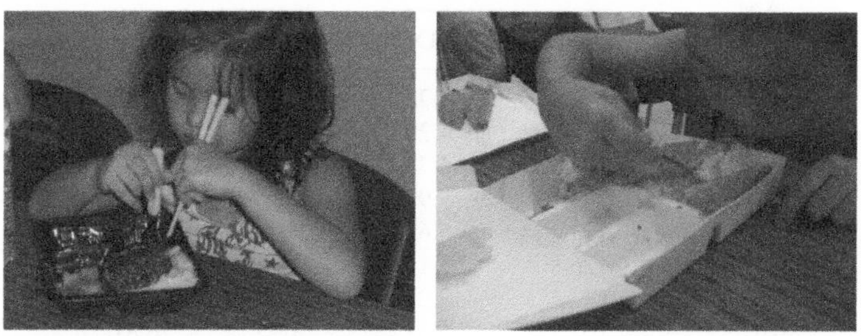

Figure 2. Dining behaviors of children by spoon.

### 3.2 *Effect of age on dimensions*

Mean stature and weight are presented in Table 1. All mean dimensions showed significant differences between age groups ($p < 0.001$). Further, Duncan's post hoc test revealed that mean stature could be divided into six subsets, in ascending order, subset I (3 years old; 99 cm), subset II (4–6 years old; 106–111.4 cm), subset III (7, 8 years old; 123.4–126.8 cm), subset IV (9 years old; 132.8 cm), subset V (10 years old; 137.7 cm) and subset VI (11–12 years old; 148–149 cm).

Duncan's post hoc test revealed that mean weight could be divided into five subsets, in ascending order, subset I (3–4 years old; 15–16.8 kg), subset II (4–6 years old; 16.8–19.6 kg), subset III (7–8 years old; 24.3–25.7 kg), subset IV (9–10 years old; 29.8–32.5 kg), and subset V (11–12 years old; 41.8–44.1 kg).

Mean length of hand could be divided into five subsets, in ascending order, subset I (3–4 years old; 106.4–106.8 mm), subset II (5, 6, 8, 9 years old; 120.1–132 cm), subset III (7, 8, 9, 10 years old; 130.1–137.2 mm), subset IV (7, 10, 12 years old; 136.1–146.6 mm) and subset V (11–12 years old; 146.6–157.6 mm).

Mean breadth of hand could be divided into seven subsets, in ascending order, subset I (3, 4 years old; 49.66–52.29 mm), subset II (4, 5, 6 years old; 52.3–57.2 mm), subset III (7, 8, 9 years old; 63.4–70.5 mm), subset IV (10, 11 years old; 75.8–76.4 mm), subset V (10, 11 years old; 75.8–76.4 mm) and subset VI (12 years old; 98.2 mm).

### 3.3 *Spoon design*

The present study provides product designers with the anthropometric dimensions of hands for children and recommend appropriate solutions for design. In addition, hand dimensions for 3–12 age children are shown in Table 2.

For Japanese data of hand length, these are about 100 mm, 110 mm, 120 mm, 130 mm 140 mm, 150 mm, 160 mm for 2 years old, 3 years old, 4 years old, 5 years old, first-second grade elementary school, third-fourth grade elementary school, and fifth-sixth grade elementary school respectively. The length of chopsticks was determined to be 30 mm greater than hand length. Thus, Japanese chopsticks could be divided into seven various sizes for children. These are about 130 mm, 140 mm, 150 mm, 160 mm 170 mm, 180 mm, 190 mm for 2 years old, 3 years old, 4 years old, 5 years old, first-second grade elementary school, third-fourth grade elementary school, and fifth-sixth grade elementary school respectively (Hashiseiwa Company). These are the general principles for applying anthropometric data to specific design problems. Certain features of equipment or facilities can be designed so they can be adjusted to the individuals who use them. Thus, the length of spoon should be adjustable from 136 mm to 187 mm. At least five levels could be set on adjustable length. In addition, a window could show the indicator of length (Figure 3).

Table 1. Stature and weight by age groups.

| Age groups | Stature (cm) | | Weight (kg) | |
|---|---|---|---|---|
| | Mean | SD | Mean | SD |
| 3 | 99.39 | 4.25 | 15.1 | 1.55 |
| 4 | 106.0 | 4.29 | 16.8 | 2.41 |
| 5 | 111.4 | 6.78 | 19.1 | 3.17 |
| 6 | 109.7 | 7.92 | 19.7 | 3.46 |
| 7 | 123.4 | 5.62 | 24.3 | 5.56 |
| 8 | 126.8 | 5.16 | 25.7 | 5.70 |
| 9 | 132.8 | 7.20 | 29.8 | 6.75 |
| 10 | 137.7 | 6.51 | 32.5 | 6.81 |
| 11 | 149.1 | 7.80 | 41.8 | 7.58 |
| 12 | 148.2 | 5.33 | 44.1 | 8.92 |

Table 2. Hand dimensions by age groups.

| | Dimensions (mm) | | | | | |
|---|---|---|---|---|---|---|
| | Length of hand | | Length between thumb and index finger | | Breadth of hand | |
| Age groups | Mean | SD | Mean | SD | Mean | SD |
| 3 | 106.4 | 7.76 | 93.2 | 11.3 | 49.7 | 3.60 |
| 4 | 106.8 | 14.7 | 97.6 | 10.4 | 52.3 | 3.57 |
| 5 | 121.8 | 9.31 | 102.3 | 14.1 | 57.2 | 4.32 |
| 6 | 120.2 | 9.38 | 98.9 | 12.6 | 56.4 | 3.91 |
| 7 | 136.2 | 6.43 | 118.2 | 9.08 | 63.4 | 5.78 |
| 8 | 130.2 | 10.72 | 112.8 | 9.44 | 70.5 | 11.7 |
| 9 | 132.1 | 17.4 | 106.3 | 13.9 | 67.3 | 6.33 |
| 10 | 137.2 | 17.4 | 123.8 | 10.2 | 75.8 | 11.1 |
| 11 | 157.6 | 6.96 | 127.1 | 3.61 | 76.4 | 12.6 |
| 12 | 146.7 | 8.86 | 131.5 | 7.03 | 98.2 | 17.6 |

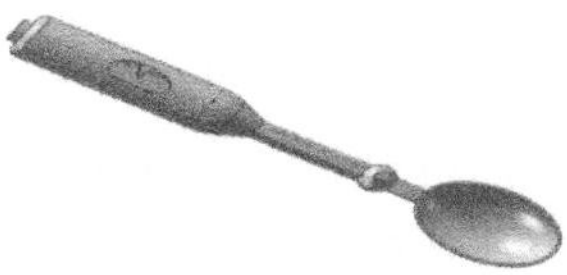

Figure 3. The length of spoon can also be adjusted from 136 mm to 176 mm by rotating the back button.

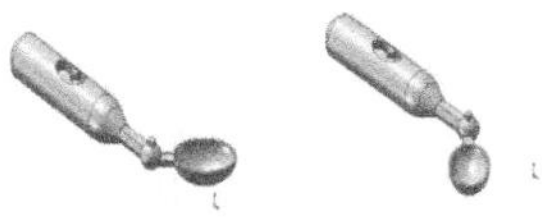

Figure 4. New spoon could turn the shallow bowl to right and left side.

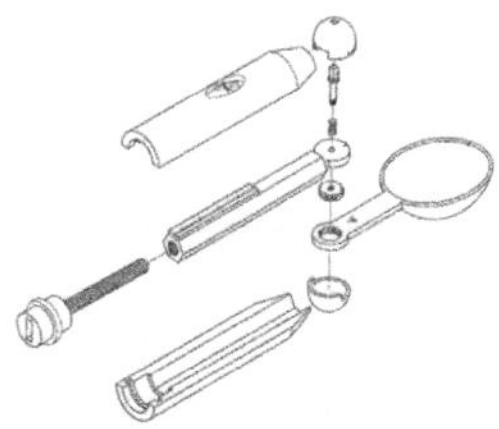

Figure 5. Product explosive chart for new spoon.

According the principles of hand tool and device design (Fraser, 1980; Freivalds 1987), tools should be designed to be used in the operator's preferred hand. For example, the handle could be moved for left-handed use, if a threaded fastener were provided in the right side of the drill housing (Greenberg & Chaffin, 1977). So, spoon design for right-hand or left-hand user should be considered for convenience. Thus, the prototype of a new spoon for children is shown in Figure 4. The shallow bowl of the spoon could be turned to the right or left and could decrease the wrist radius deviation. Figure 5 presents the product explosive chart for the new spoon. The new spoon has a window on the body to indicate the length of spoon.

## 4 DISCUSSION

Tool design can affect the user because the interface of the user with the tool will determine what the upper extremity and neck posture will be. Tools that create a need to abduct the elbow or shoulder to do a task will contribute to static muscle fatigue and limit the time the task can be sustained (Eastman Kodak, 1986).

A spoon is a utensil consisting of a small shallow bowl, oval or round, at the end of a handle. It is used primarily for serving and eating liquid or semisolid food (sometimes called "spoon-meat"), and solid foods such as rice and cereal which cannot easily be lifted with a fork. Of course, eating utensils are also hand tools that require more precision or more force than a person's hand can safely sustain.

Results of a questionnaire survey showed that users preferred buying the eating utensils through wholesale (51.4%), stores (26.9%), department stores (15.7%), and via Internet e-stores (1.2%). In addition, results of the survey reported that average buying price per each eating utensil was under 200 dollars (62.4%), 201–400 dollars (26.6%), 401–600 dollars

(6.7%), more than 601 dollars (4.2%). Further, ranking of the descriptors based on the mean ranks of the rating scored from not related (1 point) to very closely related (5 points). The factors of material, size, fragility, heatproof, certification, cleanness, easily carried, and fitness for hand are the ones most related to choice for spoon. By contrast, color, or pictures of the spoon are of less concern to customers. These results can be of help in the design of spoons (Liu, Tseng, Wu & Peng, 2009).

Hsu and Wu (1991) investigated the effects of the length of chopsticks on the food-servicing performance. Their results showed that the food-pinching performance was significantly affected by the length of the chopsticks, and that chopsticks of about 240 and 180 mm long were optimal for adults and pupils, respectively. Chen, Liu and Tseng (2009) examined the effects of age, chopstick characteristics and pinching tasks on performance and subjective ratings. Thirty elementary school students were recruited in the present study and divided into three groups (8, 10 and 12 years old). Eight types of chopsticks had been evaluated on pinching two objects (corn snacks and chocolate balls). Results of ANOVA showed that the pinching performance had a significant difference between ages. In addition, the pinching performance was better in applying the hexagon chopstick of a particular length. By contrast, the stainless steel chopstick was the worse case.

## 5 CONCLUSION

The present study was divided into three stages. The first stage is the observational study. The second stage an anthropometric survey of hand for children aged 3–12 years old was conducted. In the final stage, a new spoon for children was designed. Results of behavior analysis revealed that applied postures of chopstick and spoons were different between subjects. Some children held the middle of chopstick for dining (Figure 1). Some subjects always took the tail position of spoon and so more flexion occurred on the wrist while ladling out rice. In addition, two hundred and thirty-five participants who age ranged from 3 to 12 years old were separated into eight stratifications for analysis. Three dimensions of hand were measured by digital caliper. Results of ANOVA showed that all hand dimensions had a significant age effect. Further, post test revealed that mean length of hand could be divided into five subsets. Manufacturers should make products available in various sizes to accommodate different users at least five sizes for children and chopsticks of various sizes were based on hand length plus 30 mm. Spoon design for right-hand or left-hand user should be considered for convenience.

## ACKNOWLEDGMENT

This study was supported by a grant from the National Science Council of Taiwan (NSC96-2815-C-129-001-E). Authors would like to thank all participants for their patience during collecting the data.

## REFERENCES

Carruth, B.R. & Skinner, J.D. (2002). Feeding behaviors and other motor development in healthy children (2–24 months). *Journal of the American College of Nutrition*, 21(2), 88–96.

Chen, C.C., Liu, B.S. & Tseng, H.Y. (2009). Effects of shape and length of chopstick on pinching performance of children, *Proceedings of conference on Innovation Industry management* (PCIM), 32–37. (In Chinese)

Chen, Y.L. (1998). Effects of shape and operation of chopsticks on food-serving performance, *Applied Ergonomics*, 29(4), 233–238.

Chen, Y.L., Deng, S.Z., Huang, P.H., Chen, C.F. & Chen, K.Y. (1996). A questionnaire survey of types and pinching operations of chopsticks, Mingchi Technical Report, 1996.

Churcher, E., Egan, M., Walop, W., Huang, P.P., Booth, A. & Roseman, G. (1993). Fine motor development of high-risk infants at 3, 6, 12 and 24 months. *Physical & Occupational Therapy in Pediatrics*, *13*, 19–37.

Fraser, T. (1980). Ergonomic principles in the design of hand tools, Occupational Safety and Health Series no. 44. Geneva, Switzerland: International Labour Office.

Freivalds, A. (1987). The ergonomics of tools. In D. Oborne, (ed.), International reviews of ergonomics, vol. 1. London: Taylor & Francis.

Greenberg, L. & Chaffin, D. (1977). Workers and their tools. Midland, MI: Pendell Publishing.

Hsu, S.H. & Wu, S.P. (1991). An investigation for determining the optimum length of chopsticks, *Applied Ergonomics*, 22(6), 395–400.

Kunimoto, K. (2006). Cutlery design development for modern gastronomic culture, Annual design review of JSSD, 12(12), 6–9. (In Japanese)

Liu, B.S., Tseng, H.Y., Wu, C.C. & Peng, K.L. (2009). Observation of eating behavior and spoon design for 3–10 year-old children. The Proceedings of the 17th World Congress on Ergonomics, Beijing, China.

Sanders, M.S. & McCormick, E.J. (1993). *Human Factors in Engineering and Design*, 7th ed., New York: McGraw-Hill.

Sandra, J., Donna, J. & Jenna, D. (2002). Developmental and functional hand grasps: precision grasps. *Thorofare: Slack*, 108–109.

Shigetoshi, T. & Shigeji, M. (2009). A study on tableware used in school lunches—From the viewpoint of food education, Kagawa Nutrition University, 40, 79–86. (In Japanese)

Shigetoshi, T., Nobuko, M., Shigeji, M. (2010). A study on design and hospitality of tableware—From the viewpoint of food education, Kagawa Nutrition University, 41, 95–104. (In Japanese)

Skinner, J., Carruth, B.R., Houck, K., Moran, III J. Reed, A., Coletta, F. & Ott, D. (1998). Mealtime communication patterns of infants from 2 to 24 months of age. *Journal of Nutrition Education*, *30*, 8–16.

Wong, S., Chan, K., Wong, V. & Wong, W. (2002). Use of chopsticks in Chinese children. *Child: Care, Health & Development*, *28*(2), 157-161.

Wu, S.P. (1995). Effects of the handle diameter and tip angle of chopsticks on the food-serving performance of male subjects, *Applied Ergonomics*, 26(6), 379–385.

*Ergonomics in Asia – Shih & Liang (eds)*
*© 2012 Taylor & Francis Group, London, ISBN 978-0-415-68414-9*

# A proposal for a standard value of muscle cross-sectional area of the thigh in Japanese adults

Manami Ishiuchi
*Graduate School of Design, Kyushu University, Fukuoka, Japan*

Kiyotaka Fukumoto
*Faculty of Engineering, Shizuoka University, Shizuoka, Japan*

Satoshi Muraki
*Faculty of Design, Kyushu University, Fukuoka, Japan*

Osamu Fukuda
*National Institute of Advanced Industrial Science and Technology, Saga, Japan*

ABSTRACT: The muscle cross-sectional area in the thigh decreases with aging, which lowers the capacity to carry out activities of daily living (ADL). In the present study, we tried to make a standard value for evaluating the muscle area of the thigh, considering sex, age and body size. Muscle cross-sectional areas of the thigh in 606 Japanese adults (257 males and 349 females) were measured using an ultrasound evaluation system. Because height and weight were related to age and muscle area, we proposed measured muscle area ($cm^2$) × height (m)/weight (kg) as an index for evaluating muscle area. In both sexes, this index decreased with aging, and especially in females, the reduction was accelerated beyond 50 years old. These standard values should be useful for middle-aged and elderly people to prevent them from exhibiting a decrease in capacity to carry out ADL and becoming bedridden.

*Keywords*: muscle area, thigh, ultrasound, aging, care prevention

## 1 INTRODUCTION

Japanese society has been aging rapidly, and the issue of the increase in the number of bedridden elderly people is becoming serious. Evaluating the capacity to carry out activities of daily living (ADL), especially activities concerning muscles in the lower limbs, is very important for preventing subjects from becoming bedridden. Muscle strength is an index for evaluating the capacity to carry out ADL (Rantanen et al., 2002). However, it is sometimes risky to measure the strength of elderly people because a maximal muscular contraction might be a factor for increasing blood pressure, arthritis and osteoarthritis (Inoue et al., 1999), and elderly people have little opportunity to find out their own muscle strength (Fukumoto et al., 2011). To overcome this problem, in this study, we focused on muscle area of the thigh, which has a strong relationship to muscle strength (Akima et al., 2001; Kanehisa et al., 1994). Therefore, the aims of this study were to examine the change of the muscle cross-sectional area of the thigh with aging, and to make an index and a standard value for evaluating the muscle area of the thigh, considering age, sex and body size.

## 2 METHOD

### 2.1 *Participants*

Six hundred and six Japanese adults (257 males and 349 females), who were community-dwelling young, middle-aged and elderly people, participated in this study. Their ages ranged between 16 and 90 years old. All participants had lived as physically independent. Table 1 presents the number of participants by each generation.

### 2.2 *Measurement items*

We measured the height, weight, circumference of the thigh and muscle cross-sectional area of the thigh of the participants. Muscle area of the thigh was measured using the ultrasound evaluation system that was developed by Fukumoto et al. The position of measuring circumference and muscle area of the thigh was at 50% of the thigh length from the trochanter major point.

### 2.3 *Ultrasound evaluation system*

The ultrasound evaluation system consists of two parts, a measurement unit that includes the ultrasound system and a mechanical arm (Figure 1), and image composition software to compose a cross-sectional image from fragmental images (Figure 2). In the system, an ultrasound probe is installed on the tip of a mechanical arm and flexibly scans along the body surface. We can acquire cross-sectional images of the thigh by touching the probe to the thigh, and moving along it (Figure 3). This system is safe and is particularly suitable

Table 1. The number of participants.

| Age (yr) | Male (n) | Female (n) |
|---|---|---|
| –19 | 11 | 11 |
| 20–29 | 77 | 37 |
| 30–39 | 17 | 5 |
| 40–49 | 7 | 25 |
| 50–59 | 24 | 50 |
| 60–69 | 68 | 135 |
| 70–79 | 46 | 70 |
| 80–89 | 6 | 16 |
| 90– | 1 | 0 |
| Total | 257 | 349 |

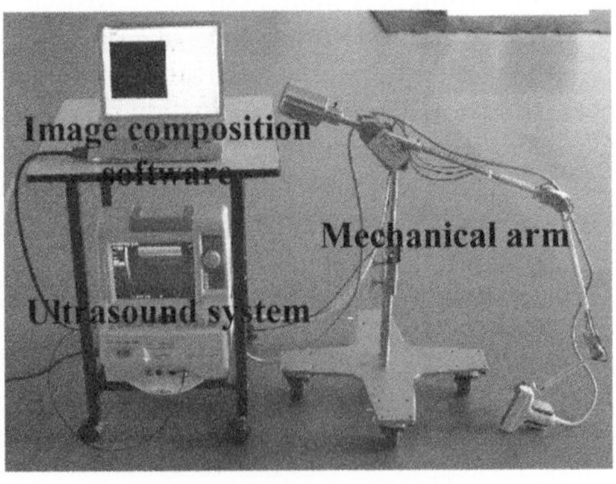


Figure 1. Ultrasound evaluation system.

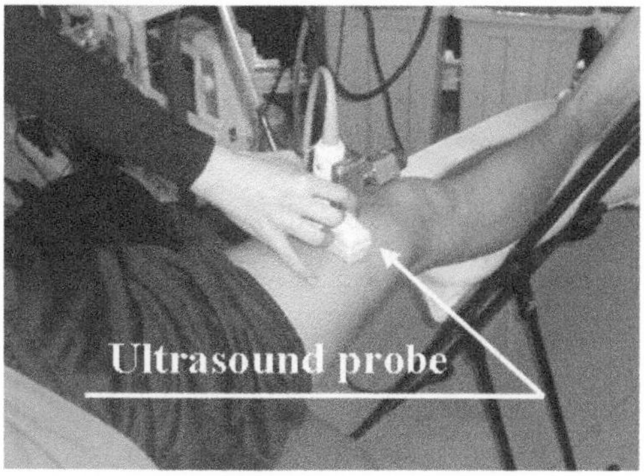


Figure 2. Measurement position.

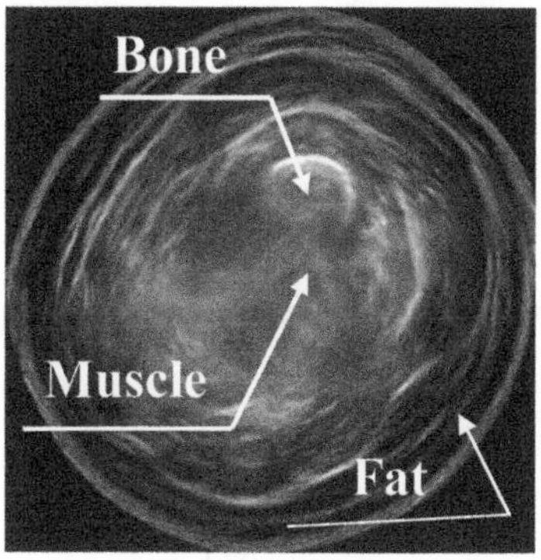


Figure 3. Cross-sectional image.

for the elderly because the participants are not required to perform any unnatural postures during measurement.

### 2.4 *Making an evaluating index*

We made scatter diagrams by age and revised muscle area, added quadratic approximation curves to them, and defined the curves as formulas for calculating a standard value.

## 3 RESULT AND DISCUSSION

Figure 4 depicts correlation diagrams between age and muscle area of the thigh. In both sexes, the muscle area decreased with aging. However, the muscle area was influenced by the subjects' body size, especially weight. Figures 5 and 6 present relationships of body height and weight to the muscle area of the thigh. Body weight showed a strong relationship to the area in both sexes. Increased body weight is associated with increased muscle area. On the other hand, the figures, especially height, decreased with aging as indicated by Figure 7. Therefore, an index for evaluating the muscle area should be considered to be revised according to body height and weight. On the basis of these relationships, we defined the following values as an index for evaluating muscle area.

Index for evaluating muscle area (RMA)
= measured muscle area ($cm^2$) · height (m)/weight (kg)

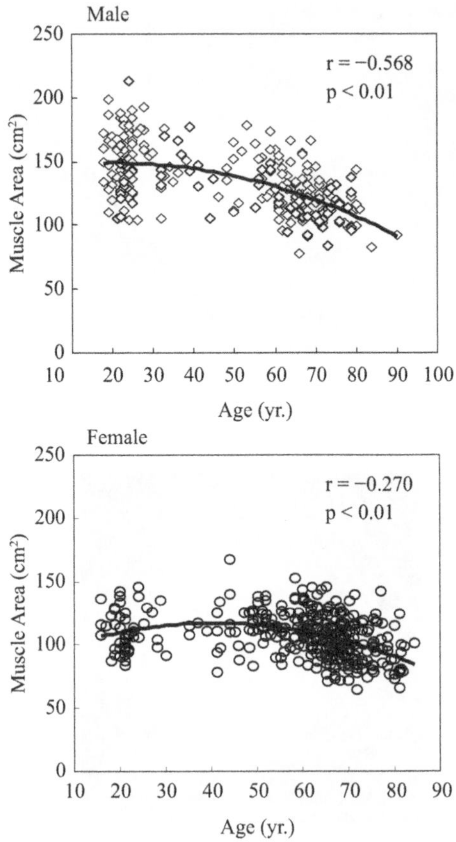


Figure 4. Correlations between age and muscle area.

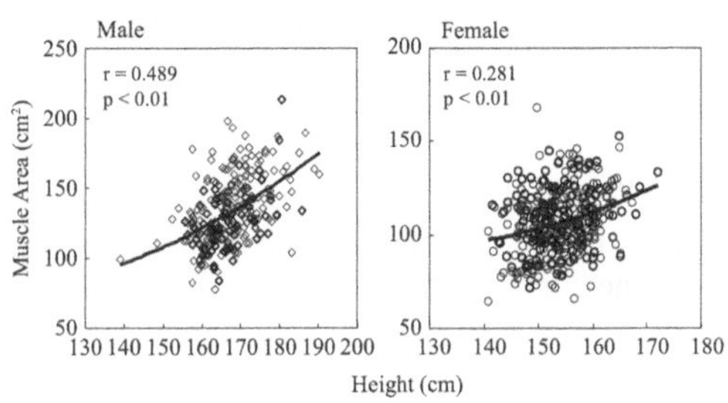


Figure 5. Relationships of body height to muscle area.

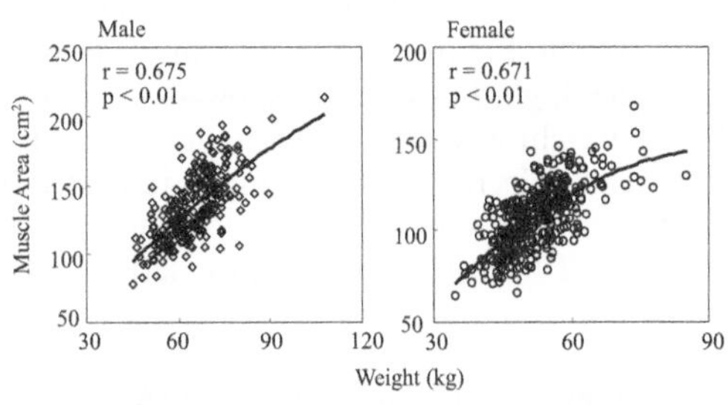


Figure 6. Relationships of body weight to muscle area.

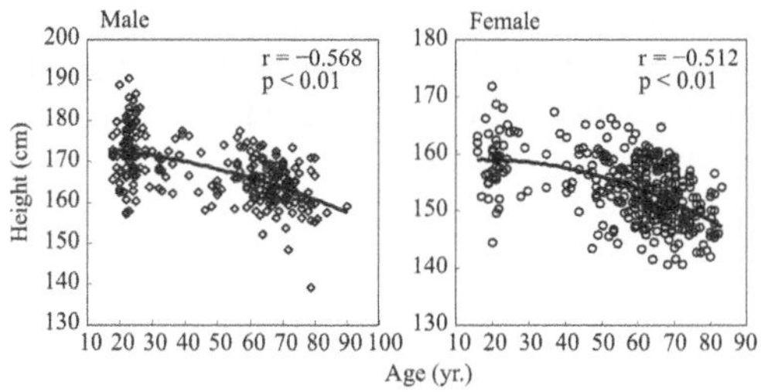


Figure 7. Relationships of age to body height.

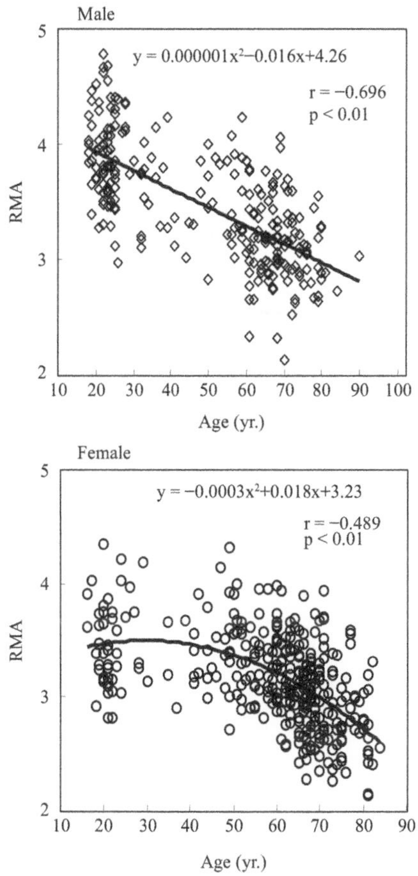


Figure 8. Correlations between age and RMA.

The muscle volume is decided by muscle area multiplied by muscle length. By substituting height for muscle length, the RMA becomes a substitute value of muscle volume per weight.

Figure 8 shows correlation diagrams between age and RMA, and we defined these quadratic approximation curves as a standard value. Their formulas were $y = 1E\text{-}06x^2 - 0.016x + 4.26$ in males, $y = -3E\text{-}04x^2 + 0.018x + 3.23$ in females. In both sexes, the muscle area decreased with aging, and especially in females, the reduction of the RMA was accelerated beyond 50 years old. This RMA would be useful for middle-aged and elderly persons to understand the present condition of their own muscle volume in the thigh, which should motivate them to prevent a decrease in ADL and becoming bedridden.

We consider that there are essential muscle areas for maintaining ADL, such as those involved in standing, walking, and climbing stairs, among others. In the future, to clarify their relationships, we should focus on the muscle area in elderly people who cannot perform ADL, in other words, elderly in need of care.

## 4 CONCLUSION

In this study, we proposed an index and a standard value for evaluating the cross-sectional muscle area of the thigh using an ultrasound evaluation system. This standard value should be useful for middle-aged and elderly people to prevent them from exhibiting decreases in the capacity for ADL and becoming bedridden.

## REFERENCES

Akima, H., Kano, Y., Enomoto, Y., Ishizu, M., Okada, M., Oishi, Y., Katsuta, S. & Kuno, S. 2001. Muscle function in 164 men and women aged 20–84 yr. *Medicine and Science in Sports and Exercise* 33(2): 220–226.

Fukumoto, K., Fukuda, O., Tsubai, M. & Muraki, S. 2011. Development of a flexible system for measuring muscle area using ultrasonography. *IEEE. Transactions on Biomedical Engineering* 58(5): 1147–1155.

Inoue, K., Watanabe, S. & Nishimoto, T. 1999. Blood pressure response to isometric contractions in healthy young males. *Kawasaki Journal of Medical Welfare* 5(1): 1–5.

Kanehisa, H., Ikegawa, S. & Fukunaga, T. 1994. Comparison of muscle cross-sectional area and strength between untrained women and men. *European Journal of Applied Physiology and Occupational Physiology* 68: 148–154.

Rantanen, T., Avlund, K., Suominen, H., Schroll, M., Frandin, K. & Pertti, K. 2002. Muscle strength as a predictor of ADL dependence in people aged 75 years. *Aging Clinical and Experimental Research* 3: 10–15.

*Ergonomics in Asia – Shih & Liang (eds)*
*© 2012 Taylor & Francis Group, London, ISBN 978-0-415-68414-9*

# Changing body posture and working system improves workers performance

I. Wayan Surata
*Mechanical Engineering, Udayana University, Denpasar, Bali, Indonesia*

A. Manuaba, N. Adiputra & D.P. Sutjana
*Medical Faculty, Udayana University, Denpasar, Bali, Indonesia*

ABSTRACT: Seaweed is one of Indonesia's potential marine commodities that may contribute to the national revenue and a new source of income for the local community, as the cultivation of seaweed is much easier and cheaper than paddy crop, because neither pesticides nor fertilizer would be required. The local practice of drying the seaweed is by spreading it over a plastic sheet on the ground and exposing it to the sun. The working posture during drying is squatting and stooping to flatten and spread evenly the seaweed. This study was conducted to examine the effects of changing the workers' working posture with the aim to improve workers' performance. This is an experimental study using a two-period cross-over design on 20 farmers of seaweed cultivation. The results of the study showed that changing the workers' working posture improved the performance of seaweed farmers.

*Keywords*: dryer redesign, working posture, seaweed drying, work performance

## 1 INTRODUCTION

Seaweed is one of Indonesia's potential marine commodities that may contribute to the national revenue and a new source of income for the local community. The villages in the district of Nusa Penida which have a beach have successfully developed seaweed cultivation. There are 2 types of seaweed that are suitable to be cultivated in Nusa Penida, namely *euchema spinosum* and *euchema cottonii*. Seaweed cultivation nowadays is a major job for the coastal community, due to the demand for seaweed to meet the export market being considerable. Nusa Penida seaweed production for the last two years was as follows: (1) in 2007 production was 105,015 tons, with a value more than Rp. 42 billion (Statistik Perikanan Budidaya Propinsi Bali, 2007); and (2) in 2008 production was 101,210 tons with a value of Rp. 85 billion (Statistik Perikanan Budidaya Propinsi Bali, 2008). Dried seaweed is exported to the countries of destination such as Japan, China, Australia, USA, UK, and other countries. Seaweed is widely used for food and medicine. Seaweed extract which is a hydrocolloid such as gelatin, carrageenan, and alginates are also much needed in various industries. Alginate is used in food, beverages, paint, graphic photo paper, textiles, and pharmaceuticals (Anggadiredja et al., 2006; Poncomulyo et al., 2006). The cultivation of seaweed is much easier and cheaper than paddy crop, because neither pesticides nor fertilizer are required. Other advantages of cultivating seaweed are the fact that it can be carried out throughout the year, and its relatively shorter duration to reach harvesting time. The local practice of drying the seaweed is by spreading it over a plastic sheet on the ground and exposing it to the sun. The working posture during drying is squatting and stooping to flatten and spread evenly the seaweed. The stooping work posture is a physiological posture that will cause a reaction in the form of musculoskeletal complaints and fatigue (Pheasant, 1991; Grandjean, 1993).

During activity the workers are exposed to direct sun, without regular rest, and have an excessive lifting load. Such working conditions are not physiological, and the workers have fatigue and musculoskeletal complaints (Aasa et al., 2006; Richardson et al., 2006). Drying methods for handling agricultural products by utilizing solar energy have been performed traditionally for a long time. Solar energy is categorized as renewable energy that is suitable to be developed in rural areas where the people lack education, have low skill levels, and a lack of economic capability. Based on these conditions, appropriate technology will be suitable to be developed and used in rural areas. There are many types of solar dryers such as tents, crates, racks, and houses. Solar dryers are designed based on the application of appropriate technology that meet the following criteria (1) technically to simplify and to speed up the work; (2) economically efficient in natural resources and affordable in cost; (3) ergonomically to improve physical and mental heath, prevent illness and injury caused by work; (4) socio-cultural considering work organization, work habits, norms, values, and trust of workers; (5) saving energy; and (6) environmentally friendly.

This study was conducted to examine the effects of changing the workers' working posture and limiting lift weight at 23 kg, and introduced a 5-minute break after an hour's work with the aim of improving workers' performance. Workers' performance consists of (1) reduction of musculoskeletal complaints, (2) decrease in fatigue, and (3) increase in productivity.

## 2 METHODS

This was an experimental study using a two-period cross-over design. The advantages of this design were that biological variability inter-subject can be controlled, and the number of samples were only half of the parallel design (Bakta, 2000; Daniel, 1999; Steven, 2005). In the cross-over design, the interval between treatments was used as a washing out period, to eliminate the residual effect of treatments that have been implemented. Subjects involved in this study were 20 farmers of seaweed cultivation in the Ped Village of Nusa Penida.

Before redesign, drying was carried out by putting the seaweed on a plastic sheet on the ground, and the working posture during drying was squatting and stooping, as shown in Figure 1.

After redesign, drying was carried out by using a dryer as high as the elbow in standing position, as shown in Figure 2. Limitation of lifting weight was 23 kg; 5 minutes break was provided for every hour worked, and availability of drinking water was improved.

The tools used in this study include anthropometer, anemometer, luxmeter, sling thermometer, globe thermometer, psychrometric chart, Nordic Body Map, and a 30-item fatigue questionnaire.

Figure 1. Drying seaweed by using plastic sheet.

Figure 2. Drying seaweed by using dryer as high as elbow.

Table 1. Mean difference of work performance.

| Variable | Before improvement | After improvement | P |
|---|---|---|---|
| Score musculoskeletal | | | |
| Before working | 29.38 | 29.15 | 0.119 |
| After working | 50.97 | 38.62 | 0.000 |
| Difference | 21.60 | 9.47 | 0.000 |
| Score general fatigue | | | |
| Before working | 31.05 | 30.80 | 0.128 |
| After working | 51.15 | 40.67 | 0.000 |
| Difference | 20.10 | 9.87 | 0.000 |
| Heart rate (pulses/min) | | | |
| Before working | 76.10 | 75.85 | 0.411 |
| After working | 96.01 | 91.06 | 0.000 |
| Workload | 19.90 | 15.21 | 0.000 |
| CVL (%) | 33.11 | 25.30 | 0.000 |
| Productivity | | | |
| Drying rate (kg/h) | 17.00 | 17.98 | 0.000 |
| Work productivity | 0.87 | 1.20 | 0.000 |

## 3 RESULTS

The farmer performances are described in Table 1. There were no significant differences of musculoskeletal, general fatigue, and heart rate before start of working ($p > 0.05$). However, after finishing working there were significant differences of musculoskeletal complaints, general fatigue, heart rate, and productivity ($p < 0.05$).

## 4 DISCUSSION

### 4.1 *Musculoskeletal complaints*

The average score of musculoskeletal complaints before working with conditions before redesign was $29.38 \pm 0.79$, and conditions after redesign were $29.15 \pm 0.81$. The results showed that the score of musculoskeletal complaints was not significantly different ($p > 0.05$).

The test results showed that the score of musculoskeletal complaints after working was significantly different ($p < 0.05$), with a score of $50.97 \pm 1.81$ for condition before redesign, and 38.62 ± 1.21 for the condition after redesign. These results indicated a decline of musculoskeletal score complaints by 24.22%. Thereby it can be assumed that the decrease in musculoskeletal complaints due to the redesign of dryer and working system.

The differences of musculoskeletal complaints before and after working for both treatments were compared, and the results were significantly different ($p < 0.05$). The mean difference score of musculoskeletal complaints of workers before redesign 21.60 ± 1.72, and after redesign of 9.47 ± 1.29, a decrease of 56.15%. Musculoskeletal complaints in the process of drying seaweed caused by squatting and stooping work posture when drying used a plastic sheet, and excessive lifting weight. Musculoskeletal complaints can occur in almost any type of work, with light category, moderate, heavy, and especially very heavy (Susila, 2002). Musculoskeletal complaints are caused by many factors (Marras et al., 2009) such as psychosocial, individual, workplace, and work organization. In the process of drying seaweed, workplace factors and work organization have been improved, resulting in a decrease in musculoskeletal complaints.

The decrease in musculoskeletal complaints in the seaweed drying process was caused by redesign of the seaweed dryer with elevation in accordance with the elbow height of the workers in standing position. Redesign of dryer has caused a change of posture from the squatting to standing position. Standing work posture produced a low level of activation of muscles in the lumbar erector spinae, the intra-discal pressure was low, and is known as an efficient position (Pheasant, 1991). Dynamic standing posture was better than static standing posture. Drying seaweed done with a dynamic standing posture, where the subject moves around the dryer to flatten the seaweed, which means the blood circulates more smoothly so that the absorption of oxygen and the transport of wastes of metabolic proceed more smoothly.

An improvement of work station on Balinese gamelan craftsmen from sitting on the floor was changed to using a table and sitting in a chair was proven to reduce the work load and musculoskeletal complaints (Tirtayasa et al., 2003). An improvement on a small industrial scale by redesigning the layout of work station, redesign of lifting equipment, design of table with the appropriate height, job rotation, giving a break and snack, can reduce workload, reduce muscle work, and increase productivity (Kogi et al., 2003).

### 4.2 *Fatigue*

The average score of general fatigue before working with conditions before redesign was 31.05 ± 0.84, and with conditions after redesign was 30.80 ± 0.70. The fatigue score before working occurred in the category of physical fatigue, where workers felt stiff in the shoulder, and had pain in the back. The results showed no significant difference ($p > 0.05$), so the decrease in fatigue is assumed to be caused by the redesign of the dryer and working system.

The average score of general fatigue after working with conditions before redesign was 51.15 ± 1.87, and conditions after redesign was 40.68 ± 1.26. The results showed a significant difference ($p < 0.05$), with a decrease in fatigue scores by 20.46%. The difference of fatigue between before and after working were compared, and the results showed there were significant differences ($p < 0.05$). The differences of fatigue score before redesign was 20.10 ± 1.63, and after redesign 9.87 ± 1.36, a decline in fatigue score of 50.84%. When the difference was analyzed by category, it showed a weakening in activity score from 8.75 to 3.70, or a reduction of by 57.71%, a improvement in motivation scores from 4.60 to 2.30, or a reduction of by 50%, and reduced physical fatigue from score of 6.75 to 3.95, or down 41.48%.

The decrease of fatigue was caused by redesign of the seaweed dryer with elevation in accordance with elbow height of the workers in standing position, the height set at the 5 percentile elbow height or 93 cm, so that the working posture changed from squatting to standing. Drying seaweed was done with a dynamic standing posture, where the subject moved around the dryer to flatten the seaweed. Lifting and carrying was limited to 23 kg weight in accordance with the recommendations of NIOSH; this caused muscle work to be lighter, and workers can do the job during work time 8 hours per day. Giving a short 5-minute break every hour worked, enabled workers to quickly restore energy to previous levels, because the subject has not experienced excessive fatigue.

### 4.3 *Workload*

The mean heart rate before working for the two groups were not significantly different ($p > 0.05$), with a mean value of 76.10 ± 3.58 ppm for the working condition before redesign, and 75.85 ± 4.00 ppm for the working condition after redesign. These data included the category of light workload that was in the range 75–100 ppm (Grandjean, 1993; Adiputra, 1998). Thus it can be assumed that the decrease of workload is caused by the redesign of dryer and working system.

Statistical analysis of heart rate after working for two groups was significantly different ($p < 0.05$), with an average of 96.01 ± 4.81 ppm for working condition before redesign, and 91.06 ± 4.51 ppm for working condition after redesign. The working heart rate included the categories of light work load that was in the range 75–100 ppm (Grandjean, 1993; Adiputra, 1998; Santosa et al., 2006). The results also showed increased heart rate after working before redesign of 26.15%, and after redesign of 20.05% from the initial conditions. The allowable increasing resting heart rate to the working heart rate 35 ppm for men, so that work can performed for 8 hours continuously (Grandjean, 1993). Based on these limits, both groups were under the required conditions, so they can carry out the work 8 hours continuously.

Similarly, the results of workload analysis show the average workload was significantly different ($p < 0.05$), with average of 19.90 ± 2.66 ppm for the working conditions before redesign and 15.21 ± 2.02 ppm for working conditions after redesign. There was a decrease of 4.69 ppm workload or 23.56%. The decrease of workload was caused by the redesign of seaweed dryer in accordance with the 5 percentile of elbow height in standing position or 93 cm, so that the work posture change from squatting to standing.

Drying seaweed was done with a dynamic standing posture, where the subject moves around the dryer to flatten the seaweed. Lifting weight was limited 23 kg as recommended by NIOSH, and lifted freely (free style), causing the muscle work to be lighter, and the subject can work 8 hours per day. The freestyle posture was the best posture for lifting loads with the smallest fatigue (Ayoub and Mital, 1989). Giving a short 5-minute break every hour worked, and improving the availability of drinking water, the worker can quickly recover, so that fatigue was reduced (Louhevaara and Kilbom, 2005).

Working with new working tools and systems which relate to aspects of anthropometry was also shown to reduce cardiovascular load (CVL) significantly ($p < 0.05$). This is evident from the decrease in % CVL, before redesign the average was 33.11 ± 3.93% and after redesign the average was 25.30 ± 2.88%. Working conditions before redesign were categorized as moderate workload, which requires 75% work and 25% of rest for every hour. After redesign a decline % CVL and changed the work conditions into the category of light work, so that workers can perform 8 hours of continuous activity.

### 4.4 *Productivity*

Analysis of drying rate for both groups showed a significant difference ($p < 0.05$), with average of 17.00 ± 0.66 kg/hour for the use plastic sheet and 17.98 ± 0.43 kg/hour for using the *bedeg (dryer)*, or an increase of 5.76%. It meant that the drying process was faster on the dryer as high as 93 cm, with a *bedeg* surface that has slope. The success of drying was determined by three factors: the heat to remove water vapor, dry air to absorb the water vapor, and circulation to carry the water vapor (Whitfield, 2000). Based on this formula, the decline in water content was caused by a higher evaporation rate on the *bedeg* surface, due to higher wind velocity. The drying process was found to be faster, needed only 2 days when the weather in sunny conditions, while drying by using plastic sheet needed at least 3 days.

Productivity analysis on both groups showed a significant difference ($p < 0.05$), with an average value of 0.87 ± 0.12 kg/(hour.ppm) for working conditions before redesign and 1.20 ± 0.17 kg/(hour.ppm) for working conditions after redesign or increase productivity by 37.93%. The increase of productivity was due to an increase in drying rate and decreasing workload.

## 5 CONCLUSION

Based on the analysis and discussion in this study, the following results can be concluded.

1. Changing body posture and working system decreased musculoskeletal complaints by 56.15%.
2. Changing body posture and working system decreased general fatigue by 50.84%, followed by an increase in activity by 57.71%, increased motivation by 50%, and reduced physical fatigue by 41.48%.
3. Changing body posture and working system increased productivity by 37.93%.

Suggestion: The redesign of the seaweed dryer, limitation on lifting load, and giving short breaks have been proven to reduce musculoskeletal complaints, reduce fatigue, and improve productivity and should begin to be implemented.

## REFERENCES


Aasa,U., Bergkvist, M.B., Axel, K. & Brulin, C. 2006. Relationships between work-related factors and disorders in the neck-shounder and low-back region among Female and male ambulance personnel. *J. Occup. Health*. November; 47(6): 481–489.

Adiputra, N. 1998. *Metodologi Ergonomi*. Program Magister Ergonomi-Fisiologi Kerja. Program Pascasarjana Universitas Udayna.

Anggadiredja, J.T., Zatnika, A., Purwoto, H. & Istini, S. 2006. *Rumput Laut*. Jakarta: Penebar Swadaya.

Ayoub, M.M. & Mital, A. 1989. *Manual Materials Handling*. London: Taylor & Francis Publishers.

Bakta, M. 2000. Uji Klinik, *J. of Internal Medicine*. Mei; 1(2): 99–107.

Daniel, W.W. 1999. *Biostatistics: A Foundation for Analysis in the Health Sciences*. 7th ed. New York: John Wiley & Sons, Inc.

Grandjean, E. 1993. *Fitting the Task to the Man, a Textbook of Occupational Ergonomics*, 4th ed. London: Taylor & Francis.

Kogi, K., Kawakami, T., Itani, T. & Batino, J.M. 2003. Low-cost improvements that can reduce the risk of musculoskeletal disorders. *International Journal of Industrial Ergonomics*. March; 31(3): 179–184.

Louhevaara, V, Kilbom, A. 2005. Dynamic Work Assessment. Evaluation of Human Work 3rd ed. London: Taylor & Francis, pp. 429–451.

Marras, W., Cutlip, R.G., Burt, S.E. & Waters, T.R. 2009. National Occupational research agenda (NORA) future direction in occupational musculoskeletal disorder health research. *Journal of Applied Ergonomics*. 40: 15–22.

Pheasant, S. 1991. *Ergonomics, Work and Health*. London: MacMillan Academic Professional Ltd.

Poncomulyo, T., Maryani, H. & dan Kristiani, L. 2006. *Budi Daya dan Pengolahan Rumput Laut*. Jakarta Agro Media Pustaka.

Richardson, G.E., Jenkins, P.L., Strogatz, D., Bell, E.M. & May, J.J. 2006. Development and initial assessment of objective fatigue measures for apple harvest work. *Journal Applied Ergonomics*. November; 37(6): 719–727.

Santosa, Azrifirwan, Putra, K. 2006. Identifikasi Antropometri Tenaga Kerja Pencangkul Sawah di Kabupaten Pasaman Barat Propinsi Sumatra Barat. *Jurnal Sains dan Teknologi*. Juni, Vol. 5., No. 1, pp. 14–24.

Statistik Perikanan Budidaya Propinsi Bali, 2007. Pemerintah Propinsi Bali Dinas Kelautan dan Perikanan.

Statistik Perikanan Budidaya Propinsi Bali, 2008. Pemerintah Propinsi Bali Dinas Kelautan dan Perikanan.

Steven, P. 2005. *Crossover Design. In Clinical Trials*: A Methodologic Perspective. 2nd ed. Hobaken, NJ: John Wiley and Sons, Inc.

Susila, I.G.N. 2002. Gangguan Muskuloskeletal. *Majalah Kedokteran Udayana (Udayana Medical Journal)*. 33(116): 78–83.

Tirtayasa, K., Adiputra, N. & Djestawana, G.G. 2003. The change of working posture in *manggur* decreases cardiovascular load and musculoskeletal complaints among Balinese gamelan craftsmen. *J. Human Ergol.*, 32: 71–76.

Whitfield, D.E. 2000. Solar Dryer Systems and the Internet: Important resources to improve food preparation. Paper International Conference on Solar Cooking, Kimberly—South Africa, 26–29 Nov.

# *Part V: Occupational safety and health*

*Ergonomics in Asia – Shih & Liang (eds)*
*© 2012 Taylor & Francis Group, London, ISBN 978-0-415-68414-9*

# Evaluation of anti-smoking measures at a Japanese factory

Hiromi Ariyoshi, Toshiko Nakamura & Rieko Yamashita
*Institute of Nursing, Faculty of Medicine, Saga University, Saga-Pref., Japan*

Yoshika Suzaki
*The Japanese Red Cross Kyushu International College of Nursing, Fukuoka-Pref., Japan*

Naoko Takayama
*Yokkaichi Nursing and Medical Care University, Mie-Pref., Japan*

Tatsuya Ishitake
*Kurume University Fukuoka-Pref., Japan*

ABSTRACT: The purpose of this research was to execute the spatial separation of smoking areas and to perform and evaluate smoking cessation support and anti-smoking education at factory "A", where sweeping anti-smoking measures have not been enforced. Health education was introduced, and through the cooperation of the Health and Safety Committee, a program for the separation of smoking areas was proposed. Through smoking cessation support and anti-smoking education as a part of general health education, and the creation of a designated smoking area, the workplace environment was improved. With the Health and Safety Committee playing a central role, it can be suggested that through the spatial separation of smoking areas and the execution of anti-smoking education, the workplace environment improved and the importance of anti-smoking was better understood by workers, which led to the creation of a comfortable workplace environment based on the ideology of health promotion.

*Keywords*: smoking areas, health promotion law, health and safety committee, health education, workplace

## 1 INTRODUCTION

The Health Promotion Law (enacted May 2003) was designed to improve national health, and involved the comprehensive promotion of the basic items necessary as a measure to improve national health. In Chapter 5, Section 2 of this law, the prevention of passive smoking is stated, and places the responsibility of the damage done by passive smoking on the management of the facility rather than on the smoker.

As social interest in smoking and its effects on health increases, designated smoking areas, completely smoke-free environments, or air filtering as methods of anti-smoking measures have been enforced in public facilities or large-scale businesses. But the Health Promotion Law is enforcing neither smoking cessation nor separation of smoking areas under a company's responsibility, and the law is not putting any fines, etc., at all on offenders. Because of this, smoking cessation and anti-smoking measures haven't been promoted at Japanese factories. Anti-smoking measures at Worksite "A" were not enforced prior to this research, therefore through cooperation with the Health and Safety Committee, this research aimed to report on the results of the spatial separation of smoking areas, smoking cessation support, and anti-smoking education.

## 2 METHODS

### 2.1 *Subjects*

The subject of this research, Worksite "A", is a printing factory for a regional urban newspaper press, which prints approximately 1.2 million copies a day. Worksite "A" is composed of the 4 major departments of printing, shipping, technology, and management (Figure 1), with an emphasis on printing and shipping, and work hours are irregular with day and night shifts. The worksite is an all male workplace with 98 employees with an average age of 46.8 ± 9.9 (as of April 2006). In 1982, the four-storied Worksite "A" was built in a suburban factory district as an asset to the newspaper company. Prior to the introduction of anti-smoking measures, smoking was permitted anywhere in the factory except the workshop, however, only during breaks.

Employee health is managed by a full time occupational health nurse and a part time industrial physician (one afternoon two times a month).

### 2.2 *Process*

1. Application of the Health and Safety Committee

12 members compose the Health and Safety Committee, including the chairman, vice-chairman, speaker, secretariat, and 1 elected member from both the company side and labor union side to represent each workplace. The Health and Safety Committee meeting was held once a month, and the issue of health and safety at Worksite "A" was discussed at these meetings. The industrial physician and the occupational health nurse requested that the Health and Safety Committee consider measures for the separation of smoking areas. Through this request, the members of the committee conducted an anonymous questionnaire on the 98 workers of Worksite "A" and distributed and collected the surveys. The pre anti-smoking measure survey contents included a figure on the number of smokers and the level of discomfort due to indoor smoke. The post anti-smoking measure survey inquired about the interest in future smoking cessation.

2. Relevant information

In order for the occupational health nurse to develop anti-smoking measures, as determined by the Health and Safety Committee, it was necessary to collect the following information and to clarify the problematic points.

a. Smoking environment

   Visitational rounds are conducted by the industrial physician and the occupational health nurse to review the room layout, position of ashtrays, state of smoking conditions, and the operation of the air filtering system.

b. Reviewing the conditions at other companies

   With written consent, the industrial physician, occupational health nurse, and members of the Health and Safety Committee toured 4 other companies in the same industry to survey the conditions of smoking.

3. Smoking cessation support and anti-smoking education

Individualized smoking cessation support and group oriented anti-smoking education was conducted by the industrial physician and the occupation health nurse.

### 2.3 *Anti-smoking measures and their evaluation*

1. Present state of smoking

The author, using the data from the simple survey conducted by the Health and Safety Committee, compared the results of pre and post spatial separation of smoking areas, conducted on April 2004 and April 2006, respectively. The proportional comparison of pre and post anti-smoking measures was calculated using the SPSS11.5J software, and the significance level of the chi-square test was set at $p < 0.05$.

2. Results of environmental measurements
A working environment measurement expert visited Worksite "A" both pre (April 2004) and post (April 2006) anti-smoking measures, and used the Kitakawa gas detector tube (Komyo Rikagaku Kogyo) to measure the levels of carbon monoxide and carbon dioxide concentrations according to the "Ordinance on Health Standards in the Office." A digital dust gauge was also used to measure the concentrations of air borne dust particles at both testings. Pre and post measurements were both taken in the same spot within the worksite, and all testing occurred during work hours on weekdays, with the average taken from 3 measurements. SPSS11.5J was used to statistically analyze the comparisons, with the Wilcoxon signed rank test significance level set at $p < 0.05$.

### 2.4 *Ethical consideration*

The author obtained written approval for the use of data for this study, through the Health and Safety Committee.

## 3 RESULTS

### 3.1 *Smoking environment*

Results of the visitational rounds conducted by the industrial physician and the occupational health nurse showed that the separation of smoking areas was not being followed other than in the workshop area. In the break room, the table area was designated as a smoking space and equipped with a ventilation machine, however the machine was not suitable in size to appropriately filter the air. In addition, it was found that workers in the office areas were smoking near paper documents and at their own desks using ashtrays, rather than smoking in the designated smoking area near ventilation. Due to the inadequacy of the ventilation machine and the lack of usage of such facilities by workers, the building was filled with the smell of smoke.

### 3.2 *The state of conditions at other companies*

Two of the four companies visited by the industrial physician, occupational health nurse, and members of the Health and Safety Committee were found to have specific and working measures for the spatial separation of smoking areas.

### 3.3 *Application of the Health and Safety Committee*

Discussions concerning anti-smoking measures were held at the monthly Health and Safety Committee meeting. In order to further the committee members' understanding of the situation, the Health Promotion Law, "Guidelines for Measures on Smoking in the Workplace", and social tendencies were described by the industrial physician and occupational health nurse. In addition, the members reported their findings of visitations to other companies to their respective work areas, and through these efforts, the members as well as company employees were able to correctly understand the necessity for the separation of smoking areas. The Health and Safety Committee board members also reported on the results of the simple survey conducted.

Among limited resources, all options were considered. Worksite "A", an older building not suitable for the construction of new facilities, had little space within to create a designated smoking area, and because over 60% of the workers were smokers, time was also needed to gain their understanding of new policies. To resolve the situation, the industrial physician found the least frequently used women's lavatory positioned between the workshop and break room and suggested turning this second floor space into a designated smoking area with only one entrance/exit door.

A ventilation fan was mounted both on the ceiling and on the window and air circulation was achieved. In addition, a 30 cm long vinyl sheet was hung from the entrance/exit doorway to prevent smoke from leaking into the hallway. The renovations, conducted in April 2005, were completed for 800,000 yen. When considering the cost to maintain tabletop air filtering machines, the renovations were comparatively more economical when considered long term.

The securing of and renovations for a designated smoking area, paying particular attention to the exterior appearance, was achieved through cooperation by the company management. In addition, one outside covered area was set up as a designated smoking area to support the worksite-wide spatial separation of smoking areas.

### 3.4 *Smoking cessation support and anti-smoking education*

The industrial physician and occupational health nurse developed a health education program based on Orem's self-care guidelines. Three of the 98 workers expressed a desire to stop smoking, and through the smoking cessation support program, which began assisting them in May 2003 through the use of nicotine patches and gum, they were able to successfully continue their abstinence from smoking. In addition, the industrial physician and occupational health nurse, using a PowerPoint tool to explain important topics, conducted group health education programs. These 60-minute programs, sponsored by the Health and Safety Committee, were given to groups of 50 workers at a time and were conducted once to twice a year. The program content included lectures, discussions with smoking cessation expert advisors, and demonstrations using the "smokerlizer". The industrial physician and occupational health nurse also conducted anti-smoking education through intra-office news and bulletins boards, and informative campaigns were continued. As a result, health education participants, members of the Health and Safety Committee, and non-smokers, expressed to the industrial physician and occupational health nurse, their desire for more aggressive programs in the future.

### 3.5 *Measures for the spatial separation of smoking areas and their evaluation*

The evaluation of the measures for the spatial separation of smoking areas can be seen in Tables 1 and 2.

The present state of smoking:

The results of the simple survey (effective return rate of 100%), given by the Health and Safety Committee to all 98 workers at Worksite "A", show that in April 2004, or pre spatial

Table 1. Evaluation of spatial separation of smoking areas.

| Unit | 2004 | 2006 | |
|---|---|---|---|
| Smoking rate % difference | 61.3 | 58.3 | Not significant |
| Discomfort from indoor % environment | 60.2 | 49.5 | $p < 0.001$ |
| Carbon monoxide ppm concentration difference | 1.00 | 1.00 | Not significant |
| Carbon dioxide ppm concentration | 532.4 | 490.4 | $p < 0.01$ |
| Airborne dust mg/m$^3$ Particle concentrations | 0.03 | 0.02 | $p < 0.01$ |

*Based on the business worksite prevention regulations, above each item was measured once a day at noon when the most of the people are said to smoke for 3 days. The figures are averages.

separation of smoking areas, 60 workers (61.3%) were smokers and 38 workers (38.7%) were non-smokers, and in April 2006, or post spatial separation of smoking areas, 56 workers (58.3%) were smokers and 42 workers (41.7%) were non-smokers (no significant difference). When smokers were asked post separation of smoking areas (April 2006), if they had plans to stop smoking in the future, 24 workers (23.0%) replied "yes", 47 (48.7%) replied "no", and 27 (28.3%) replied "not certain". When asked pre separation of smoking areas (April 2004), about the existence of indoor environmental discomfort, 54 workers (60.2%) replied "yes", compared to post separation of smoking areas (April 2006) when 49 workers (49.5%) replied "yes" ($p < 0.001$). In addition, 48 out of 56 smokers (86.0%) responded that they felt there was indoor environmental discomfort post separation of smoking areas, compared to 42 all non-smokers (100%) who replied that they felt no discomfort.

Results of environmental measurements:

In April 2004, the carbon monoxide measurement was found to be 1.00 ppm compared to 1.00 ppm in April 2006 (no significant difference). The carbon dioxide levels pre separation of smoking areas (April 2004) were 532.4 ppm compared to post separation of smoking areas (April 2006) when the levels were measured at 490.4 ppm ($p < 0.1$). Airborne dust particle concentrations were 0.03 mg/m$^3$ in April 2004 and 0.02 mg/m$^3$ in April 2006 ($p < 0.1$).

## 4 DISCUSSION

Tobacco use not only affects the smokers themselves, but also poses a health risk for those around the smoker who are unwillingly exposed to secondhand smoke. The enactment of the Health Promotion Law placed the responsibility of the development of lung cancer in non-smokers who were exposed to long-term secondhand smoke in offices and break rooms, on the shoulders of the business executives and/or industrial physician. These and other negative influences posed by tobacco use have been and are becoming a great social problem.

In larger enterprises and public facilities, there is a tendency towards completely non-smoking areas. However, in smaller and middle sized enterprises, the movement towards completely non-smoking areas is sluggish due to the lack of understanding on the part of the management and employees, and as well as to economical circumstances, older buildings, and lack of resources. It was difficult to implement such understanding with the subject of this research, a small scaled worksite with less than 100 workers, however due to the activities of the Health and Safety Committee, the spatial separation of smoking areas was achieved.

In accordance with the "Ordinance on Health Standards in the Office" and the "Tobacco Action Plan Study Group Report", it was deemed necessary to utilize company funds for ventilation machines and the designation of smoking areas, not only for the sake of the company, but also to maintain appropriate human relations between smokers and non-smokers. This research, with the Health and Safety Committee playing a central role, conducted a questionnaire survey and considered various costs as well as the opinions of smokers and non-smokers when creating measures for the spatial separation of smoking areas. Through the creation of this space, human relations among company workers were maintained. In addition, a key characteristic of this research is its bottom-up style approach to the creation of a comfortable workplace environment based on the concepts of health promotion.

Previously existing research shows an example where a ceiling to floor vinyl sheet partitioned off a corner area of an office in order to secure a designated smoking area. In this study, a women's lavatory was transformed into a designated smoking area. The "Report on the Effects of Passive Smoking on the Respiratory System: Lung Cancer and Other Related Diseases" states that non-smokers prefer areas where smoke can not leak out as opposed to smoking corners, thus emphasizing the need for smoking areas sealed off from non-smoking areas. In addition, the "Guidelines for Measures on Smoking in the Workplace"also states that tobacco smoke should not be allowed to leak into non-smoking areas, as well as stating the necessity for proper ventilation in the designated smoking areas.

For the effective spatial separation of smoking areas, the opening between the designated smoking area and non-smoking area should not be a large area, as with smoking corners, but should be limited to only the entrance/exit door, as with a designated smoking room. In this study, the windows were sealed shut, ventilation fans were built into the ceiling and windows each to provide proper air circulation to the outside, and the only opening was at the entrance/exit door, where a 30 cm long vinyl sheet was hung to prevent smoke from leaking into the non-smoking area. This study was effective and successful in creating and securing a designated smoking area through the remodeling of the women's lavatory, with regards to the exterior look and fire prevention.

The results of the environmental measurements showed that the levels of carbon monoxide and carbon dioxide concentrations, and the concentrations of airborne dust particles both pre and post measures for the separation of smoking areas were within the normal range set by the "Ordinance on Health Standards in the Office". The results of the survey showed a decrease in the level of indoor environmental discomfort after the implementation of the spatial separation of smoking areas. In particular, the discomfort level of non-smokers was completely eliminated. Although there is no barometer to confirm the existence of the tobacco smell chronically stained into the walls or the smoggy air as a result of the tobacco smoke pre-anti-smoking measures, it can be inferred that the separation of smoking areas, which led to the disappearance of the tobacco smell and smoke, played a key role in the restoration of a normal air environment. On the other hand, smokers, who were forced to smoke only in the designated areas, found the environment uncomfortable. In the future, environmental measurements will be taken in accordance with the "Guidelines for Measures on Smoking in the Workplace", and it will be necessary to measure the air currents and wind velocity in the area between the designated smoking area and non-smoking areas. Concurrently, it will be necessary to take into consideration the smoking area itself in order to ensure appropriate spatial separation as well as to maintain a comfortable workplace environment.

Pre anti-smoking measures, the smoking rate was 61.3% (April 2004), compared to 58.3% (April 2006) post anti-smoking measures, although no significant difference was shown. It is thought that the reason for the lack of decline in smoking rates at Worksite "A" is attributed to the fact that pre separation of smoking areas, workers were only permitted to smoke during breaks and only outside of the factory, and by creating an additional space, within the building, where workers would be permitted to smoke, the smokers may have found it practical rather than inconvenient. In addition, it was found that post spatial separation of smoking areas, 23% of smokers were interested in future tobacco cessation. It is thought that this interest can be attributed to the tobacco cessation support and anti-smoking education, which was provided as a part of the health education program in order to promote self care efficiency among workers. In addition, it is thought that the interest in tobacco cessation also stems from the reality that smokers faced in only being able to smoke in the designated smoking area, which was secured with the Health and Safety Committee playing a central role.

As stated in "Healthy Japan 21: Health Promotion and Fitness", it is important to promote smoking cessation not only for the sake of the smokers themselves, but also for the health of non-smokers, and the separation of smoking areas is necessary. In the future, with the Health and Safety Committee continuing to play a central role, it will be crucial to continue to maintain the separation of smoking areas as well as to promote and provide smoking cessation support.

The limitations of this study were that for both pre and post spatial separation of smoking areas, the behaviors or details of smoking, such as the acquisition of tobacco or the number of cigarettes smoked daily were not assessed at the pre or post health education opportunities. It was especially necessary to conduct awareness surveys on smokers prior to the spatial separation of smoking areas. These details will be left for future studies.

Worksite "A" does keep the measure of spatial separation of smoking areas at present, but is not still the same level as large-scale worksites or public facilities where smoking is prohibited 100%. Also, in Japan, many middle and small-scale worksites are not even taking any measures for spatial separation of smoking areas. This reality signifies the necessity to keep evaluating the research.

## 5 CONCLUSION

In this study, anti-smoking measures were re-evaluated at factory "A", a small-scale worksite.

1. With the Health and Safety Committee playing a central role, cooperation was obtained from both the management side and employee side, and the women's lavatory was transformed into a designated smoking area.
2. Through smoking cessation support and anti-smoking education provided through health education opportunities, the level of awareness of the need for the separation of smoking areas was promoted among employees.
3. With the Health and Safety Committee playing a central role, systemic anti-smoking measures with a foundation in health promotion contributed to the creation of a comfortable workplace environment.

## REFERENCES


Fujiuchi, S. Otawa Constitution and health promotion. Public Health. 1997; 61: 636–641.

Guidelines for anti-smoking measures in the workplace. H.17 Labor and Health memos. Ministry of Health, Labor, and Welfare, the Labor Standards Bureau, Tokyo: Sanshusha, 2006: 267–275.

H.15 the Guide of Labor's Hygiene, the Ministry of Health. Labor, and Welfare, Edited by the Labor Standard Bereau. Tokyo: Sannsyusya, 2003: 77–78.

H.15 the Guide of Labor's Hygiene, the Ministry of Health. Labor, and Welfare, Edited by the Labor Standard Bureau. Tokyo: Sannsyusya, 2003: 218–223.

Handbook of the Safety-Hygiene Laws. Central Industrial Accidents Prevention Assoc. 2006: 712–716. http://www.cuc.ac.jp/^yasuhiro/env4.html

Kasuya, Y. The concept of self-care and practical nursing. Tokyo: Health Publications, 1994; 19–38.

Kawakami, K. & Haratani, R. Reforming the workplace environment. Occupational Health Journal. 2000; 23: 45–49.

Masuda, K. Deciding on the rules for smoking cessation and the separation of smoking areas through group meetings. Psychiatric Mental Health Nursing. 2004; 31: 14–19.

Nakamura, M. Anti-smoking measures and risk communication. Public Health. 2004; 67: 524–528.

Odate, J. Health Promotion Law Legislation. Life Hygiene. 2003; 47: 40–41.

Ogawa, H. Health Promotion Law and Japan's measures against tobacco. Physical Science. 2004; 237: 80–84.

Otsuka, Y. Tobacco cessation guidance for staff and patients: results from efforts by the committee board. Psychiatric Mental Health Nursing. 2004; 31: 14–19.

Shimauchi, K. Health promotion and the metro health culture. Japanese Journal for Public Health Nurse. 1999; 55: 276–286.

Yamato, H. Measures against passive smoking in the workplace. Physical Science 2004; 237: 62–67.

Yamato, H., Seto, T., Morimoto, Y. et al. Considerations for highly effective measures for the spatial separation of smoking areas in offices. Japan Society for Occupational Health. 2000; 42: 1–5.

Yamato, H., Akiyama, I., Ogami, A. et al. Installing effective smoking areas and the usefulness of real-time monitoring of dust density. Japan Society for Occupational Health. 2004; 46: 55–60.

Zanetti, F., Gambi, A., Bergamachi, A. et al. Smoking habits, exposure to passive smoking and attitudes to a non-smoking policy among hospital staff. Public Health 1998; 112: 57–62.

*Ergonomics in Asia – Shih & Liang (eds)*
*© 2012 Taylor & Francis Group, London, ISBN 978-0-415-68414-9*

# Analysis of productivity with an ergo-mechanical approach for making banten elements

I. Wayan Bandem Adnyana
*Mechanical Engineering, Udayana University, Denpasar, Bali, Indonesia*

ABSTRACT: One activity that can be found in the lives of Bali people is making *banten* elements such as *tumpeng* and *penek*. *Tumpeng* and *penek* are done in home industry and are used in religious ceremonies. Manufacturing processes are done manually sitting on the floor. Production process is sped up by gas-fired dryer to make the products ready to be marketed. One of the goals of industrial activities is productivity. How do our efforts increase productivity for making *banten* elements? Increasing the productivity of the production process will be achieved through an ergo-mechanical approach, because the ergo-mechanical can be a simultaneous problem solving approach. The ergonomics aspects are oriented to human performance and the mechanical engineering aspects are oriented to the production process of dryer performance.

*Keywords*: productivity, ergo-mechanical, *banten* elements

## 1 INTRODUCTION

Bali is an unique area in respect of religious ceremonies, works of art, a variety of foods and other things. Religious ceremonies are offered as banten form and are very important in the life of Hindus in Bali. They are always available every day, there are not many days without a religious ceremony. The first ceremony can be created when a baby's birth is greeted with pemagpag rare ceremony, breaking the umbilical cord is the keles kambuhan ceremony, then the three months ceremony, otonan ceremony (six months), haircut ceremony and on to adulthood ceremony (Swastika, 2010).

Components of the bantens are tumpeng and penek. Tumpeng is shaped like a cone as the Mountain symbol. Mountain is a symbol of Purusa (male) prosperity and also as a symbol of Sang Hyang Akasa as protection or guidance. Penek is rather flat-shaped symbol of ocean (sea) or a lake. Sea or lake is a symbol of Predhana (female/Sang Hyang Ibu Pertiwi) that gives immediate (existence) of life (Wijayananda, 2003).

The tumpeng and penek are the result of home industry activities created manually by hand and dryer tools. One of the goals in industry activity is continuity of productivity. Productivity generally implies comparison between the results achieved (outputs) with all the resources used (inputs). Competitive business can't just rely on cheap human resources to increase production, but should optimize existing resources and infrastructure to support production activities.

How do our efforts increase the productivity of the banten elements industry? One of the efforts to increase productivity can be conducted through an analytical approach using an ergo-mechanical method. In the ergo-mechanical method of solving the problem, ergonomics is mechanical oriented to the human process and simultaneously, engineering is oriented to the production process of dryer tools.

## 2 PRODUCTION PROCESS

The production processes of tumpeng and penek are done manually by hand and the drying process by dryer tool. Rice is cleaned, soaked and then steamed. Rice that has been steamed is mixed with warm water, and starch is added as an adhesive, then the resulting dough is the raw material for manufacturing the tumpeng and penek. The dough has been cold-formed by mold manually to fit the desired shape of either the tumpeng or penek. The result of the production process of forming just manually has not been marketed yet. What is needed further is a simple process to dry using a dryer which has been marketed already. The production process can be shown as in Figures 1 to 4 below.

## 3 DISCUSSION

Production process of tumpeng or penek will be influenced by the performance of workers and the dryer performance. To know the factors that affect worker and dryer tool performance an analysis approach by ergo-mechanical method is employed. The ergo-mechanical approach is solving a simultaneous problem, with ergonomics oriented to human processes and mechanical engineering oriented to the production process which aids drying.

In the ergonomics of performance a person will be influenced by the balance between the demands of the task and the capabilities and limitations of humans (Manuaba, 2006). Performance of a person can be seen from changes in productivity. One of the variables that can affect productivity is the workload. The workload is greatly determined by the working posture at the time of doing the activity. Working posture in Figures 1 and 2 in the production process is not natural or not physiological, so it needs an ergonomic intervention. These interventions are hoped to result in changes to work posture which becomes physiologically natural so it can reduce the workload. This includes external workload (stressor) and internal workload (strain) (Manuaba, 1996 and Adiputra, 1998).

### 3.1 *External workload (stressor)*

Stressor is the workload that comes from outside the body such as tasks, organization, and environment. Tasks include static and dynamic muscle activity, frequency and speed of use of assistive devices, as well as quantity and quality of production. Organization involves working

Figure 1. Sitting with one leg crossed and asymmetric position.

Figure 2. Sitting with both legs folded.

Figure 3. Penek and tumpeng in dryer conditions.

Figure 4. Walls dryer without insulator.

together (team work), turn (shift), and scheduled work breaks. The working environment is associated with physical barriers, microclimate, lighting, noise, aspects of anthropometry, range, high, and low working facilities.

### 3.2 *Internal workload (strain)*

Strain is the workload that comes from inside the human body, including: somatic factors and psychological factors. Somatic factors include: gender, age, body size, health condition, nutritional status and others. Psychological factors include: motivation, perception, desire, emotion, satisfaction, confidence, self esteem, responsibility and others.

The workload can be known based on changes in resting pulse rate with beating pulse rate per minute of work. When the workload is increasing, the pulse per minute rate is calculated based on the difference between working and resting pulse rate. The pulse rate after working on a job making offerings was measured by ten beats. The resting pulse rate before the job is also measured by the method of ten beats. The method of ten pulses is the method involving palpation of the radial artery on the left hand, and is calculated based on how long it takes between the first pulse and the eleventh, the results given in seconds.

Some of the indicators for determining the severity of the work will be the pulse rate at work, oxygen consumption, and the person's energy needs. Pulse relationship to workload, oxygen consumption and energy expenditure can be seen in Table 1 below.

According to Sutjana (2000) productivity is the ratio between the amount of output with input which can be calculated on the formula:

$$P = \frac{O}{I_1 \times t_1} \tag{1}$$

where $P$ = productivity of workers; $O$ = (output) the average weight banten element (tumpeng and penek) in kg; $I_1$ = (input) the average energy expenditure for activity doing banten element workloads (beats/min); and $t_1$ = (time) the average time expenditure to make banten element (hours).

Equation 1 above can be converted to inputs (I) based on oxygen consumption and energy expenditure (cal/min) with helping of appropriate methods interpolation (Table 1). From the equation above the productivity of workers will be affected by the production, energy and time used for the production process.

Mechanical engineering, especially the energy conversion performance of dryers, is a better indicator of productivity performance, and is also a more productive process. The drying production process used an LPG gas-fired dryer. Performance of the dryer is determined by the effectiveness of the process of drying and the utilization efficiency of energy contained in the fuel. The amount of energy that is used in the drying process will be determined by the balance of energy in the combustion process such as equation 2 as follows:

$$q_s = NKA - X_{H_2O} \cdot LH_{H_2O} - q_b - q_{ub} - q_r \tag{2}$$

Table 1. Pulse relationship workload, oxygen consumption and energy expenditure.

| Workload | Oxygen consumption (liter/min) | Energy expenditure (cal/min) | Heart rate during work (Beats/min) |
|---|---|---|---|
| Light | 0.5–1.0 | 2.5–5.0 | 60–100 |
| Moderate | 1.0–1.5 | 5.0–7.5 | 100–125 |
| Heavy | 1.5–2.0 | 7.5–10.0 | 125–150 |
| Very heavy | 2.0–2.5 | 10.0–12.5 | 150–175 |

where $q_s$ = useful heat (kcal/kg); $NKA$ = calorific value of fuel (kcal/kg); $X_{H_20}$ = total mass of $H_2O$ produced during combustion (kg $H_2O$/kg); $LH_{H_20}$ = latent heat of vaporization of $H_2O$ (kcal/kg $H_2O$); $q_{gb}$ = heat carried by the exhaust gas (kcal/kg); $q_{ub}$ = heat fuels which are not burned (kcal/kg); $q_r$ = heat exposed to the outside of the system (kcal/kg).

An energy balance such as the above to dry tumpeng and penek can be achieved by minimizing the energy lost with isolation wall dryers so that the wall temperature is approximately equal to the air temperature working environment. Isolation is expected to reduce the heat energy that is exposed to the outside of the system. Lesser energy exposed to the outside of the system will improve the performance of dryers so that the drying rate is faster and the rate of fuel consumption decreases and heat lost to the environment is reduced. This situation will also be able to reduce the additional workload due to thermal effects of environmental influences.

The drying process does not only depend on the amount of energy that is available but also on how the phenomenon of water vapor existing in the offerings moves into the air so that the drying process becomes effective. The process of drying involves heat and mass transfer phenomena simultaneously. Heat transfer occurs when there is a difference in temperature while the mass movement occurs when there are differences in concentration. This process can work well when there is air movement. The air movement can be improved if a natural draft is added on the dryer.

The drying process will depend on the process above. The combustion process will depend on the effectiveness of the fuel while drying and will affect the drying time. If both these variables can be minimized then productivity will be better. The productivity of the dryers can be estimated as follows:

$$P = \frac{O}{I_2 \times t_2} \tag{3}$$

where $P$ = productivity of dryer tool working; $O$ = (output) the weight of banten element (tumpeng and penek) in kg; $I_2$ = the total energy spent in fuel during the drying process kg/hour (cal/min); $t_2$ = the total time used to dry the banten elements (hours).

Analysis of the productivity process by the ergo-mechanical approach can be formulated as follows:

$$P = \frac{O}{(I_1 \times t_1)(I_2 \times t_2)} \tag{4}$$

where $P$ = productivity of production process; $O$ = (output) the weight of banten elements that can be made weight (kg); $I_1$ = the total energy expenditure for activity in making banten element workloads (cal/min); $t_1$ = the total time expenditure to make banten element (hours); $I_2$ = the total energy which is spent as fuel during the drying process for banten elements kg/hour (cal/min) and $t_2$ = the total time used to dry the banten elements (hours).

## 4 CONCLUSIONS

1. Increased productivity of production process for making *banten* elements is achieved to through the ergo-mechanical approach.
2. The equation used in the ergo-mechanical approach is as in equation 4 above:

$$P = \frac{O}{(I_1 \times t_1)(I_2 \times t_2)}$$

## REFERENCES


Adiputra, N. 1998. Ergonomic Methodology. Master-Work Programs Ergonomic Physiological. London: Postgraduate Courses Ergonomic and Physiology Exercise Physiology University of Udayana.

Manuaba, A. 1996. Ergonomics and Exercise Physiology for Human Development and Whole Community Indonesia. Denpasar: Udayana University Graduate Program.

Manuaba, A. 2006. Total Approach Is A Must For Small And Medium Enterprises To Attain Sustainable Working Conditions And Environment, With Special Reference To Bali, Indonesia. Industrial Health. January 2006; 44(1): 22–26, National Center for Biotechnologi Information (NCBI). U.S. National Library Of Medicine And National Institute For Health. [Cited 2007 April 21]. Available From: URL: http://141.99.140.157.d/aws/index.gov

Sutjana, I.D.P. 2000. Application of Ergonomics Improve Productivity and Livelihoods of Communities. Scientific Orations Inauguration Udayana University Professor of Physiology, 11 November.

Swastika, P.F.M. 2010. Mepandes (Cut Teeth), the First Matter. Denpasar.Kayumas Court.

Wijayananda, M.J. 2003. Tetandingan lan Sorohan Banten. Surabaya: Paramita.

*Ergonomics in Asia – Shih & Liang (eds)*
*© 2012 Taylor & Francis Group, London, ISBN 978-0-415-68414-9*

# Intervention in the stamping process has improved work quality, satisfaction and efficiency at PT ADM Jakarta

Titin Isna Oesman
*Industrial Engineering Department, Institute of Science & Technology, Akprind Yogyakarta, Indonesia*

Sudarsono
*Mechanical Engineering Department, Institute of Science & Technology, Akprind Yogyakarta, Indonesia*

I. Putu Gede Adiatmika
*Ergonomic Postgraduate, Udayana University, Indonesia*

ABSTRACT: The body component of a car is the finished product component that is mainly produced by a large press machine. In the production process of body components of a car in the stamping plant division, the manual task is performed by two operators feeding metal sheets into the press machine simultaneously. After the knobs were pressed, the press machine started to stamp the material sheet and the outcoming product is taken by the next operator. This task is performed repeatedly until the last sheet. It is a manual and comprehensive task that is combined with speed in operating the big press machine involving muscular tension which at the end could cause muscular complaints and premature fatigue, and which could decrease production and productivity. The current study of ergonomic intervention in the stamping process of body components of cars was conducted in order to determine whether ergonomic intervention could improve the work quality, satisfaction and time efficiency in the stamping process. The subjects of study were 10 persons who were taken randomly. It was designed experimentally with treatment by subject design. The method of data collection was by questionnaire distribution and direct measurement of the subject's condition, time of stamping process and work environment before and after intervention. The Shapiro Wilk test showed that all of the data had normal distribution and was followed by t-paired for data on work quality, work satisfaction, time efficiency, and the environment. The significance level is set at 5%. The research result showed that work quality was improved (muscular complaints were decreased by 6.65%, fatigue was decreased by 5.47% and work boredom was decreased by 5.87%), work satisfaction was improved by up to 6.43%, time efficiency of stamping process was improved by up to 10.7%, production was improved by up to 2.59% and productivity was improved by up to 32.65%. Hence, the study concludes that ergonomic intervention in the production process of car body components in the stamping plant division could improve work quality, work satisfaction and time efficiency.

*Keywords*: ergonomic intervention, work quality, work satisfaction, time efficiency

## 1 INTRODUCTION

In the free trade era, tight competition, complexity and fast change are challenges in the industrial world. The emerging problems and demands are very complex, and must be solved by a total approach. Management ability and individual quality should be improved, with added-value and with changes to the mindset, work culture and organizational structure (Manuaba, 2005).

Competitive capacity could be improved when the human resources are tough, ready to work hard, be collaborative and work according to the smart principle. A very serious attention along with distinct and direct action is needed as an empowerment process, so, work performance is achieved and knowledge can be continually improved (Manuaba, 2005; Sukapto, 2005).

Some of the growing industries in Indonesia are the automotive, financial service, agribusiness and heavy duty equipment businesses. PT ADM is part of the PT X group. The production capacity of PT ADM in the year 2005 was 10,000 cars per month. The production is increasing continuously based on market demand (Anonim, 2004).

One output of the stamping plant is the automobile body as a finished part that is mostly procuced by a big press machine. The production process at the stamping plant is dominated by manual work and this means that the human factor has an important role in the production process. Manual tasks may lead to work accidents and could cause occupational diseases (Agustin et al., 2003; Kroemer & Grandjean, 2000).

In the production process, the operators are involved in taking, lifting and putting the metal sheets into the big press machine repetitively. Using arms and hands involved in repetitive manual tasks continuously and combined with work speed in using the industrial equipment may cause muscle tension that could produce muscular complaints and premature fatigue (Manuaba, 2005; Kroemer & Grandjean, 2000).

One of the protection efforts for the operators from the risk factor in working is improvement of working conditions through ergonomic intervention. The ergonomic principle was applied specifically through application of a total ergonomic approach. The total ergonomic approach was conducted integratedly, by combining the SHIP (Systemic, Holistic, Interdisciplinary and Participatory) approach and Appropriate Technology (Manuaba, 2005).

Based on the principle of fitting the task to the man, a harmony between human, product, machine, method and environment suitable for the ability, skill and limitations of humans will create an optimum product quality (Manuaba, 2005). In order to create a healthy, safe, comfortable and efficient working condition and environment as well as high productivity, a functional maximal and optimal utilization of the human body is required (Kroemer & Grandjean, 2000).

Problem identification based on 8 ergonomic problems in the 2 A Production Line showed that nutritional intake while working was not yet optimally implemented. 73.3% of operators reported being thirsty after working and their weight declined 0.71 kg after work. The upper part of the body was not in a natural position for the task assigned; while the tasks were repetitive and monotonous with taking, lifting and placing the metal sheets with a weight of 4–8 kg. The work position was not varied, as legs and feet supported the body statically and the work was done by standing for a long time. The temperature of the environment was 31.5°C–33°C, relative humidity (Rh) was 56%, and noise was 97.5 dBA.

The schedule of short break or active break time had not been implemented. The social condition of operators showed a lack of opportunity for social interaction among colleagues. The information condition was not well arranged, and interaction of human-machine/tool occurred when pushing the push button knob that was located behind the operator's body. These data were strengthened by secondary data from the PT ADM clinic which reported that musculoskeletal disorders were the most common among operators who seek treatment in the clinic. Indeed, the above eight ergonomic problems need a holistic evaluation (Manuaba, 2005; Djestawana et al., 2002).

It is suspected that the boredom that is felt by operators in the stamping plant division was caused by monotonous and repetitive tasks. If this condition could not be solved, this could bring a negative impact on the work quality and work satisfaction. In addition, time efficiency was not optimal due to the impact from musculoskeletal complaints and fatigue.

Therefore, the current study was conducted to determine the contribution of ergonomic intervention toward operators' work performance in the division of the stamping plant of PT.

ADM Jakarta. It is expected that the bad work body posture and the static/continuous muscle contraction could be corrected; by so doing the work quality, life quality, and the work productivity could be improved.

## 2 MATERIALS AND METHODS

The study was an experimental one, using treatment by subject design. It was held at 2 A Line of the Stamping Plant Division in PT ADM in Jakarta in March–April 2008. The sample was 10 persons who were randomly chosen from the big machine operators.

The research object was a push button on the 2 A Line. The object was redesigned because its position was not appropriate and its size did not fit with operator's anthropometric. The redesigning was followed by changing operator's work position on die shifting, arranging active break times and adding nutrition intake. The shape and dimension of the former push button (the old one) was rectangular with a length of 500 mm, width 140 mm, thickness 100 mm and height was adjustable (Figure 1). The modified push button (the new one) was U-shaped, the inner length was 610 mm and outer length was 810 mm, width was 200 mm as well as 100 mm, thickness was 85 mm and the height was adjustable. The other redesigning change was in the position of the emergency stop button (Figure 1). The old emergency stop button was located in the middle while the new one was located on the right side just behind the push button.

The chosen sample worked as they used to (before applying ergonomic intervention) and then worked with ergonomic intervention. Work quality was assessed based on the workload (resting heart rate, working heart rate), musculoskeletal complaints (Nordic Body Map questionnaire), fatigue (Scale Rating Questionnaire of 30 items), and work boredom (work boredom questionnaire). Work satisfaction was rated by using 20 items of Minnesota Satisfaction Questionnaire (MSQ) and the time efficiency was a comparison of the process time taken to produce one body component both before and after applying ergonomic intervention, measured using a stop–watch. Work environment conditions were recorded in terms of noise with a sound level meter, temperature with a WBGT–meter, relative humidity also with a WBGT- meter, and illumination intensity with a lux meter. The collected data were tested for normality by using a Shapiro Wilk analysis and t-paired test for data of quality, work satisfaction and time efficiency, while the environment data were analysed by using t-group with a significance level of 5%.

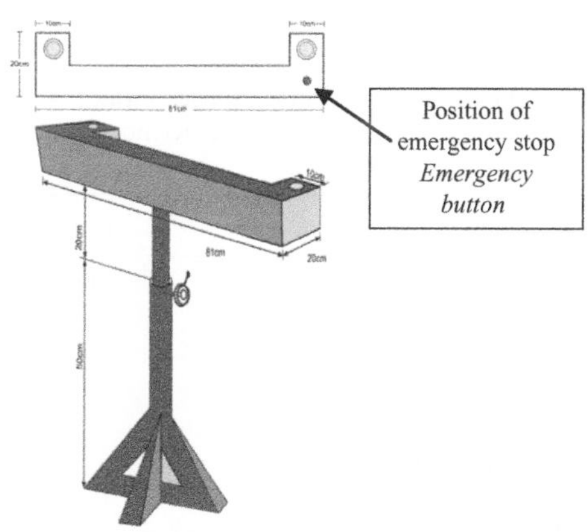


Figure 1. The new one (after redesign).

## 3 RESULTS

### 3.1 *Subject characteristic*

At the average age of the sample group, a person has optimum muscular and physical strength capacity to work and a productive age that supported the study (Table 1). Operators were in a healthy condition to do their physical work as their average blood pressure was in normal condition. A person's physical fitness level can be shown by his resting heart rate.

*Operator's Work Condition Before and After Ergonomic Intervention.*

Measurement results on the operator's work condition without and with ergonomic intervention and its economic analysis is shown in Table 2.

As shown in Table 2, the average of working heart rate and working pulse showed no significant difference before and after intervention ($p > 0.05$). However, the average of work boredom score, average of fatigue score differences, musculoskeletal complaints score differences, work satisfaction score, process time differences and productivity before and after intervention were all significantly different ($p < 0.05$).

*Operator's Environment Condition Before and After Ergonomic Intervention.*

The operator's environmental condition without and with ergonomic intervention in the study is presented in Table 3.

As shown in Table 3, the average level of noise before and after intervention was not significantly different ($p > 0.05$). However, the average of WBGT, air temperature, relative humidity and illumination before and after intervention was significantly different ($p < 0.05$).

Table 1. Operator data at 2 A line of stamping plant division PT ADM Jakarta ($n = 10$).

| No | Parameter | Average ± SB |
|---|---|---|
| 1 | Age (years) | 21.9 ± 2.23 |
| 2 | Systolic blood pressure (mmHg) | 107.0 ± 6.75 |
| 3 | Dyastolic blood pressure (mmHg) | 68.0 ± 6.32 |
| 4 | Resting heart rate (bpm) | 78.0 ± 11.43 |
| 5 | Body weight (kg) | 58.6 ± 4.62 |
| 6 | Body height (cm) | 171.7 ± 4.43 |
| 7 | Hip + wrist width (cm) | 47.25 ± 2.03 |
| 8 | Standing elbow height (cm) | 106.3 ± 3.29 |
| 9 | Body mass index (BMI) | 19.88 ± 1.52 |
| 10 | Work experience (years) | 1.83 ± 1.61 |

Table 2. Working heart rate, working pulse, work boredom, fatigue, musculoskeletal complaints, work satisfaction, efficiency of time and economic analysis.

| No. | Parameter | t | p |
|---|---|---|---|
| 1 | Working heart rate (bpm) | 1.96 | 0.08 |
| 2 | Working pulse (bpm) | 2.11 | 0.06 |
| 3 | Work boredom | 2.17 | 0.02 |
| 4 | Fatigue | 2.334 | 0.044 |
| 5 | Musculoskeletal complaints | 2.330 | 0.045 |
| 6 | Work satisfaction | −3.00 | 0.01 |
| 7 | Process time (minute) | 10.33 | 0.00 |
| 8 | a. Productivity | 3.00 | 0.00 |
| | b. Profit (Rp) | | |
| | c. BCR | | |

Table 3. Work environment conditions at 2A Line of stamping plant division in PT ADM Jakarta.

| | Before | After | |
|---|---|---|---|
| Parameter | Mean ± SD | Mean ± SD | p |
| Noise (dBA) | 99.72 ± 4.18 | 100.84 ± 2.24 | 0.71 |
| Air temperature (°C) | 26.19 ± 0.48 | 27.11 ± 0.47 | 0.00 |
| | 28.68 ± 0.52 | 30.80 ± 0.69 | 0.02 |
| Relative humidity (°C) | 68.48 ± 2.36 | 59.28 ± 4.33 | 0.04 |
| Illumination (lux) | 195.30 ± 39.30 | 271.27 ± 31.47 | 0.03 |

Table 4. Contribution of work environment to NBM, fatigue, work boredom, work satisfaction and processing Time at 2 A line of stamping plant division in PT ADM Jakarta.

| Parameter | $R^2$ Value |
|---|---|
| NBM | 0.126 |
| Fatigue | 0.030 |
| Work boredom | 0.331 |
| Work satisfaction | 0.272 |

The regression analysis showed that work environment contributed to the NBM by 12.6%, fatigue was 3%, work boredom was 33.1% and work satisfaction was 27.2%.

## 4 DISCUSSION

### 4.1 *The effects of ergonomic intervention*

Redesigning the push button resulted in a change in work posture that previously was not in-natural position. The operator's upper extremity position was in abduction position, away from the body with fore-arm and back-arm making a 111° angle (Figure 2b). After using the new design, the operator's upper extremity posture was in-natural position, because the redesigned push button is located just to the left and right side of the operator's body. The change permits the fore-arm and back-arm to make a wider angle, i.e., 151° (Figure 2a) or nearly parallel with body axis. Indeed, this posture reduced the load on the shoulder joint and the muscles while operating the redesigned push button.

The emergency stop button was positioned at the bottom right side of the push button because most of the operators were using their right hand (right handed). This improvement had significantly reduced the musculoskeletal complaints, fatigue and work boredom ($p < 0.05$), hence, this means that the work quality was improved. This result was followed a reduction of musculoskeletal complaints by 6.65%, fatigue by 5.47% and work boredom by 5.87% which is also supported by secondary data from the clinic in PT ADM which reported an 80% reduction in complaints of musculoskeletal disorders among the operators who visited the clinic.

This is in compliance with the previous study (Sukapto, 2005; Anonim, 2004) which stated that the usage of a hand wheel with a vertical handle on the pressing tool of a coconut grating machine could reduce the musculoskeletal complaints to 23.22%.

However, the operator's workload (working heart rate and working pulse) is used as an indicator of a person's physical load, or the workload that affects the cardiovascular system, and that was not significantly different ($p > 0.05$). It is predicted that this condition originated from the light category of the workload, and the work mechanism before and after

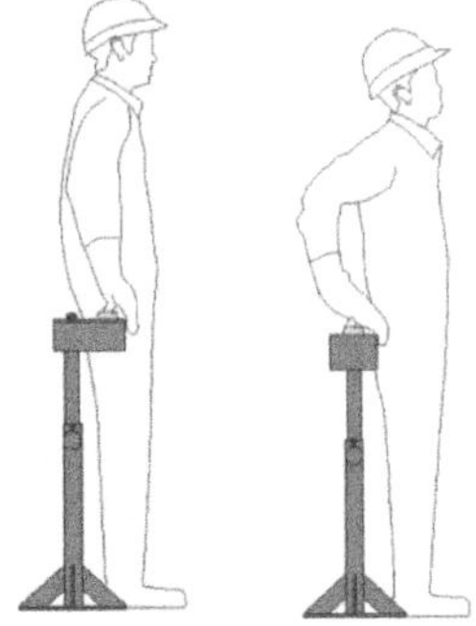

Figure 2a. Work position with the new one, 2b. Work position with the old one.

intervention was the same from the fact that manual tasks were improved and yet the machine tasks were still the same.

Musculoskeletal complaints, fatigue and work boredom were reduced because of the new improved work system. Ergonomic intervention was also applied by: (1) arranging the work position. Arranging operator's work position during work time was carried out by balancing workload between the left and the right body sides. Work position management could reduce musculoskeletal complaints and fatigue, and could reduced the boredom among the workers; (2) arranging work organization. Arranging work organization influenced operators' work. Arranging work organization was carried out by changing dies in the active break time for 5 minutes because changing dies was carried out by another operator. Active break time was used to have a break (drinking and eating snacks) that are prepared by the company. Indeed, it was found that activity changing reduced the boredom, particularly for long duration working time.

This result was in accordance with a study that stated that adding active short breaks and Balinese pop songs could reduce the muscular complaints and boredom as well as increase work productivity (Djestawana et al., 2002).

The operators at Line 2A of Stamping Plant division in PT. ADM Jakarta had been exposed to good experience, which was manifested by an improvement in work satisfaction while working with the modified push button. The such modified push button has decreased the musculoskeletal complaints and fatigue and finally influenced work satisfaction.

Ergonomic intervention that used a total approach permits all operators to participate in the process of push button redesigning. Thus, the opportunity had a great influence on operators' satisfaction.

### 4.2 *The working environment conditions*

The working environment conditions that were considered in the study were noise intensity, WBGT, air temperature, relative humidity, ball temperature and lighting.

The result of the study showed that there was no significant difference in noise while working with the old and the new design push button ($p > 0.05$). Consequently, an ear protection program was implemented by using ear plugs that coud reduce the noise to about 25 dBA.

The measurement result of air temperature, humidity (before and after intervention) was in the tolerable category. Accordingly, work in the light work category was in an air temperature of between 33°C to 35°C and the moderate work category had an air temperature range between 29°C and 31°C. Air temperature during working with the former push button and modified push button was still in the tolerable level that could be accepted by operators without experiencing health complaints. It is generally accepted that Indonesians are able to acclimatize well in an air temperature between 29–30°C with air humidity between 85–95%.

Measurement results of illumination levels (before and after intervention) was good compared previous researcher's category. According to the statement that the minimal illumination needed to carry tasks well is about 200 lux, the environmental condition of illumination on the former push button and modified push button was considered in to be in the good category (Hartono, 2004; Tayyari & Smith, 1997).

Some work environmental parameters in the study had significant statistical differences and this will influence the result of the study. The regression analysis found that work environment contributed to the result of NBM of 12.6%, fatigue 3.0%, work boredom 33.1% and work satisfaction 27.2%. With such a contribution to the study result, the reduction which was related to design factors on musculosceletal complaints was 6.65%, on fatique reduction 5.47%, on work boredom reduction 5.87% and work satsifaction was increased up to 6.43%.

Such results were evoked by the fact that before intervention was applied it was the rainy season (March) and during intervention was the dry season (April). The season difference could not be controlled for its natural condition so that work environment had contribution toward the study result.

### 4.3 *Production output and productivity*

The product produced by PT ADM Jakarta was a car body component that was a work-in-process item. The study being examined was a chasis with serial W 1032 of dimensions: length = 1830 mm; width = 340 mm; thickness = 1.4 mm and weight = 6.837 kg.

Operator's work productivity in the study was seen from the length time of 1 (one) part body component, starting from placing a metal sheet for stamping using a big press machine. Process duration for each part is $2.17 \pm 0.12$ seconds (before intervention) and $1.96 \pm 0.09$ seconds (after intervention). Based on the statistical test, there was a significant difference in the time process ($p$ value $< 0.05$). This indicated that work activity using the modified push button could decrease the processing time and improve operator's work productivity by up to 32.65% and the production result was improved by up to 2.59%. The height of the push button fits the height of the hand when the operators stand up so that the arm position is straight to the push button.

These are the efforts made to make the operators healthier, safer, more satisfied, effective and efficient so that they were more productive. It is in accordance with the statement that ergonomic application could increase work productivity and people's health (Surata, 2001).

### 4.4 *Time efficiency*

The study found that the reduction in manual time processing and efficiency of time processing was 10.7%. The reduction of time processing was caused by changing the push button position and size where it is positioned at a flat surface which previously located in vertical surface behind the operator's body that needs more time to reach.

With the processing time reduction, unnecessary movements were omitted so that it was more efficient and effective and the time taken was shorter. An efficient and effective movement means that work is more productive, muscle complaints are decreased so that the operator's life quality is also improved. This is in accordance with previous research that improvement on productivity means improving efficiency of production process time.

## 5 CONCLUSIONS

Based on the results of the study, we conclude that total ergonomic intervention has improved work quality, work satisfaction, time efficiency, production and productivity, reduced processing time, and eventually increased the company's profit.

## 6 SUGGESTIONS

The current study on ergonomic intervention coupled with a total ergonomic approach should be broadened to include other variables such work environment.

## REFERENCES


Adiatmika, I.P.G. 2007. Perbaikan Kondisi Kerja Dengan Pendekatan Total Menurunkan Keluhan Muskuloskeletal dan Kelelahan Serta Meningkatkan Produktivitas Perajin Pengecatan Kerajinan Logam di Kediri—Tabanan. Disertasi. Program Doktor Ilmu Kedokteran Program Pascasarjana Universitas Udayana, Denpasar.

Agustin, E., Tjitro, B. & Prawira, Y.W. 2003. Perancangan Fasilitas Kerja yang Ergonomis Pada Departemen Produksi dan Departemen Pengerolan di CV. Jaya Plastindo Raya, Semarang. Dalam: Nora Azmi, dkk. editor. Prosiding Seminar Nasional Ergonomi 2003 "Ergonomi dalam Desain Produk dan Sistem Kerja". Jakarta, 9–10 April, halaman 175–181.

Anonim, 2004. Laba Bersih Astra Melonjak 44,6% Mencapai Rp. 2,6 trilliun Pada Semester Pertama 2004, Majalah Astra, Edisi 04 Juli-Agustus Tahun XXXIV, halaman 5–6.

Christensen, E.H. 1991. Physiology Of Work. In: Parmeggiani, L Editor Encyclopedia Of Occupational Helath And Safety. Third (revised) Edition. Geneva: ILO.

Djestawana, I.G.G., Tirtayasa, K. & Adiputra, I.N. 2002. Intervensi Ergonomics Pada Proses Manggur Mengurangi Beban Sistem Kardiovascular dan Keluhan Muskulosekeletal Perajin Gamelan Bali. Jurnal Ergonomi Imdonesia (The Indonesian Journal Of Ergonomics), Vol. 3 No. 2 Desember 2002, halaman 59–66.

Hartono, W. 2004. Hubungan Antara Sikap dan Posisi Kerja Anggota Tubuh Dengan UEWMSDs Pada Perajin Rotan di Perusahaan "X". (tesis). Jakarta: Universitas Indonesia.

Kroemer, K.H.E. & Grandjean, E. 2000. Fitting The Task to The Human Fifth Edition A Textbook of Occupational Ergonomics. U.K. Taylor & Francis.

Manuaba, A. 2005. Pendekatan Total Ergonomi Perlu Untuk Adanya Proses Produksi dan Produk Yang Manusiawi, Kompetitif dan Lestari. Makalah. Dipresentasikan pada Seminar Nasional 2005 Perancangan Produk "Collaborative Product Design" Jurusan Teknik Industri Atmajaya Yogyakarta, 16–17 Februari. Halaman 385–392

Sukapto, P. 2005. Peran Personalitas Dalam Membentuk Tim Yang Dinamis Dalam Pengembangan Produk Baru Yang Tepat Guna (Studi Kasus Di Industri Otomotif). Dalam: Isa Setiasyah Toha, Sritomo, W., Benedictus, E., Heri, S., Desi, K., Editor. Proceedings Kongres BKSTI dan Seminar Nasional Teknik Industri IV Palembang, 24–25 Juni.

Surata, I.W. 2001. Penggunaan Roda Tangan Berhandel pada Alat Pres Parutan Kelapa Mengurangi Keluhan Sistem Muskuloskeletal dan Meningkatkan Produktivitas Kerja Pembuat Minyak Kelapa. Jurnal Ergonomi Indonesia (The Indonesian Journal Of Ergonomics), Vol. 2 No. 2 Desember 2001, halaman 54–62.

Tayyari, F. & Smith, J.L. 1997. Occupational Ergonomics Principles and Applications. New York: Chapment & Hall.

Wulanyani, N.M.S. 2003. Pemberian Istirahat Pendek Aktif dan Lagu Pop Bali Menurunkan Keluhan Otot Skeletal dan Kebosanan serta Meningkatkan Produktivitas Pelinting Kertas Rokok. Jurnal Ergonomi Imdonesia (The Indonesian Journal Of Ergonomics), Vol. 4 No. 2 Desember 2003, halaman 73–77.

*Ergonomics in Asia – Shih & Liang (eds)*
*© 2012 Taylor & Francis Group, London, ISBN 978-0-415-68414-9*

# Practice of reducing ergonomic hazards in a display panel plant—experience sharing by NB CMI

Hung-Kai Huang, Pin-Hsuah Lee, Pei-Hsun Lin, Stella, Wan-Jun Yang, Cho-Fan Hsu, Chin-Lien Tsai, Alice & Yung-Fen Lin
*Chimei Innolux Corp. Tainan Site SHE Department, Taiwan*

Feng-Ren Wu, Dan-Dan Qiu & Chong-Yan Xu
*Ningbo Chi Mei Optoelectronics Ltd., China*

ABSTRACT: The TFT-LCD industry continues to increase rapidly. The assembly process is speeding up whether in Taiwan or in overseas factories. Therefore, there are a number of risk factors for ergonomic hazards that we need to consider. We established an ergonomic self-management group in our panel industry cooperation. In order to help the plants overseas adopt ergonomic hazard evaluation and improve systems, we share experiences and methods of evaluating the risk factors of ergonomic hazards and improve strategies with them. The research includes four parts: (i) Key indicators method (KIM), (ii) the program for carrying designed by IOSH, (iii) basic training for employees, (iv) improving strategies. The results of the research should not only be adopted for designing the work station and for developing self-management groups in factories overseas, but would also provide relevant high-tech electronic industries with methods of evaluating the factors of ergonomic hazards.

*Keywords*: KIM, working analysis, ergonomic hazard, self-management

## 1 INTRODUCTION

In recent years, the TFT-LCD industry has increased rapidly (PIDA, 2008). The assembly process is speeding up whether in Taiwan or overseas factories. Therefore, there are a number of risk factors for ergonomic hazards that we need to consider. Moreover, we are actively preventing our workers being injured from repeated movements at work which would cause muscle aching or lower-back pain.

CMI established an ergonomic self-management group in our panel industry cooperation, in order to help the plants overseas adopt ergonomic hazard evaluation and improve systems, and share experiences and methods of evaluating risk factors for ergonomic hazard and improve strategies with them.

## 2 OBJECTIVE

The process of the research goes from investigating the workplace; using tools to assess the underlying human factors of ergonomic hazards, and then, based on the features of the manufacture process, to providing proper improving strategies.

The objective of the research is to evaluate and improve human factors of ergonomic hazard in order to help workers in overseas plants.

Since there are limitations on space and productivity, most of the workers are standing or walking, so that means of preventing exhaustion among workers due to long periods of working is needed. Workers who keep an unnatural working position or stay in an unsafe working environment with long-time repeated work are liable to induce muscle, skeletal and nerve trauma, also called cumulative trauma disorder (CTD) or musculoskeletal disorders (MSDs). With the intention of preventing the disorders above, ergonomic engineering here is extremely valued.

NINDS in July 2003, showed that Americans spend at least $50 billion each year on treating low back pain (LBP), the most common cause of job-related disability and a leading contributor to missed work (NINDA, 2003).

According to NIOSH (1997) *Musculoskeletal Disorders and Workplace Factors* several studies suggested that both lifting and awkward postures were important contributors to the risk of low-back disorder (NIOSH, 1997).

Sean Gallagher (2008) found that severe pain may result in a severe handicap or may preclude performance of an activity altogether. Further, he said that LBP is a complex problem that involves both factors that cannot currently be controlled (such as certain effects of age) and factors that can be changed or influenced (such as the design of jobs, equipment, and work culture). We can try to reduce exposure to physical risk factors to prevent initial onset of LBP and design work to better accommodate workers with LBP (Sean Gallagher, 2008).

Due to the fact that working postures often do not conform to ergonomics, many workers suffer from waist pain and low back pain. This situation exists in every industry. It is found that improper postures, overexertion, and highly repetitive movements are the main causes for the cumulative damage cumulative trauma disorders, CTDs) (I-Fan Lin, 2003).

In recent years, symptoms of chronic occupational diseases which are due to CTDs, are taken more and more seriously. For the worker, discomforts and lower workability are minor problems, but inability of limbs' mobility could occur, and this affects effect quality of life. For the company, productivity and efficiency would be lower, with loss of working time, and moreover, the company would get a bad image from the increasing chronic occupational disease cases.

For the reasons above, we established an ergonomic self-management group in our panel industry cooperation. In the factories in Taiwan, we invited professionals to join our ergonomic improvement project. In order to help the plants overseas adopt the ergonomic hazard evaluation and improve systems, we share experiences and methods of evaluating risk factors for ergonomic hazards and improve strategies with them.

## 3 METHODS

The study team included occupational safety engineers, health engineers, nurses, the occupational medicine physician, and a physiotherapist.

This study intended to use the fastest, easiest method with the lowest cost and highest effects to investigate the relation between the wrong working postures, and also aimed to find ways to help them.

The research included five parts: (i) Key indicators method (KIM), (ii) the program for carrying designed by IOSH, (iii) the basic training for employees; we investigated the working environment for basic ergonomic hazards and high risk manufacturing process, such as measuring the working platform, analyzing angles of working position, and counting frequency and weight of panel-carrying, etc. (iv) improving strategies; which includes: (1) correcting the working position, and promoting the muscle stretch exercise in the whole company, (2) provide tools to help reduce discomfort at work and increase working efficiency, (3) change the position to reduce waist pressure. (v) survey the employees about the improvements for reducing muscle-skeletal symptoms after practicing all the improving strategies above.

### 3.1 *Research procedure*

Our ergonomic team used multiple research methods to assess the human risk factors, complete self evaluation and carry out improving methods. We also built up the system for overseas plants, held ergonomic seed camps, established an ergonomic hazard assessment mechanism for them and investigate the working environment together.

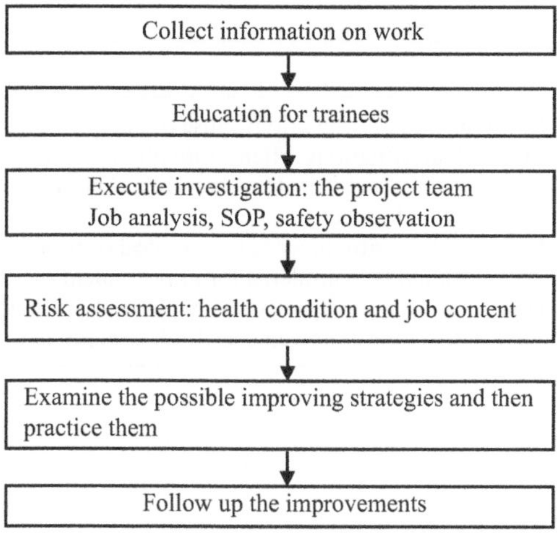


Figure 1. Research flowchart.

After having a conference with the higher level executives, we got their support to work on possible improving strategies. The research process was as follows:

### 3.2 *Subjects*

This study evaluates every part of the module manufacturing process toward the overseas plants, and we focused on investigating the employees in the materials preparation area of module B.

## 4 RESULTS

### 4.1 *KIM index*

The key indicators are quantified in individual scales. The scales correspond to the conditions encountered in practice and range from minimum/optimum to maximum/poor. The classification on these scales gives an indication of any load bottlenecks. A total can be obtained by multiplying the scale value for the daily duration by the sum of the other scale values. A risk assessment can take account of additional work-organizational and individual indicators. KIM is used to evaluate the ergonomic risk by active indexes which include the weight of object, posture, working situation and duration of work etc.

The risk of musculoskeletal disorders would increase with a rising score. The risk value would be calculated by formula (1), and the meaning would be explained with Table 1. The KIM risk assessment method (Steinberg, Behrendt & Caffier, 1994), was adopted to analyze the working situations.

$$(\text{Carrying Weight} + \text{Posture Type} + \text{Working Conditions}) \times \text{Times} = \text{Risk points} \quad (1)$$

### 4.2 *Analyzing working procedures*

To analyze workers' postures, angles, carrying weight, etc. in the interviews, the workers described their working procedures in the materials preparation area. We also adopted the operation analysis, safety observation, and KIM risk assessment. Operation analysis is broken down a work into a few steps, for individual's work, must be completed steps. Then, each operation which people are doing with risk factors, was investigated and identified (IOSH, 1998). Furthermore, operation analysis was used to analyze the operating frequency and

Table 1. Risk level.

| Risk level | Rating points | Interpretation |
|---|---|---|
| 1 | <10 | Low load situations: seldom cause physical overload. |
| 2 | 10≦<25 | Middle load situations: often cause physical overload to those who have weak recovery ability. This level should be considered in redesigning job content and tools. |
| 3 | 25≦<50 | Increased load situations: cause physical overload to a normal worker. This level should be considered for improvement of job content and tools. |
| 4 | ≧50 | High load situation: causes highly physical overload. This level should be considered for improvement of job content and tools immediately. |

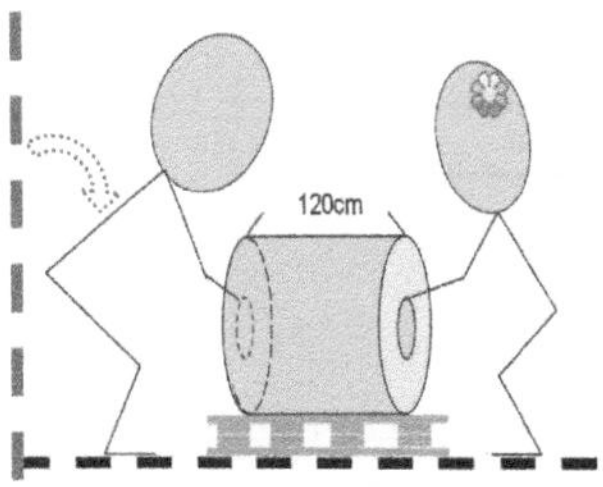


Figure 2. Before.

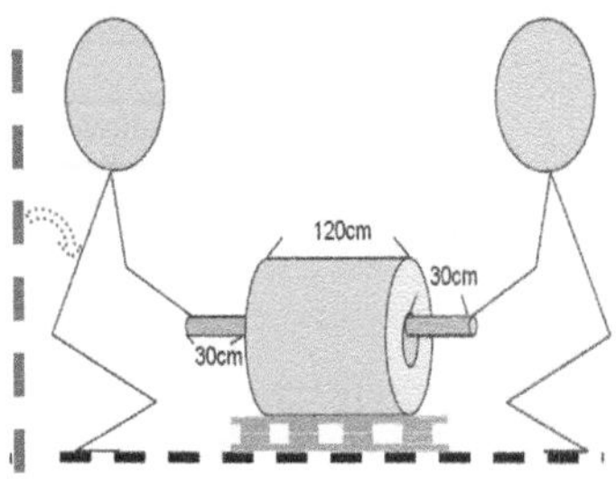


Figure 3. After.

weight for the work of carrying, cleaning, and assembling in order to identify the key risk factors for health.

Scott Cook and Terry E. McSween suggested the importance of both formal and informal leader involvement in behavioral safety initiatives (Cook and McSween, 2000). The Safety Observation method was employed, including video recording, picture taking, and data recording. The safety observation method was used to analyze operating positions and moving angles. The results are shown in Figures 2 and 3.

### 4.3 *MMH index*

We estimated the compression stresses on L5/S1 forces by IOSHMMH (Yi-Shyong Ing, 1999).

The compression stresses on L5/S1 of the employees in the materials preparation area of module B for the lifting task were larger than for other tasks, implying that the risk of musculoskeletal disorders would be higher in the lifting tasks. The MMH (manual materials handling) indexes are shown in Table 2.

### 4.4 *Health exercise*

We introduced a three-minute health exercise everyday after lunch break to extend muscles, release working stress and prevent CTDs. All workers in overseas plants worked out daily, as shown in Figure 4.

### 4.5 *Conclusions*

Based on the data above, the working position observed in this station has a higher ergonomic risk by carrying heavy weights and wider bend-over angles. After the assessment, the team could understand more of the ergonomic hazard risks of the working style. Moreover, we could improve them:

Figure 4. Health exercise in overseas plants.

Figure 5. Workers bending position.

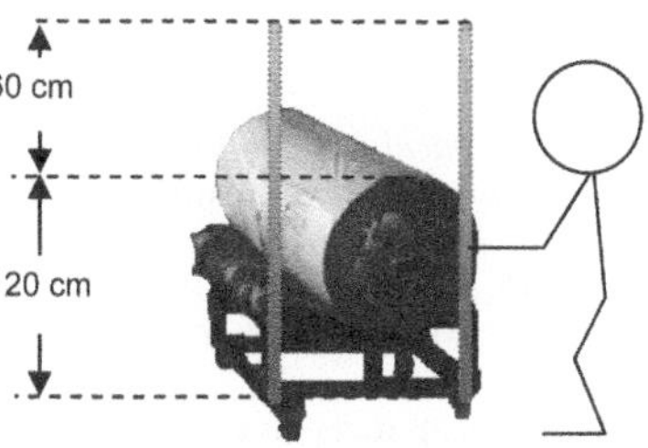


Figure 6. Fixture diagram.

Table 2. MMH index.

| | L5/S1 load (N) | | Back to Hip angles (°) | |
|---|---|---|---|---|
| Situation | Men | Women | Men | Women |
| Before | 8205 N | 7643 N | 80° | 85° |
| After | 6295 N | – | 90° | – |

*1. Hip angle means the angle between back and thigh which is observed from the side.

1. By holding training for preventing ergonomic hazards to show the workers how to identify the risks.
2. By educating workers when two are working together, taking care with the differences of height, weight, sex and body build between them and try to find the balance between them.
3. By introducing health exercises to train major muscles and release stress. At present, the exercise is followed all around the overseas plants.
4. By installing the hardware to change the carrying position which caused workers to bend over a wider angle >30° (as in Figure 5) and also prevent injury of the hands. We provided

a fixture of a wheel stand (as in Figure 6) to avoid over-bended posture of workers in the carrying procedure.

## 5 DISCUSSION

Our ergonomic self-management group followed simple, effective, low-cost, and high profit rules to investigate the risks of ergonomic hazards in the workplace. The result of the research would not only be adopted for designing the working station and developing self-management groups in factories overseas, but also provide relevant high-tech electronic industries with methods of evaluating the factors of ergonomic hazard.

Besides protecting our employee's health on both the physical and psychological levels, the research would raise the production efficiency and enhance the wellbeing of workers in the TFT-LCD industry.

## ACKNOWLEDGEMENTS

We would like to give special thanks to General Director Geng-Rong Xu, Directors of manufactories: Shou-Yi Chen, Kun-Yu Tsai, Jun-Yan Chen, Wen-Zhang Lin, Le-Nian Tan, Zheng-Zhang Guo, Zhi-Hong Huang, Cai-Mao Chen, Zong-Xian Tsai, colleagues of Safety & Health Department of Ningbo Chi Mei Optoelectronics, and all the trainees of the ergonomic seed camp as well. Because of all of your support and help, the project could proceed smoothly in the plants.

## REFERENCES

Cheng-Kung & Cheng. 1999. *Biomechanical Evaluation and Application Study—Biomechanical Model of Low Back Injury in Field Improvement*. Institute of Occupational Safety and Health. IOSH88-H123.

Chih-Ming Chen & Yay-uan Xu. 2009. *Evaluation of A Health Promotion Program for Drivers Deliverer In Domestic Home Delivery Company*. 2009. Institute of Occupational Safety and Health. IOSH97-M317.

Chih-Yong Chen, 1999. The Influence of Back Belts on Lifting Trajectory. IOSH.

Council of Labor Affairs. Labor Insurance Person-time Payment of Occupational Low Back Pain: 2001–2009. http://statdb.cla.gov.tw/statis/webproxy.aspx?sys=210&kind=21&type=1&funid=q08061&rdm=hXgjt95W.

Dipl.-Ing. Ulf Steinberg, Sylvia Behrendt, Dr. SC. med. Gustav Caffier, Research Project F 1994. *Key Indicator Method Manual Handling Operations Design and Testing of a Practical Aid for Assessing Working Conditions*, Dortmund/Berlin/Dresden 2008.

I-Fan Lin. 2003. Repetitive work and poor working postures may cause the "cumulative trauma disease" or "repeated strain injury. *Health World*, 2003 No. 210.

IOSH. 1998. Application Guidebook of Human Factor. IOSH84-T-002.

James, T. Wassell, Ph.D., Lytt I. Gardner, Ph.D., Douglas P. Landsittel, Ph.D., Janet, J. Johnston, Ph.D. Janet M. & Johnston, Ph.D. 2000. A Prospective Study of Back Belts for Prevention of Back Pain and Injury. *JAMA*, December 6, 2000 ol 284. No. 21. http://www.cdc.gov/niosh/jamapapr.html, NINDS. 2003. Low Back Pain Fact Sheet. NIH Publication No. 03–5161 http://www.ninds.nih.gov/disorders/backpain/detail_backpain.htm.

NIOSH. 1997. Musculoskeletal Disorders and Workplace Factors. No. 97–141.

NIOSH. 2007. Ergonomic Guidelines for Manual Material Handling. NO. 131.

PIDA. 2008. Taiwan Ranks No. 1 in world TFT-LCD panel production. PIDA Post. February 2008.

Scott Cook & Terry, E. McSween. OCT. 2000. The Role of Supervisors in Behavioral Safety Observations. A Case Study Examination of an Ongoing Debate *American Society of Safety Engineers*.

Sean Gallagher, Ph.D., CPE. 2008. Reducing Low Back Pain and Disability in Mining. NIOSH. IC 9507.

Sean Gallagher and Christopher A. Hamrick, 1994. NIOSH. *A Scientific Look at Back Belt*, NIOSHTIC-2 No. 10005494. P 44. http://www.cdc.gov/niosh/mining/pubs/pubreference/outputid2127.htm.

Yi-Shyong Ing, Shih-Yi Lu, Chih-Yong Chen & Cheng-Lung Lee. OCT. 1999. Development of Evaluation Diagrams for Manual Lifting Based on Biomechanical Approach.

*Ergonomics in Asia – Shih & Liang (eds)*
 *ISBN 978-0-415-68414-9*

# Human reliability and unsafe behavior factor analysis of chemicals loading process in a dock

G-H. Kao
*Graduate Institute of Occupational Safety and Health of Kaohsiung Medical University, Taiwan*

C-W. Lu
*Department of Industrial and Systems Engineering of Chung Yuan Christian University, Taiwan*

## ABSTRACT

*Introduction*: Because Taiwan is surrounded by sea, its international trade depends on air and sea transportation. Sea transportation is the most popular way to transport chemical goods. The workers who participate in this process include sailors, operators and ship members. This study aims to detect the potential human factors and calculate human reliability within this process.
*Method*: The Hierarchical Task Analysis (HTA) was used to decompose the whole work into units and to analyze the processes; Systematic Human Error Reduction and Prediction Approach (SHERPA) was used to detect the possible human factors in this process. The questionnaire also was used to investigate the workers' experience about program design and the hardware practicability. A group of 40 workers participated.
*Result*: According to HTA, the whole work decomposed into 7 parts: arrivals, loading preparation, pre-loading, loading, pro-loading, leaving preparation, and departures. SHERPA found the most popular error types were action error (49%) and check error (44%). And the questionnaire analysis of the workers' experience about the program design and hardware practicability showed that the workers' knowledge and attitude had a negative relationship.
*Discussion*: The results of HTA and SHERPA indicate that the potential human factors play a huge proportion in this process. The suggestions are that program design could be improved and the workers need reeducation.

## 1 INTRODUCTION

### 1.1 *Background*

International trade in Taiwan depends on air and marine transportation. Marine transportation is the main way to ship chemicals, like petroleum, gasoline, natural gas, and methylbenzene. The GHS stipulate some rules for these chemicals transportation and their physical and chemical characteristics. The transportation process needs staff to operate many kinds of interfaces. These interfaces, like human-software and human-hardware, have some potential disadvantages. The human factors in these disadvantages are action error, management oversight, misjudgment, improper use of protective equipment, error of machine operation, safety inspection insecurity before work, and error of process operation. Most of these factors refer to unsafe movement or behavior.

### 1.2 *Aim*

The chemical industry is an important industry in Taiwan. It plays an important role in other industries and in daily use for transportation, refining, and distribution. There are many kinds of complex and dangerous chemical substances, like raw materials, finished and half-finished products. These toxic, corrosive, inflammable, or explosive substances could cause explosion and fire during transportation and use. Careless manipulation faults during operations in the chemical industry could cause a serious accident. According to the previous serious accidents analysis, we could determine some prevention and protection methods to decrease the risk of accidents and financial loss (Kuo, 2004).

The port of unloading for the Southern Petroleum Company is this study's object. Its main products are gasoline, natural gas, and inflammable materials. These could cause accidents easily when staff load and unload cargos inappropriately. This study's aim is to use reliability methods to analyze human error in modes of loading and unloading processes and utilize these results to determine the staffs' reliability and unsafe behaviors. Furthermore the aim is to establish a database for long-term tracking and to calculate failure rate or reliability analysis. These would be contributive to improving the process's safety and quality. This study used field survey, job analysis, motion analysis, questionnaire survey, and error analysis to determine error type and failure action level, and calculate its failure rate. These would be help to improve human error in the chemical industry and its human reliability validation.

### 1.3 *Subjects*

The subjects in this study were staff and sailors who worked in the unloading port of the Southern Petroleum Company. Their major work was to transport chemicals, like gasoline, natural gas, and alkenes. This study's aim is to determine the potential human error risks during their work and be help to plan the best process.

## 2 METHODS

### 2.1 *Human reliability analysis*

#### 2.1.1 *Hierarchical Task Analysis, HTA*

HTA (Annett, 2004) is the most used method in industrial safety, and can be used to analyze the workers actions. HTA can completely describe the whole work activity. It can be used in many ways, and not only in ergonomics. For example, it also be used in the nuclear emergency system. HTA was aimed at the structure of the operations and to develop a useful hypothesis to solve the efficiency problem. Figure 1 is the functional flow diagram of HTA.

#### 2.1.2 *Systematic Human Error Reduction and Prediction Approach, SHERPA*

SHERPA includes the error mode and the behavior mode, and it can stratify types of operating analysis and classify the operations, it can find the potential human errors or design

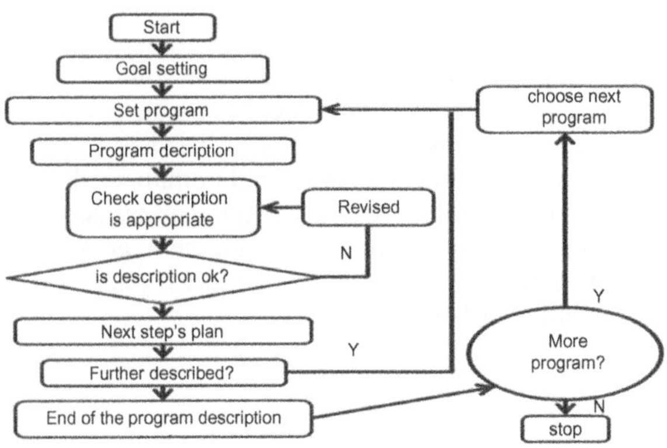


Figure 1. HTA's flow chart.

errors. Due to the references, the precision of the forecast error in SHERPA is the most accurate for the identification of human errors. Figure 2 is the flow chart of SHERPA.

Analysis steps:

1. Hierarchical Task Analysis, HTA.
   Analyze every action step. We collected the information by interviewing the workers, and observed the operation activity.
2. Task classification.
   Classify the operation as "Action" "Retrieval" "Check" "Selection" and "Communication" after having analyzed the operation.
3. Human error identification.
   After having classified the operation, we analyzed the possible error of every act, and described the error actions, and classified the possible error actions by using SHERPA (Figure 3).
4. Consequence analysis.
   Describe the consequence of all the possible error actions.

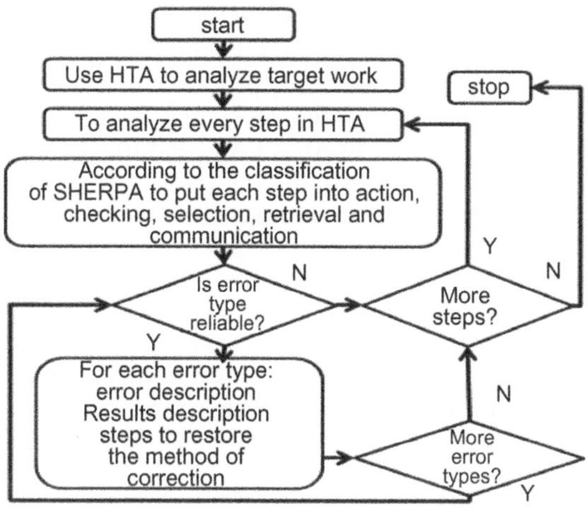


Figure 2. SHERPA's flow chart.

| The error model | Code | Error mode |
|---|---|---|
| Action errors | A1 | Act too long time or too short time |
| | A2 | Act out time |
| | A3 | Act wrong side |
| | A4 | Act too low or too much |
| | A5 | Match error |
| | A6 | Right action on the wrong event |
| | A7 | Wrong action on the right event |
| | A8 | Action omit |
| | A9 | Uncompleted action |
| | A10 | Wrong action on the wrong event |
| Checking errors | C1 | Check omit |
| | C2 | Uncompleted check |
| | C3 | Right check on wrong event |
| | C4 | Wrong check on right event |
| | C5 | Checking out time |
| | C6 | Wrong check on wrong event |
| Retrieval errors | R1 | Misinformation |
| | R2 | Retrieved wrong information |
| | R3 | Uncompleted retrieved information |
| Communication errors | I1 | Information not passed on |
| | I2 | Pass the wrong information |
| | I3 | Uncompleted pass the information |
| Selection errors | S1 | Selection omit |
| | S2 | Wrong selection |

Figure 3. SHERPA error mode.

#### 2.1.3 *SHEL model*

It is developed from the traditional system "human-machine-environment". It can find out the relationship between the workers and another interface, for example: human-hardware interface. After that, we can modify the interface to decrease the weak point of the action.

### 2.2 *The case of the study*

The company is part of the petrochemical business and also is the biggest import and export port of the south part of Taiwan. The workers, including the ship pilots and dock workers, total forty individuals. The study was performed by operate observation, and questionnaire.

## 3 RESULTS

### 3.1 *Basic data*

The basic data showed the workers age, work age and education (Figure 4).

The result from the HTA separated the oil tank loading process into seven parts: 1. Ship into the dock 2. Preloading process 3. Preloading process action 4. Start loading 5. Loading complete 6. Prepare to leave 7. Ship leave the dock (Figure 5). The result from SHERPA showed the error code of the workers: A4.A6.A7.A8.A9.A10.C1.C2.R1.I1, and A8.C1.C2 took a large percentage of the error (Figure 6). The error code classified to "A" for action error, "C" for checking error, "R" for retrieval error, "I" for communication error, and action errors and checking errors took the largest percentage of the error model (Figure 7), and also showed the numbers of the errors (Figure 8).

The result from SHEL model showed that the knowledge of the SOP helped the workers to do the correct action, but the attitude of the workers could lead to negative relations in the operation. The organization helps the workers to get the knowledge, but it got negative reaction to directing the workers operation. The hardware showed the negative rection of to the workers. The behavior of some foreign ship members led to the workers failing to accomplish the operation.

## 4 DISCUSSION AND CONCLUSION

In the whole operation, tasks must be done by all kinds of workers at the dock, including the ship pilot and the dock workers. To ensure their safety and reduce the weak points and increase human reliability there are some points which have been noticed. At first the

| Ratio data | Mean | SD |
|---|---|---|
| Age | 49.53 | 9.11 |
| Work time(year) | 19.62 | 11.99 |
| Category data | numbers(N) | percentage(%) |
| Gender | | |
| Male | 40 | 100 |
| Female | 0 | 0 |
| Education | | |
| Junior high | 0 | 0 |
| Senior high | 25 | 64.1 |
| College | 12 | 30.2 |
| Institute | 2 | 5.1 |

Figure 4. Basic data.

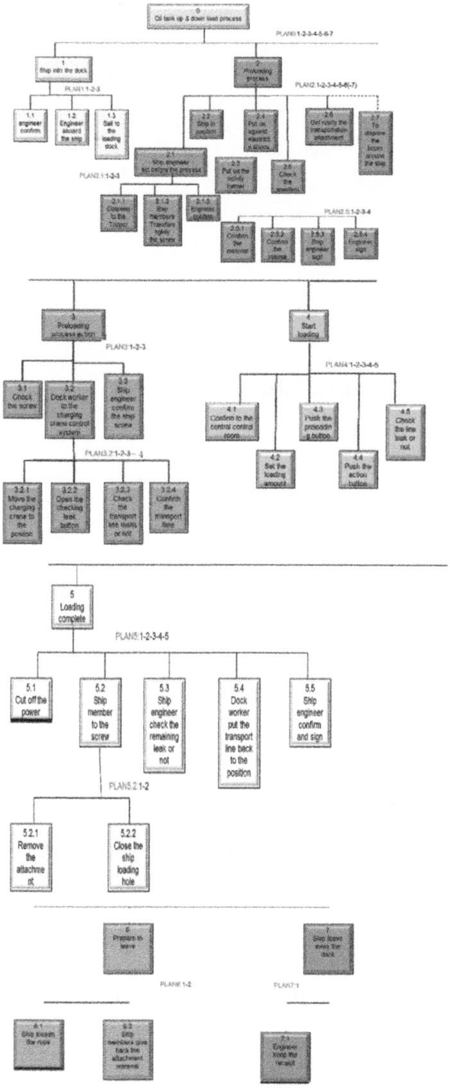


Figure 5. The result from HTA.

| The error code of the workers | Number of error | percentage |
|---|---|---|
| A4.act too low or too much | 55 | 8.6% |
| A6.right action on the wrong event | 10 | 1.6% |
| A7 wrong action on the right event | 34 | 5.2% |
| A8 action omit | 127 | 19.8% |
| A9 uncompleted action | 56 | 8.7% |
| A10.wrong action on the wrong event | 31 | 4.8% |
| C1 check omit | 200 | 31.1% |
| C2 uncompleted check | 80 | 12.4% |
| R1.Misinformation | 23 | 3.6% |
| I1.Information not passed on | 27 | 4.2% |
| Total | 643 | 100% |

Figure 6. The error code of SHERPA.

| Error model code | number | percentage |
|---|---|---|
| Action Errors | 313 | 49% |
| Checking Errors | 280 | 44% |
| Selection Errors | 23 | 3% |
| Retrieval Errors | 27 | 4% |
| Communication Errors | 0 | 0% |
| Total | 643 | 100% |

Figure 7. The error classification model of SHERPA.

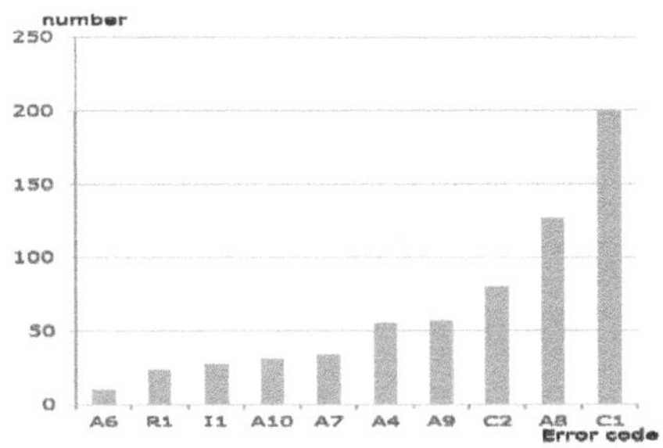


Figure 8. The numbers of errors.

supervisor should remind the operator of the correct action and retrain the workers on a regular basis. Also the hardware at the dock should be checked regularly and some of the hardware should be upgraded to decrease the weak points.

Due to the international business, the operation must be done in cooperation with foreign members, but when the members don't follow the rules this can be the biggest problem in the operation. Our company should advise the foreign company to manage their members appropriately.

## REFERENCES


Annett, J. (2004). Hierarchical task analysis. In: Diaper, D., Stanton, N.A. (Eds.), The Handbook of Task Analysis for Human–Computer Interaction. Lawrence Erlbaum Associates, Mahwah, NJ, pp. 67–82.

Baber, C. & Stanton, N.A. (1996). Human error identification techniques applied to public technology: predictions compared with observed use. *Applied Ergonomics*, Vol. 27, pp. 119–131.

Bertolini, M. (2007). Assessment of human reliability factors: A fuzzy cognitive maps approach. *International Journal of Industrial Ergonomics*, Vol. 37, No. 5, pp. 405–413.

Chang, Y-H. & Yeh, C-H. (2010). Human performance interfaces in air traffic control. *Applied Ergonomics*, Vol. 41, pp. 123–129.

Christensen, J.M., Topmiller, D.A. & Gill, R.T. (1988). Human factors definitions revisited. *Human Factors Society Bulletin*, Vol. 31, pp. 7–8.

Kirwan, B., Scannalit, S. & Robinson, L. (1996). A case study of a human reliability assessment for an existing nuclear power plant. *Applied Ergonomics*, Vol. 27, No. 5, pp. 211–302.

Lasala, K.P. (1998). Human Performance Reliability: A Historical Perspective, *IEEE Transactions on Reliability*, Vol. 47, No. 3, pp. 365–371.

Ruckart, P.Z. & Burgess, P.A. (2007). Human error and time of occurrence in hazardous material events in mining and manufacturing. *Journal of Hazardous Materials*, Vol. 142, No. 3, pp. 747–753.

*Ergonomics in Asia – Shih & Liang (eds)*
*© 2012 Taylor & Francis Group, London, ISBN 978-0-415-68414-9*

# Ergonomics redesign minimized unsafe actions in a wood working workshop

L. Sudiajeng, N. Sutapa, I.G. Wahyu A. & Ngr. Sanjaya
*Civil Engineering Department Baly State Polytechnic, Bali, Indonesia*

I.P.G. Adiatmika
*Department of Physiology, Faculty of Medicine, Udayana University, Bali, Indonesia*

T.I. Oesman
*Department of Industrial Engineering, Akprin-Jogjakarta, Indonesia*

ABSTRACT: Woodworking workshops are full of high technology equipment and are very hazardous. A preliminary study showed that there were inappropriate working conditions, and therefore unsafe actions by workers. An ergonomic redesign was carried out, and research was conducted to assess the value of the new working conditions. Results showed that it had minimized the unsafe actions of the workers and increased their health status. Workers who wore ear plugs, masks, and gloves increased by 63.74%, 85.9%, and 95.66% respectively. The use of bending and squatting postures decreased by 92.24% and 92.51%. Similarly, loss of body weight, working heart rates, MSDs, and general fatigue score decreased by 78.72%, 17.93%, 24.78%, and 22.45% respectively. In conclusion, the workers' behavior was greatly improved, and so was their health status. A program arising from this result should be widely implemented, especially for small-medium scale woodworking industries so as to build a sustainable ergonomics working culture.

## 1 INTRODUCTION

Woodworking workshops are full of high technology equipment and are very hazardous, especially when used improperly. Stock can get stuck in a blade and pull the operator's hand into machine. Employees can be injured when the machine is not properly adjusted or maintained. Besides electrical and mechanical hazards, the physical environment is also dangerous. There are high levels of noise, vibration, and wood dust exposure. The U.S. Department of Labor (1997) reported that there were 393.1 cases among 10,000 full-time workers with 180.5 caused by contact with objects, 89.9 cases of overexertion and other event or exposure leading to injury or illness in lumber and wood product industries. Those hazards are greater if there are unsafe actions by the workers, poor organization, and lack of preventive action. Varonen & Mattila (2000) reported that organization and preventive action have a strong correlation with the accident rate in woodworking industries. Developing an ergonomics working culture is one of the curative actions that could minimize the unsafe actions, accident rate and occupational illnesses and eventually increase the health status and productivity of the workforce (Manuaba, 2000).

## 2 MATERIAL AND METHOD

### 2.1 *Materials*

Research was conducted in the woodworking workshop at Bali State Polytechnic. Participants were 20 students in the Civil Engineering Department, male, 18–24 years old, and with body mass index in the normal category (18.5–24.9).

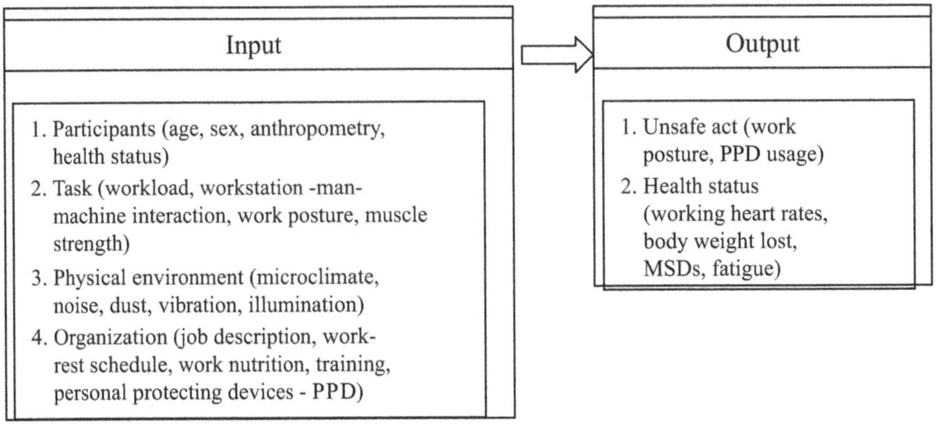


Figure 1. Conceptual framework.

### 2.2 *Method*

The methodology of this research was built based on the conceptual framework as described in Figure 1.

Research was conducted with pre- and post-test control group design. Participants were divided into 2 groups, the first group as control group (CG) performed the task in the original working conditions and the other one as treatment group (TG) in new conditions. The main task for both CG and TG was making a door frame (0.85 cm width and 200 cm height. But CG performed the task in the original work conditions, while TG in an ergonomics one.

The radial saw, drilling machine and the fabrication working table were modified based on the workers' anthropometry, while the redesigned organization was done by giving a short break for about 5 minutes after 1.5 hours working for drinking mineral water, and providing intensive training in occupational safety and health.

The unsafe actions were performed through the working posture (bending, squatting, and sitting on the floor in the fabrication process). The awareness of workers in using personal protecting devices (mask, ear-plug, and gloves) was recorded every 15 minutes. The health status was measured through the working heart rate, MSDs and general fatigue. The assessment of health status was done by the ten pulse method for heart rate, Nordic Body Map for MSDs, and general fatigue questionnaire for general fatigue.

## 3 RESULT AND DISCUSSION

### 3.1 *Participants*

Table 1 shows that there were no significant differences of age and body mass index between control and treatment groups and it means that both groups were in equal condition before they started working ($p > 0.05$). The average age of the participants was between 18.50–18.60 years old and the BMI were 20.26–21.43. At that age, the relative capability and productivity level increases gradually and reaches the maximum value between the early 20s and the end of the 30s (Avolio et al., 1994 and Skirbe, 2003). Besides, people with BMI > 25 have higher risk of hypertension, and people with BMI > 29 (Obesity) have 2.5 higher risk of musculoskeletal disorders (MSDs) compare with those with BMI < 20 (Droyvold et al., 2005). Furthermore, the result of the analyses showed that there was no significant difference of BMI within the two groups ($p > 0.05$). It means that the characteristics of subjects in all groups were comparable and did not influence the result of the study since the subjects were all at the same level of relative capability, productivity and health.

Table 1. Participants characteristics data.

| Variables | No. of participants | Control group | Treatment group | p |
|---|---|---|---|---|
| Age (years) | 10 | 18.60 ± 0.52 | 18.50 ± 0.71 | 0.317 |
| Body mass index | 10 | 20.26 ± 2.63 | 21.43 ± 3.08 | 0.380 |

Table 2. Physical environment data.

| Variables | No. of assessments | Control group | Treatment group | p |
|---|---|---|---|---|
| Noise (dBA) | 6 | 78.99 ± 3.74 | 80.40 ± 3.49 | 0.587 |
| Humidity (%) | 6 | 75.33 ± 16.59 | 86.80 ± 3.90 | 0.218 |
| WBGT index (Co) | 6 | 27.39 ± 1.04 | 26.95 ± 0.30 | 0.228 |
| Lighting (Lux) | 6 | 886 ± 204.79 | 916 ± 326.71 | 0.197 |

### 3.2 *Physical environment*

From those indicators of the physical environment, only the WBGT index was not adequate. The average of WBGT indexes in CG and TG were 27.39 ± 1.04 and 27.39 ± 1.04°C, while the Indonesian National Standard recommended WBGT index for 8 hours continuous working is 26.7°C for moderate workload category and 25°C for heavy workload category (BSN, 2004). Another indicator that needed preventive action was the level of noise. There were 78.99 ± 3.74 dB(A) in CG and 80.40 ± 3.49 dB(A) in TG and both of them nearly in the upper limit (85 dBA). A level of noise more than 85 dBA will increase the blood tension and heart rate, cause hearing loss, early fatigue, and finally decreased productivity. Meanwhile, a level of noise more than 80 dB A is not suitable for conversation (Grantham, 1992). However, statistic analyses showed that there were no significant differences of humidity, WBGT Index, noise and lighting intensity between the groups ($p > 0.05$). This means that the physical work environment in both groups were comparable and will not influence the result of the study, since each group received the same effect from the physical work environment.

### 3.3 *Unsafe actions*

Tables 3 and 4 showed that the ergonomics redesign minimized the unsafe actions. The intensive training in Occupational Safety and Health (OSH) increased the awareness of participants concerning the standard operating procedure properly, including using personal protection devices (PPD). The bending posture decreased by 77.88%, while the squatting postured was 92.51% less, and there were no more participants who were sitting on the floor in the fabrication process. Furthermore, the participants who wore masks increased by 220.15%, and less than 10% of participants were not using ear-plugs and gloves during working hours. It proved that the working culture could improve by training.

### 3.4 *Health status*

Health status was described through the indicators of workload, body weight lost, MSDs, and fatigue.

#### 3.4.1 *Workload*

The workload was examined through the value of the resting heart rate (RHR) and working heart rate (WHR) and the results are shown in Table 5.

Table 5 showed that there was no significant difference of RHR ($p > 0.05$), but the WHR was ($p < 0.05$). The WHR in CG was in the heavy workload category and was moderate in TG. The WHR in TG decreased by about 17.93% from CG. The decreasing of workload in

Table 3. The working posture.

| Posture | No. of assessments | Control group | Treatment group | p |
|---|---|---|---|---|
| Bending/15 minutes (%) | 15 | 49.15 ± 2.55 | 10.87 ± 3.46 | 0.000 |
| Squatting/15 minutes (%) | 15 | 13.76 ± 2.19 | 1.03 ± 1.03 | 0.000 |
| Sitting on the floor/15 minutes (%) | 15 | 15.64 ± 3.50 | 0 | – |

Table 4. The usage of PPD.

| PPE | No. of assessments | Control group | Treatment group | p |
|---|---|---|---|---|
| Masks/15 minutes (%) | 15 | 29.18 ± 11.1 | 93.42 ± 5.48 | 0.000 |
| Ear-plugs/15 minutes (%) | 15 | 0 | 95.81 ± 4.14 | 0.000 |
| Gloves/15 minutes (%) | 15 | 0 | 92.38 ± 8.50 | 0.000 |

Table 5. Heart rates.

| Variables | No. of participants | Control group | Treatment group | p |
|---|---|---|---|---|
| RHR/(pulse/minutes) | 10 | 81.20 ± 5.00 | 82.63 ± 3.63 | 0.473 |
| WHR/(pulse/minutes) | 10 | 113.21 ± 3.04 | 92.91 ± 2.93 | 0.000* |

Table 6. Body weight lost.

| Variables | No. of participants | Control group | Treatment group | p |
|---|---|---|---|---|
| Body weight | 10 | | | |
| Body weight lost (kg) | 10 | 0.94 ± 0.34 | 0.20 ± 0.20 | 0.000 |

TG was caused by the improvement of organization and workstations. Working in a natural posture avoids the over-exertion, maintains the blood flow and the supply of oxygen to the body muscles, avoids muscle fatigue and additional workload (Grandjean, 1993).

### 3.4.2 *Body weight lost*

Two glasses of drinking water (240 cc each) improved the work nutrition compared that shown by the reduction of the weight lost as described in Table 6.

The weight lost in CG was 1.72% which was more than the recommended limit (< 1.5%). In contrast, the body weight loss in TG significantly decreased compared to CG ($p < 0.05$). A body weight loss more than 1.50% could cause dehydration, early fatigue and decreased health status (Genady, 1996).

### 3.4.3 *Musculoskeletal Disorders (MSDs)*

The major hazards in woodworking activities were lifting loads and awkward postures caused by the facilities, and working tools that were not suitable with workers' anthropometry. Results showed that there were significant differences of the MSDs score after working within groups ($p < 0.5$). In addition, the daily analyses showed that MSDs score inclined gradually during manual working processes, and then declined when they used woodworking machines. Furthermore, it was increasing again during fabrication and finishing process in CG, but constant in TG. This proved that the workstation that suited workers' anthropometry in TG decreased the MSDs effectively. Ergonomic intervention in organization was due to the understanding of students about how to behave appropriately, while the anthropometric work station minimized the awkward posture. In addition, the comprehensive ergonomic intervention shortened the working period from six days to five days. Participants could perform their tasks in more convenient and healthy working conditions that decreased the MSDs, increased their vigilance, quickened the working process and avoided early fatigue (Grandjean, 1993). Further analyses showed that ergonomics intervention in workstations minimized the bending, squatting, and sitting postures, even eliminating the sitting posture in TG. This data is in accordance with the study in some small industries in the Philippines during 1994–1996 and the results showed that the improving of workstations minimized the tension on the body muscles and decreased the MSDs. (Kogi et al., 2003).

### 3.4.4 *General fatigue*

Results showed that general fatigue in CG and TG was significantly different ($p < 0.05$). The previous study reported that there was a very strong correlation between general fatigue and the increase in heart rate that indicated the workload category (Wijesuiya et al., 2007). In line with the MSDs, the general fatigue score increased gradually during manual working processes, and then declined when woodworking machines were used. Furthermore, it was increasing again during fabrication and finishing processes in CG but constant in TG. This proved that the workstation that suited participants' anthropometry in TG decreased the general fatigue effectively. Moreover, the analyses showed that the daily general fatigue score after working in TG kept constant from the first to the sixth day.

## REFERENCES


Avolio, B.J. & Waldman, D.A. (1994). Variations in Cognitive, Perceptual, and Psychomotor Abilities Across the Working Life Span: Examining the Effects of Race, Sex, Experience, Education, and Occupational Type. *Psychology and Aging*, Vol. 9 No. 3, pp. 430–442.

Badan Standarisasi Nasional. (2004). Nilai Ambang Batas Iklim Kerja (Panas), Kebisingan, Getaran tangan-lengan dan Radiasi Sinar Ultra Ungu di Tempat Kerja. Jakarta.

Drøyvold, W.B., Midthjell, K., Nilsen, T.I.L. & Holmen, J. (2005). Change in body mass index and its impact on blood pressure: a prospective population study. *International Journal of Obesity*, Vol. 29, No. 5, pp. 650–655.

Genady, A.M. (1996). Physical Work Capacity. In Bhattacharya A. & McGlothlin D. 1996. Occupational Ergonomics—Theory and Application. Marcel Dekkers Inc. New York: pp. 219–234.

Grandjean, E. (1993). *Fitting the task to the man*, 4th ed. London: Taylor & Francis Inc.

Grantham, D. (1992). *Occupational Health & Hygiene. Guidebook for the WHSO*. Australia: Merino Lithographics Moorooka Queensland, pp. 52–94.

Kogi, K., Kawakamia, T., Itanib, T. & Batinoc, J.M. (2003). Low-cost work improvements that can reduce the risk of musculoskeletal disorders. *International Journal of Industrial Ergonomics*, Vol. 31, pp. 179–184.

Skirbekk, V. (2003). Age and Individual Productivity: A Literature Survey. Max Planck Institute for Demographic Study. Germany.

Wijesuriya, N., Tran, Y. & dan Ashley. 2007. The psychop-hysiological determinants of fatigue. *International Journal of Psycophysiology*. Vol. 63, No. 1, pp. 77–86.

*Ergonomics in Asia – Shih & Liang (eds)*
 *ISBN 978-0-415-68414-9*

# Study of quality of life (QOL) in home oxygen therapy patients and related factors of actual life situation

Nanae Shintani & Tomoko Morimoto
*Gifu University of Medical Science, Seki, Japan*

ABSTRACT: Home oxygen therapy is the most general of many kinds of home healthcare and its effects are regarded as not only extending patients' vital prognosis but also guaranteeing their conditions of life and quality of life (QOL). However, it was reported that the social activities of the patients were restricted because of social, mental and physical troubles, and an acute exacerbation was brought about due to depression by shutting themselves away. Therefore, the related factors of actual life situation affecting their QOL will be examined in order to clarify the relation between the actual life situation and QOL.

The followings are examined in this survey; 1) "basic attributes"—age, sex, underlying disease and the extent of impediment 2) "actual life situation"—quantity of oxygen inhalation, inhalation times per day, understanding of chest physical therapy, frequency and relations with their caregiver 3) "Health Related QOL (SF-36v2)". As a result, it is revealed that the vitality and mental health of patients relate to the understanding of chest physical therapy significantly. The patients who have stable understanding of chest physical therapy have high cognitive faculty and can take care of themselves rigorously, so they can accept their disease positively and maintain their vitality. Therefore, it is suggested that repeated education for patients is required. Furthermore, understanding of chest physical therapy influences the mental health of patients significantly. Dyspnea seen in the patients who practice home oxygen therapy not only drives them into a critical state and creates the fear of a sudden occurrence of the symptoms but also engenders an anxiety that the patient will encounter the same situations unexpectedly. As they have felt great anxiety, securing patient safety is a core aim for QOL of the patients on home oxygen therapy. For patients, increasing awareness of their disease and improving their capability for self-management such as preventing increasing the symptoms and avoiding aggravation by acquiring the knowledge of chest physical therapy, is connected with not only decreasing the symptoms, anxiety and stress of the disease but also with getting a sense of security which is important for the patients.

*Keywords*: home oxygen therapy, chest physical therapy, QOL, SF-36

## 1 INTRODUCTION

In Japan, the number of patients on home oxygen therapy was only about 500 in 1985. However, the number was estimated to be 13 thousand in 2008 due to improving the supply of oxygen equipment and application of health insurance, and the number increases by 4 to 5 thousand every year (Kawakami, 1997). An increase in of the number of patients who started to practice home oxygen therapy was related to air pollution, smoking and aging. Therefore, introduction of home oxygen therapy is presumed to be likely to continue to increase in the future. Home oxygen therapy is the most general of many kinds of home healthcare and the effects of it are regarded as not only extending the patients' vital prognosis but also guaranteeing their conditions of life and quality of life (QOL). However, it was reported that the patients were restricted in their social activities because of social, mental and physical troubles, and brought about an acute exacerbation due to fall depression by shutting themselves away (Nagata et al., 1987). Preventing the acute exacerbation and learning how to improve the QOL is important in home oxygen therapy. In addition, it is necessary to clarify the relationship between the actual life situation and QOL for extending the life under medical treatment of the patients with chronic pulmonary insufficiency who have irreversible pathological damage to lungs. According to the study by Dr. Ishizaki that has analyzed the QOL from the point of view of the actual life situation, it is elucidated that presence of mind connects with the improvement of the QOL regardless of the extent of dyspnea (Ishizaki et al., 2000) Therefore, we will examine related factors of the actual life situation that affects their QOL to clarify the relation between the actual life situation and QOL.

## 2 METHODS

### 2.1 *Subjects*

We conducted a questionnaire by semi-structured questionnaire form and an interview with 30 pairs (patients—caregivers). The subjects for this investigation were the patients on home oxygen therapy (male:female = 8:2) which consisted of the male group (average age $77.5 \pm 9.4$) and the female group (average age $66.4 \pm 13.2$) in the home nursing situation in Hiroshima prefecture, T city and in the general hospital in H city. Furthermore, we limited the subjects to over 40 year-old patients who had applied for Long-Term Care Insurance item (ii) because the chronic obstructive pulmonary disease which accounts for a high percentage of the basic diseases of the patients of on home oxygen therapy was certificated as an intractable disease on Long-Term Care Insurance.

### 2.2 *Contents of questionnaire and collection term of the data*

We defined the concept of QOL as "individual comfort and satisfaction in life" and decided the contents of questionnaire which consisted of 1) "basic attributes" 2) "actual life situation" 3) "Health Related QOL (SF-36v2)". Each component is shown below in concrete terms.

1. Basic attributes
We ask the examinee the questions—age, sex, underlying disease, the extent of the impediment (respiratory functional disorder, cardiac disturbance), the term of the disease, experience of home oxygen therapy, other diseases, monthly income, family relationship to caregiver, whether living with caregiver, family members (living together).
2. Actual life situation
We also ask the examinee the questions—quantity of oxygen inhalation (resting, going out), times of inhalation a day, frequency of exercise a month, whether you have kept the doctors instructions, understanding of chest physical therapy, the frequency of going out, interactions with other people, relationship to caregiver, existence or nonexistence of someone to turn to for advice about worries, feelings about the care, livelihood in the future.
3. Health Related QOL (SF-36) of the patients
We used "SF-36" as the measurement scale about health-related QOL (SF-36v2). SF-36 is composed of 8 concepts of health—(1) Physical functioning (2) Physical role (3) Emotional role (4) General health perceptions (5) Social functioning (6) Bodily pain (7) Vitality (8) Mental health. We created 36 questions including these 8 concepts of health and kept the score. Then we transformed it to 0–100 points based on established methods.

The reasons why we selected "SF-36"as a measure of the Health Related QOL was—(1) It can measure the QOL of not only the patients who have intractable disease but also the patients on home oxygen therapy or even healthy people. (2) You can evaluate the QOL of targets and compare them to national standard values. (3) SF-36 is widely known as the measurement scale for health-related QOL (SF-36v2).

### 2.3 *Analytical method*

We practiced the Spearman's rank correlation analysis by using "Health Related QOL (SF-36)", "basic attributes" and "actual life situation". We used SPSS Ver16.0 in the statistical analysis.

### 2.4 *Ethical consideration*

We consider the ethical issues such as (1) The questionnaires were filled out anonymously. (2) The patients can refuse or interrupt the treatment and avoid sustaining disadvantages according to their will. (3) Practicing the statistical treatment so as not to identify the patients' personal information. (4) The data were used only for the purpose of the study. (5) We gave the subjects previous notice of (1) to (4) through announcements orally and in writing.

This survey was inspected and approved by the ethics committee at Hiroshima International University, School of Nursing.

## 3 RESULTS

Basic attributes are shown in Table 1, and actual life situation is shown in Table 2.

### 3.1 *Results of analysis*

The categories which connected with the QOL related to the patients' health ($p$ value $< 0.05$); on the basic attributes and actual life situation was understanding of chest physical therapy (Table 3). As you can see from the Table 3, understanding of chest physical therapy influences the mental health of patients significantly.

## 4 CONSIDERATION

The understanding of chest physical therapy of patients related to their vitality significantly. The category of patients vitality included these questions such as "Are you hale and hearty?", "Do you have exuberant liveliness?" and "Are you tired?"

On prior research, it is found that the physical, social, mental and psychological QOL of the patients who participated in the class of respiratory rehabilitation was much better than the patients who did not join it, but there are few researches referring to understandings of chest physical therapy (respiratory rehabilitation) in the group guidance. Moreover, it was suggested that the effects of comprehensive respiratory rehabilitation included educational guidance, recreation and mental care which were evaluated by POMS, and STAI brought about a rise of the patients vitality and a decrease in their anxiety (Yamamoto et al., 2002). Other studies showed that the effects of respiratory rehabilitation was expected to even decrease or discontinue the quantity of oxygen inhalation not just the aspects of exercise tolerance, ADL and QOL (Hiramatsu et al., 2003). Further, Dr. Ueki mentioned that the patients demanded to get knowledge and skills for improving their self-management such as dealing with their increment according to the questionnaire of home respiratory care, so the education for patients played an important role for improving the patients QOL (Ueki, 2006). This survey has not evaluated the effects before and after the chest physical therapy, but inquired into the understanding of chest physical therapy, to be sure, the patients who have stable understanding of chest physical therapy have high competency for affiliation and can take care of themselves rigorously, so they can accept their disease positively and maintain their vitality. In a similar way to Dr. Ueki, we also found that it is necessary to spread and perform the education for patients for getting the skill of self-management repeatedly. Besides, the interviews indicated that for the patients taking a large amount of oxygen inhalation do not correlate to the vitality which is felt by the patients, and mental calmness is related to the patients happiness and energy.

Table 1. Basic attributes.

| Parameter | n = 30<br>N | ratio (%)<br>% |
|---|---|---|
| Sex | | |
| Male | 25 | 83% |
| Female | 5 | 17% |
| Age | | |
| 40 s | 1 | 3% |
| 50 s | 2 | 7% |
| 60 s | 4 | 13% |
| 70 s | 9 | 30% |
| 80 s | 12 | 40% |
| 90 s | 2 | 7% |
| Average age (male) | 77.5 | |
| Maximum age (male) | 93 | |
| Minimum age (male) | 55 | |
| Standard deviation (male) | 9.4 | |
| Average age (female) | 66.4 | |
| Maximum age (female) | 88 | |
| Minimum age (female) | 49 | |
| Standard deviation (female) | 13.2 | |
| Underlying disease | | |
| Chronic pulmonary emphysema | 17 | 57% |
| Chronic bronchitis | 3 | 10% |
| Fibrosis of lung | 3 | 10% |
| Pulmonary tuberculosis sequelae | 2 | 7% |
| Bronchitic asthma | 1 | 3% |
| Lung cancer, metastatic lung cancer | 1 | 3% |
| Diffuse bronchitis | 0 | 0% |
| Bronchiectasia | 0 | 0% |
| Pulmonary thromboembolism | 0 | 0% |
| Essential pulmonary hypertension | 0 | 0% |
| Others | 2 | 7% |
| Unknown | 1 | 3% |
| Other diseases | | |
| High blood pressure | 4 | 13% |
| Diabetes | 5 | 17% |
| Cataract | 3 | 10% |
| Gastrointestinal disorder | 1 | 3% |
| Liver | 0 | 0% |
| Heart | 6 | 20% |
| Kidney | 1 | 3% |
| Lumbago | 1 | 3% |
| Others | 1 | 3% |
| None | 8 | 27% |
| Existence or nonexistence of living with caregiver | | |
| Yes | 28 | 93% |
| No | 2 | 7% |

| Parameter | n = 30<br>N | ratio (%)<br>% |
|---|---|---|
| Cardiac disturbance | | |
| None | 25 | 83% |
| First rank | 0 | 0% |
| Third rank | 1 | 3% |
| Fourth rank | 1 | 3% |
| In the middle of application | 0 | 0% |
| Unknown | 3 | 10% |
| Respiratory functional disorder | | |
| None | 10 | 33% |
| First rank | 2 | 7% |
| Third rank | 9 | 30% |
| Fourth rank | 5 | 17% |
| In the middle of application | 2 | 7% |
| Unknown | 2 | 7% |
| The term of the disease | | |
| 12–60 months | 6 | 20% |
| 61–120 months | 11 | 37% |
| 121–180 months | 3 | 10% |
| 181–240 months | 3 | 10% |
| 241–420 months | 3 | 10% |
| Over 421 months | 1 | 3% |
| Unknown | 3 | 10% |
| Experience of home oxygen therapy | | |
| 4–12 months | 5 | 17% |
| 13–24 months | 3 | 10% |
| 25–36 months | 7 | 23% |
| 37–60 months | 3 | 10% |
| 61–84 months | 6 | 20% |
| 85–120 months | 4 | 13% |
| 121–240 months | 2 | 7% |
| Unknown | 0 | 0% |
| Monthly income(yen) | | |
| Over 400000 | 8 | 27% |
| 200000–400000 | 12 | 40% |
| 100000–200000 | 4 | 13% |
| 50000–100000 | 2 | 7% |
| Under 50000 | 2 | 7% |
| Unknown | 2 | 7% |
| Family relationship to caregiver | | |
| Wife | 19 | 63% |
| Husband | 2 | 7% |
| Daughter-in-law | 3 | 10% |
| Daughter | 4 | 13% |
| Son | 1 | 3% |
| Brother, Sister | 0 | 0% |
| Others | 1 | 3% |

Table 2. Actual life situation of the subjects.

| | n = 30 | ratio (%) |
|---|---|---|
| Parameter | N | % |
| Quantity of oxygen inhalation | | |
| Maximum | | |
| Resting | 3 L | |
| Going out | 5 L | |
| Minimum | | |
| Resting | 0.5 L | |
| Going out | 0.5 L | |
| Average | | |
| Resting | 1.54 L | |
| Going out | 1.94 L | |
| Time of inhalation a day | | |
| Throughout the day | 14 | |
| Not throughout the day | 5 | |
| Unknown | 11 | |
| Frequency of excercise a month | | |
| 1 time a month | 20 | 67% |
| 2 times a month | 3 | 10% |
| 4 times a month | 4 | 13% |
| Unknown | 3 | 10% |
| Whether you have kept the doctors instructions | | |
| Always do it | 17 | 57% |
| Generally do it | 9 | 30% |
| Not either | 1 | 3% |
| Hardly do it | 2 | 1% |
| Not do it | 0 | 0% |
| Unknown | 1 | 3% |
| Understanding of chest physical therapy | | |
| Fully understand it | 5 | 17% |
| Reasonably understand it | 9 | 30% |
| Not either | 5 | 17% |
| Hardly understand it | 7 | 23% |
| Not understand it | 3 | 10% |
| Unknown | 1 | 3% |

| | n = 30 | ratio (%) |
|---|---|---|
| Parameter | N | % |
| How often do you refrain from going out? | | |
| None | 5 | 17% |
| Seldom | 7 | 23% |
| Not either | 3 | 10% |
| Sometimes | 7 | 23% |
| Always | 6 | 20% |
| Unknown | 2 | 7% |
| Interactions with other people | | |
| everyday | 4 | 13% |
| 1 time a week | 8 | 27% |
| 1 to 2 times a month | 5 | 17% |
| Several times a year | 5 | 17% |
| None | 6 | 20% |
| Unknown | 2 | 7% |
| Existence or nonexistence of someone to turn to for advice about worries | | |
| Exist | 26 | 87% |
| Not exist | 3 | 10% |
| Unknown | 1 | 3% |
| Feelings about the care | | |
| Always appreciate it | 22 | 73% |
| Sometimes appreciate it | 3 | 10% |
| Not either | 4 | 13% |
| Sometime do just as you want to | 0 | 0% |
| Always do just as you want to | 0 | 0% |
| Unknown | 1 | 3% |
| Relationship to caregiver | | |
| Very good | 21 | 70% |
| Good | 5 | 17% |
| Not either | 0 | 0% |
| Poor | 2 | 7% |
| Bad | 1 | 3% |
| Unknown | 1 | 3% |
| Livelihood in the future | | |
| Live in your home all the time | 24 | 80% |
| Mostly live in your home and occasionally go to the institutions or hospitals | 4 | 13% |
| Live in your home and go to the institutions or hospitals alternately | 0 | 0% |
| Mostly go to the institutions or hospitals and occasionally live in your home | 0 | 0% |
| Live in the institutions or hospitals | 1 | 3% |
| Unknown | 1 | 3% |

Table 3. The results of spearman's rank correlation analysis correlation analysis between health related QOL and actual life situation of the patients.

| Health related QOL (SF-36) | | Understanding of chest physical therapy |
|---|---|---|
| Physical functioning | | –0.114 |
| Physical role | | –0.217 |
| Bodily pain | | –0.152 |
| General health perceptions | | –0.125 |
| Vitality | ** | –0.55 |
| Social functioning | | –0.291 |
| Emotional role | | –0.166 |
| Mental health | ** | –0.502 |

**$p < 0.01$.

Furthermore, understanding of chest physical therapy influences the mental health of patients significantly. The category of mental health included this question such as "Have you been feeling calm, happy and peaceful?" Dyspnea seen in the patients who practice home oxygen therapy not only drives them into a critical state and creates the fear of a sudden occurrence of the symptoms but also engenders an anxiety that the patient may encounter the same situations unexpectedly.

As they have felt great anxiety, securing the patient's safety is a core aim for QOL of the patients on home oxygen therapy. Dr. Tsuda also reported that the patients were rationed in their social activities because of dyspnea and brought about the decreasing of their ADL levels and an acute exacerbation due to depression by shutting themselves away, so it is important for home oxygen therapy to prevent the acute exacerbation and to improve the QOL (Tsuda et al., 2006). For patients, increasing awareness of their disease and improving their capablility of self-management such as preventing progression of the symptoms and avoiding aggravation by acquiring the knowledge of chest physical therapy correlates with not only decreasing the symptoms, anxiety and stress of the disease but also getting a sense of security which is important for the patients.

## 5 CONCLUSION

We clarified that the QOL of the patients correlated with what concerned the actual life situation of the patients and caregivers. As a result, it is suggested that studying chest physical therapy plays an important role for the patients' mental health.

## REFERENCES


Fukuhara, S. & Suzukamo, Y. (2001). "Health Related QOL SF-36 Japanese Manual" Public Health Research Center, 2001.

Hiramatsu, T., et al. (2003). "dyspnea kan to rehabilitation" gairai kokyuu rehabilitation program 51(2), 2003, 211–217.

Ishizaki, T., et al. (2000). "A study on Home Oxygen Therapy Patients and their family: Patients' Quality of Life and family burden" Journal of Fukui Medical University 1–3(2000): 523–534.

Kawakami, Y. (1997). "wagakuni ni okeru home oxygen therapy no rekisi to genjyou" The Journal of the Japan Medical Association 117–5(1997): 663–667.

Nagata, H., et al. (1987). "mansei kokyuu fuzen kanjya ni mirareru utsu to fuan ni tsuite" 3(2), 1987, 32–35.

Tsuda, T., et al. (2006). "home oxygen therapy" kokuu 25(9), 2006, 890–897.

Tsuji, K., et al. (2000). "home oxygen therapy kanjya no QOL koujyou ni mukete: kokyuu rihabiri kyoushitsu no yuukousei" Integrated studies in nursing science 1–2(2000): 45–49.

Ueki, J. (2006). "kanjya kyouiku" nihon iji shinpou No. 4266, 2006, 21–24.

Yamamoto, R., et al. (2002). "The Roles of Clinical Psychologist in Rehabilitation for Patients on Home Oxygen Therapy" Japanese journal of psychosomatic internal medicine 16–3(2002): 141–146.

*Ergonomics in Asia – Shih & Liang (eds)*
*© 2012 Taylor & Francis Group, London, ISBN 978-0-415-68414-9*

# The effects of stretching, hot pack and massage treatments on muscle thickness and hardness

Satoshi Muraki & Makoto Ohnuma
*Faculty of Design, Kyushu University, Fukuoka, Japan*

Kiyotaka Fukumoto
*Faculty of Engineering, Shizuoka University, Japan*

Osamu Fukuda
*The National Institute of Advanced Industrial Science and Technology, Japan*

ABSTRACT: The purpose of this study is to clarify the changes in muscle hardness by three treatments (stretching, hot pack and massage) that are commonly used to relieve muscle fatigue. Ten young male adults received the treatments in their anterior thigh. The thickness and hardness of the subcutaneous fat and the muscles in the anterior thigh were measured using an elasticity-measuring instrument with an ultrasound signal. In the stretching and hot pack conditions, an index of muscle hardness significantly decreased without change in muscle thickness, which indicates softening of the muscles. In contrast, the massage condition failed to show significant changes in the muscle hardness. These findings suggested that the stretching and warming of the muscles soften them, which might reflect relief of muscle fatigue.

*Keywords*: muscle thickness, muscle hardness, aging, ultrasound device

## 1 INTRODUCTION

Advances in technology have led to the development of sedentary lifestyles and working practices, such that any continuous strains on local muscles can cause muscle fatigue. In addition, accumulation of muscle fatigue is a factor that can lead to the deterioration in the performance of athletes. Therefore, speedy recovery from muscle fatigue is important. In general, various treatments such as stretching, hot pack application and massage are used to relieve muscle fatigue (Greenberg, 1972; Howatson & Someren, 2008). However, because there are no methods to quantitatively measure muscle fatigue, it is difficult to examine the effectiveness of such treatments.

It is considered that muscle fatigue is related to hardening of the muscles (Ashina et al., 1999; Murayama et al., 2000). Accordingly, muscle hardness might be a good indicator for use to examine the effectiveness of treatment. We therefore developed a new ultrasonic hardness meter, an elasticity-measuring instrument that uses an ultrasound signal (EUS) (Fukuda et al., 2007; Tsubai et al., 2008). This device can distinguish borders between tissues layers comprising subcutaneous fat, muscle and bone using pulse echoes. The device compresses the skin surface at a constant pressure and emits ultrasonic waves simultaneously. Information about the borders before and after the compression provides values of displacement of muscle thickness, which is an index of muscle hardness.

Therefore, in the present study, we investigated the changes in muscle thickness and hardness by various treatments (stretching, hot pack and massage) using the EUS.

## 2 METHODS

### 2.1 *Participants*

Ten males, whose ages ranged from 22 to 27 years old, participated in this study. Their body height and weight were 172.7 ± 7.0 cm and 63.4 ± 4.6 kg, respectively. No subjects had a history of cardiac, metabolic, pulmonary or orthopedic diseases. Written informed consent was obtained from all of the subjects before starting the study. This study was approved by the Research Ethics Committee of the Faculty of Design at Kyushu University.

### 2.2 *Experimental procedures*

Each subject, wearing a shirt and shorts, remained in a supine position on a bed for at least 30 min. The participants on the bed then received the following four treatment conditions: 1) stretching, 2) hot pack, 3) massage and 4) control conditions. In the stretching condition, their muscle groups in the anterior thigh were stretched by bending of the knee at the dorsal position with another's support (Figure 1a). The stretching for 40 sec was repeated five times with intervening rests of 20 sec. In the hot pack condition, using a hot pack, the skin temperature on their anterior thigh was controlled between 40 and 42°C for 15 min in a supine position (Figure 1b). In the massage condition, the anterior thigh was massaged in a supine position using a commercial compact massager (MD7300, Thrive Co. Ltd., Japan), which gives a vibration at 40 Hz (Figure 1c). The massage for 30 sec was repeated five times with intervening rests of 30 sec. In the control condition, the participants simply maintained their supine position for 15 min.

The experiment was carried out in an artificial climate chamber, and ambient temperature and relative humidity were maintained at 27°C and 50%, respectively.

### 2.3 *Measurement of muscle hardness*

Tissue thickness and hardness were measured in the subcutaneous fat and the muscle group (rectus femoris muscle and vastus intermedius muscle) of the anterior thigh using the EUS, as described by Tsubai et al. (2008). In brief, the system consists of an ultrasound sensor unit, a sensor driver and a personal computer with software that controls the main unit and processes signals (Figure 2). An examiner presses the head of the ultrasonic sensor perpendicular to the body surface at a constant pressure of 10 N maintained by an internal coiled spring (Figure 3). The thicknesses of the subcutaneous fat and the muscle group are monitored from pulse echoes reflected by tissue boundaries. Changes in thickness between before and during the compression are measured for each tissue (Figure 4). The ratios of the values before compression to those during compression are calculated as an index of the hardness. Higher ratio means greater hardness of the tissue. Figure 5 shows a model of thickness and hardness measurements of soft tissues using the non-invasive EUS.

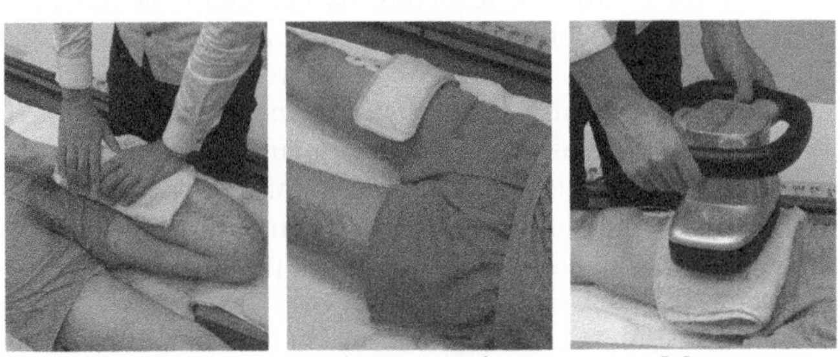

a. Stretching b. Hot pack c. Massage

Figure 1. The treatments.

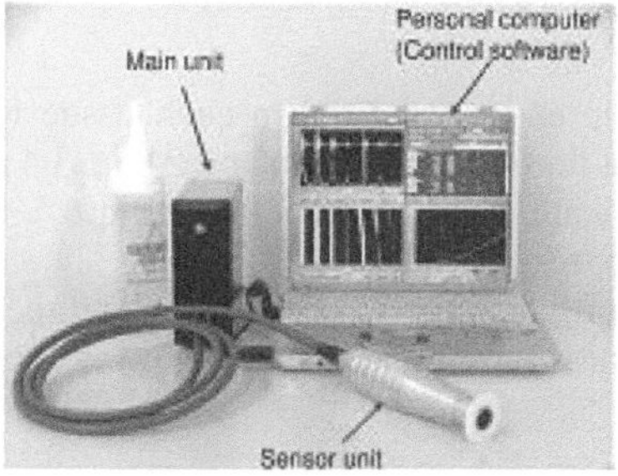


Figure 2. The elasticity-measuring instrument using an ultrasound signal.

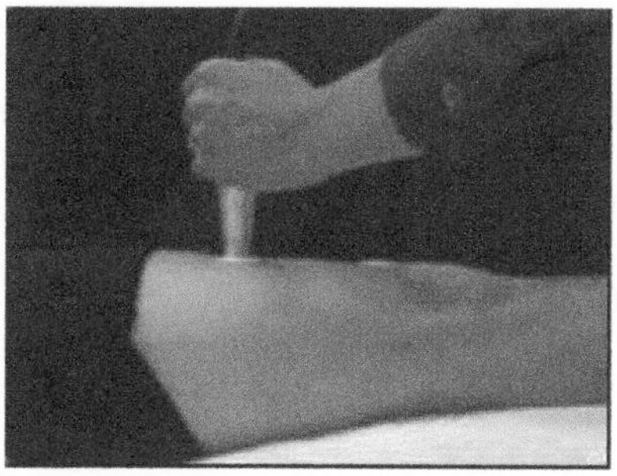

Figure 3. Measuring muscle hardness in the anterior thigh.

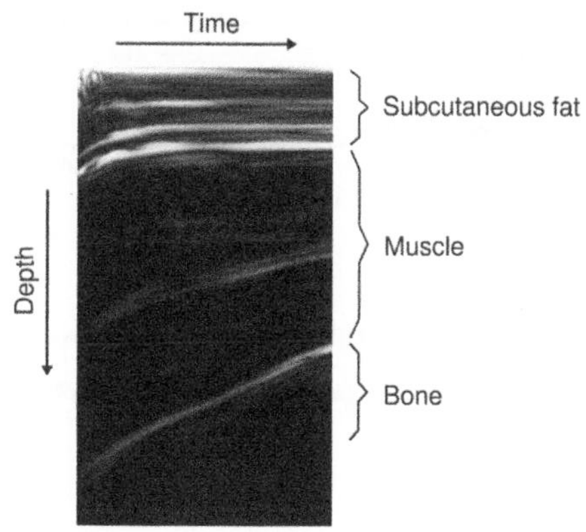


Figure 4. An image obtained using the elasticity-measuring instrument with an ultrasound signal.

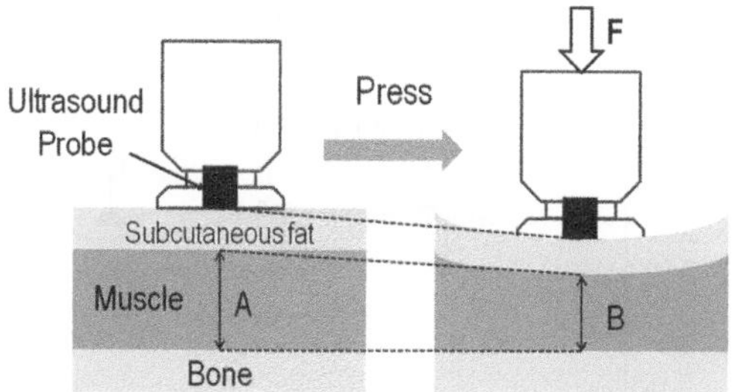


Figure 5. Mechanism for measuring hardness of individual tissues using elasticity-measuring instrument with an ultrasound signal.

These measurements were obtained before treatment, immediately after treatment, and at 15 and 30 min after treatment. At each measurement, the mean thickness and hardness were calculated from means of three determinations.

## 2.4 *Statistical analysis*

A two-way (conditions by time) analysis of variance with repeated measures was used to compare the changes in the muscle hardness. Significant F-ratios were followed by post-hoc comparisons using Dunnett's test. The level of significance was established at 5%.

## 3 RESULTS

Before treatments, thicknesses before and after compression of EUS in each tissue were 7.6 ± 2.8 mm and 7.2 ± 2.6 mm in the subcutaneous fat and 33.0 ± 4.4 mm and 24.5 ± 3.4 mm in the muscle group, respectively. These values indicate hardnesses of 95.4 ± 1.6% and 74.3 ± 2.3% in the subcutaneous fat and muscle group, respectively.

In all conditions, tissue thickness did not significantly change in both the subcutaneous fat and the muscle group (Figure 6). Although hardness of subcutaneous fat did not show

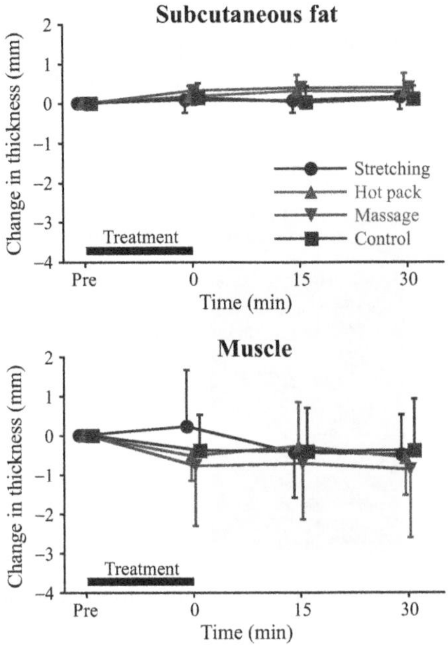


Figure 6. Changes in tissue thickness (subcutaneous fat and muscle) by four treatment conditions.

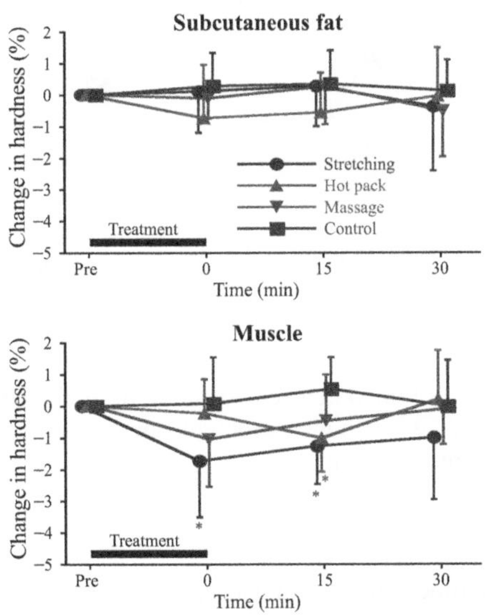


Figure 7. Changes in tissue hardness (subcutaneous fat and muscle) by four treatment conditions.

any significant changes by treatment in any conditions, that of the muscle group showed significant effect of time and an interaction effect (Figure 7). In the stretching condition, muscle hardness significantly decreased immediately after the treatment. Hot pack condition also showed a significant decrease at 15 min after the treatment. These phenomena indicate softening of the muscle group.

## 4 DISCUSSION

Some previous studies used a hardness meter to examine changes in muscle hardness (Horikawa et al., 1993, Kashima et al., 2004). Because their devices press the tissue and then measure responsive force, the obtained values reflect hardness of the whole tissue including the subcutaneous fat and the muscle group. In contrast, the EUS can detect the hardness of only muscle by observing the change in muscle thickness and eliminates this disadvantage of the existing devices.

After stretching and hot pack conditions, the change in hardness of the muscles significantly decreased without significant changes in the tissue thickness of both subcutaneous fat and muscle and hardness of the subcutaneous fat. These results suggested that the stretching and hot pack treatments softened the muscles without changing their thickness. Previous studies reported that stretching and hot pack treatments led to various physiological and anatomical changes in the muscles, such as increases in blood flow, temperature, flexibility and length of the muscles (de Weijer et al., 2003; Lee & Ng, 2008; Okada et al., 2006; O'Sullivan et al., 2009). Some of these changes would reflect the decrease in hardness of the muscles.

In contrast, the massage condition failed to show softening of the muscles. Some previous studies showed its effectiveness in removing muscle fatigue with physiological changes in the muscle (Durkin et al., 2006; Mori et al., 2004). The absence of significant changes in muscle hardness in the massage condition in this study might have been due to insufficient pressure or duration of massage to soften the muscles. In addition, the muscles in the present study were not in a state of fatigue. Ohnuma et al. (2010), using the EUS, reported that exercise of repeated knee extension caused hardening of the muscles in the anterior thigh. If we massaged the hardened muscles, softening of them might be observed using the EUS. However, Wiltshire et al. (2010) refuted the positive effect of massage with the results of muscle blood flow and lactic acid removal. Further studies should be undertaken to examine the changes in muscle hardness after exercise using the EUS.

In conclusion, these findings suggested that stretching and warming of the muscles soften them, which might reflect relief of muscle fatigue.

## REFERENCES

Ashina, M., Bendtsen, L., Jensen, R., Sakai, F. & Olesen, J. 1999. Muscle hardness in patients with chronic tension-type headache: relation to actual headache state. *Pain* 79(2–3):201–205.

de Weijer, V.C., Gorniak, G.C. & Shamus, E. 2003. The effect of static stretch and warm-up exercise on hamstring length over the course of 24 hours. *Journal of Orthopaedic and Sports Physical Therapy* 33(12):727–733.

Durkin, J.L., Harvey, A., Hughson, R.L. & Callaghan, J.P. 2006. The effects of lumbar massage on muscle fatigue, muscle oxygenation, low back discomfort, and driver performance during prolonged driving. *Ergonomics* 49(1):28–44.

Fukuda, O., Tsubai, M. & Ueno, N. 2007. Impedance estimation of soft tissue using ultrasound signal. *Proceedings of the 29th Annual International Conference of the IEEE EMBS* 3563–3568.

Greenberg, R.G. 1972. The effects of hot packs and exercise on local blood flow. *Physical Therapy* 52(3):273–276.

Horikawa, M., Ebihara, S., Sakai, F. & Akiyama, M. 1993. Non-invasive measurement method for hardness in muscular tissues. *Medical & Biological Engineering & Computing* 31(6):623–627.

Howatson, G. & van Someren, K.A. 2008. The prevention and treatment of exercise-induced muscle damage. *Sports Medicine* 38(6):483–503.

Kashima, K., Higashinaka, S., Watanabe, N., Maeda, S. & Shiba, R. 2004. Muscle hardness characteristics of the masseter muscle after repetitive muscle activation: comparison to the biceps brachii muscle. *Cranio* 22(4):276–282.

Lee, G.P. & Ng, G.Y. 2008. Effects of stretching and heat treatment on hamstring extensibility in children with severe mental retardation and hypertonia. *Clinal Rehabilitation* 22(9):771–779.

Mori, H., Ohsawa, H., Tanaka, T.H., Taniwaki, E., Leisman, G. & Nishijo, K. 2004. Effect of massage on blood flow and muscle fatigue following isometric lumbar exercise. *Medical Science Monitor* 10(5):CR173–178.

Murayama, M., Nosaka, K., Yoneda, T. & Minamitani, K. 2000. Changes in hardness of the human elbow flexor muscles after eccentric exercise. *European Journal of Applied Physiology* 82(5–6):361–367.

Ohnuma, M., Fukumoto, K., Kuroiwa, M., Muraki, S. & Fukuda, O. 2008. Effects of different exercise intensity and duration on muscle tightness by in vivo elasticity measuring instrument using an ultrasound signal. *Proceedings of the 23rd Symposium on Biological and Physiological Engineering* 173–174.

Okada, K., Yamaguchi, T., Minowa, K. & Inoue, N. 2005. The influence of hot pack therapy on the blood flow in masseter muscles. *Journal of Oral Rehabilitation* 32(7):480–486.

O'Sullivan, K., Murray, E. & Sainsbury, D. 2009. The effect of warm-up, static stretching and dynamic stretching on hamstring flexibility in previously injured subjects. *BMC Musculoskeletal Disorders* 10:37.

Tsubai, M., Fukuda, O., Ueno, N., Horie, T. & Muraki, S. 2008. Development of an ultrasound system for measuring tissue strain of lymphedema. *Proceedings of the 29th Annual International Conference of the IEEE EMBS* 5294–5297.

Wiltshire, E.V., Poitras, V., Pak, M., Hong, T., Rayner, J. & Tschakovsky, M.E. 2004. Massage impairs postexercise muscle blood flow and "lactic acid" removal. *Medicine and Science in Sports and Exercise* 42(6):1062–1071.

*Ergonomics in Asia – Shih & Liang (eds)*
*© 2012 Taylor & Francis Group, London, ISBN 978-0-415-68414-9*

# Mental health approaches by municipalities after personnel reduction in Japan

Yoshika Suzaki
*The Japanese Red Cross Kyushu International College of Nursing, Fukuoka-Pref., Japan*

Toshiko Nakamura
*Aroma & Maternity School Herb Warming, Japan*

Naoko Takayama
*Anan National of Technology, Tokushima-Pref., Japan*

Hiromi Ariyoshi
*Institute of Nursing, Faculty of Medicine, Saga University, Saga-Pref., Japan*

Shuhei Ryu
*Tokyo Wemen's Medical University School of Nursing, Japan*

ABSTRACT: Mental health measures for employees have been increasingly focused these days and a survey on the organization structural stress status in public employee's workplaces, which have been pointed out to be causing too much stress, was performed. Compared to general office employees, management employees are indicated to have higher values for "Caution" in "emotional deprivation" and "life satisfaction" and resulted in significantly higher values for the "depressive tendency". In relation to other items, it can be assumed that there is also an impact from the structural characteristics of public employees' workplaces after a personnel reduction in addition to their life environments. Not only improving the target/evaluation system but also considering the reconstruction of daily lives such as reviewing leisure activities is assumed to be effective as a measure against this.

*Keywords*: municipality, personnel reduction, multi-faceted life stress survey form

## 1 INTRODUCTION

Nowadays, mental health issues including depression and increased suicides have been focused as a crucial challenge in terms of the health management of public employees or companies. The Basic Act on Suicide Prevention was established in 2006 and comprehensive suicide measures have been pushed forward, implementing responsibilities of national/local public organizations etc. In addition, the General Principles of Comprehensive Suicide Prevention were formulated as a measure guideline in 2007; however, the number of deaths from suicides have still been as high with more than 30,000. 1) An intervention study on mental health measures targeting employees in medium sized workplaces reported that an individual approach is efficient and effective. 2) However, comprehensive individual approaches based on factual surveys have not yet been implemented, particularly in large sized workplaces. Under the current situation of mental health measures in the target municipal office, more information of mental health training is being communicated than in university hospitals, however, they lack fundamental resources for measures. Moreover, personnel reduction is currently being implemented with the Act on Promotion of Administrative Reform and organizational efforts are being expected. This study is intended to clarify the structural stress status in life and health aspects through a stress survey in public employee's workplaces, to consider the proper way of effective mental health measures in public employee's workplaces, and to provide recommendations.

Operational Definitions of Terms

Management employees—Assistant Manager or above.

## 2 TARGET AND METHOD

### 2.1 *Questionnaire research*

1. About the Target

The target was 1,000 individuals including the municipal staff of A city and the staffs of related organizations in A city of a certain prefecture (its population is approximately 46,000).

2. Survey Period

The survey was started from January 2011 and is still in progress.

3. Survey Method

A "multi-faceted life stress survey forms" 3) intended to mainly survey the multi-faceted life stress of people in employment were used. These survey forms consist of the following eleven criteria: living habits, degree of tiredness, life adaptation, depressive tendency, neurotic tendency, psychosomatic disorder tendency, alcohol dependence, emotional deprivation, over-adaptation personality, aggressive personality, and life satisfaction including life scores, and the forms provide scores of 0, 1, or 2 points for the three level answers of "No", "N/A", or "Yes" to 138 question items. For survey forms whose 95th percentile of the lie scale was at 8 points of total criteria or above, the reliability of all the answers (eleven criteria) was considered to be low and they were excluded. This means that the analysis was performed on the eleven criteria other than the lie scale. The individuals were categorized in the following four levels: "Good", "Generally good", "Some caution needed", "Caution" according to the total points of each item of the eleven criteria: living habits, degree of tiredness, life adaptation, depressive tendency, neurotic tendency, psychosomatic disorder tendency, alcohol dependence, emotional deprivation, over-adaptation personality, aggressive personality, and life satisfaction. When performing the statistical analysis, the analysis categorization was performed combining "Good" and "Generally good" as "Good" while combining "Some caution needed" and "Caution needed" as "Caution".

4. Ethical Considerations

Both verbal and written descriptions about the survey were provided for the target individuals, describing that they will not be disadvantaged even if they don't join the survey etc. and the individuals were asked to anonymously fill in the survey forms for themselves only with their free will. Their consent for this study was gained on collecting the survey forms.

5. Analysis Method

To compare between the two groups of general office and management employees, a statistics analysis $X^2$ test on the eleven criteria in the "multi-faceted life stress survey forms" was performed.

## 3 RESULT

### 3.1 *Collected result of survey forms*

The 335 answers were collected from the municipal staffs of A city and the staffs of related organizations in A city, and three of them were unclear answers. Currently, the survey is still in progress.

### 3.2 *Stress status of general office and management employees*

As for the seven items of "living habits", "depressive tendency", "neurotic tendency", "alcohol dependence", "emotional deprivation", "over-adaptation personality", and "life satisfaction", the "Caution" values were higher for management employees than for general office employees. In particular, a significant difference was observed for the "depressive tendency", in which the "Caution" percentage for management employees resulted in approximately 45%.

Table 1. Stress status of general office and management employees.

| | | | Management | General | p |
|---|---|---|---|---|---|
| Living habits | Good | n | 44 | 241 | |
| | | % | 78.6 | 87.3 | |
| | Caution | n | 12 | 35 | |
| | | % | 21.4 | 12.7 | 0.095 |
| Degree of tiredness | Good | n | 47 | 207 | |
| | | % | 83.9 | 75.0 | |
| | Caution | n | 9 | 69 | |
| | | % | 16.1 | 25.0 | 0.170 |
| Life adaptation | Good | n | 41 | 183 | |
| | | % | 73.2 | 66.3 | |
| | Caution | n | 15 | 93 | |
| | | % | 26.8 | 33.7 | 0.351 |
| Depressive tendency | Good | n | 31 | 197 | |
| | | % | 55.4 | 71.4 | |
| | Caution | n | 25 | 79 | |
| | | % | 44.6 | 28.6 | 0.026* |
| Neurotic tendency | Good | n | 41 | 215 | |
| | | % | 73.2 | 77.9 | |
| | Caution | n | 15 | 61 | |
| | | % | 26.8 | 22.1 | 0.486 |
| Psychosomatic disorder tendency | Good | n | 39 | 173 | |
| | | % | 69.6 | 62.7 | |
| | Caution | n | 17 | 103 | |
| | | % | 37.3 | 37.3 | 0.362 |
| Alcohol dependence | Good | n | 46 | 238 | |
| | | % | 82.1 | 86.2 | |
| | Caution | n | 10 | 38 | |
| | | % | 17.9 | 13.8 | 0.410 |
| Emotional deprivation | Good | n | 26 | 159 | |
| | | % | 46.4 | 57.6 | |
| | Caution | n | 30 | 117 | |
| | | % | 53.6 | 42.4 | 0.141 |
| Over-adaptation personality | Good | n | 40 | 206 | |
| | | % | 71.4 | 74.6 | |
| | Caution | n | 16 | 70 | |
| | | % | 28.6 | 25.4 | 0.618 |
| Aggressive personality | Good | n | 46 | 212 | |
| | | % | 82.1 | 76.8 | |
| | Caution | n | 10 | 64 | |
| | | % | 17.9 | 23.2 | 0.482 |
| Life satisfaction | Good | n | 23 | 145 | |
| | | % | 41.1 | 52.5 | |
| | Caution | n | 33 | 131 | |
| | | % | 58.9 | 47.5 | 0.143 |

In addition, the "Caution" percentages for the "emotional deprivation" and "life satisfaction" resulted in approximately 54% and 60% respectively, making the majority of management employees compared to general office employees.

## 4 DISCUSSION

Yamamoto (2007) states that management employees are more likely to have psychosomatic disorders or depressions. Although this survey also resulted in significantly higher values for the depressive tendency compared to general office employees, no significant difference was observed as for the psychosomatic disorder tendency.

In general, it can be assumed that companies also ask for more bottom-up type organization actions, however, management employees are mainly required to communicate instructions from their senior management to general office employees and to ensure they are performed. This difference of organization characteristics may be one of the factors causing the tendency difference in this survey.

No significant difference for the emotional deprivation and life satisfaction was present from general office employees, but their Caution percentages held the majority as for management employees, which resulted in very high values compared to other criteria. This fact may also affect the reduction of their sense of contentment or happiness in daily lives outside their workplaces in addition to the issues caused by the workflow that is specific to municipalities.

It can be assumed that reviewing the ways how management employees work or how the organization functions to implement improvement measures can be effective in municipalities so that they can gain a sense of accomplishment through their business practice.

Yamamoto (2007) also states that it is important for management employees to cherish their time outside workplaces. Although management employees may often lose their hobbies due to their age or physical limits or they may often feel a sense of loss of any response from their adolescent children at home, it is also necessary for them to try to improve their quality of life by finding a new hobby or reconstructing their family relationships so that they can feel satisfaction in their daily lives.

## 5 CONCLUSION

Management employees in an organization are driven into a situation that is difficult to improve only through their own actions. Most of the municipalities have strong aspects of being hierarchical organizations. However, in order for management employees to gain a sense of accomplishment through their business practice in such an organization, it is thought to be necessary to introduce creative changes in their ways of defining organizational goals, evaluating personnel, and assessing business results.

In addition, gaining a sense of accomplishment will become easier if they can perform organization actions based on proposals to local citizens, but controls as a municipal organization will be difficult, which requires careful considerations.

As for their life aspects, some measures are also thought to be required such as preventing management employees from working for a prolonged period of time, etc.

## POSTSCRIPT

This study is intended to clarify the stress on staff caused by the hierarchy structure of an organization from such a point of view. This survey is still an interim survey at a certain point, therefore, a continuous survey is required to figure out the dynamism.

## REFERENCES


Health, Labour and Welfare Statistics Association; "The Trend of National Health/Indicator of Health and Welfare, Health, Labour and Welfare Statistics Association, August 2009.

Kumai, S., Suzaki, Y. et al.: "About the Validity and Reliability of New Multi-faceted Life Stress Survey Forms Targeted for Workers", Journal of the Japan Health Medicine Association, 16(1), 8–16, April 2007.

Suzaki, Y., Ariyoshi, H. et al.: "Deployment and Assessment of Mental Health Measures in Mid-sized Workplaces", Journal of Ergonomics in Occupational Safety and Health, 11(1), 1–6, 2009.

Yamamoto, H. "Mental Health of Workers", (No. 2) How to Cleverly Deal with your Stress, The ROSAI Monthly, 58(12) 11–15, December 2007.

*Ergonomics in Asia – Shih & Liang (eds)*
*© 2012 Taylor & Francis Group, London, ISBN 978-0-415-68414-9*

# Effects of self stair-climbing exercise on physical fitness of clinical nurses

Ching-Wen Lien
*Department of Nursing, Taipei Veterans General Hospital, Taipei, Taiwan*

Bor-Shong Liu
*Department of Industrial Engineering and Management, St. John's University, New Taipei City, Taiwan*

Shir-Ling Lin & Chuan-Hui Yu
*Department of Nursing, Taipei Veterans General Hospital, Taipei, Taiwan*

ABSTRACT: The purpose of the present study was to provide a stair-climbing exercise program for nurses and examine the effectiveness of an intervention program on physical fitness. The 199 stair-climbing exercises (about 22 floors) was utilizing a public access staircase during the working week. Each participant needed to take the stair-climbing exercise thrice a week and a total of eighteen trials should be executed during six weeks. Results of ANOVA showed that the climbing time and physical fitness of nurses were significantly improved after intervention stair-climbing exercise. The mean climbing time for 199 steps was decreased from 159 to 138 sec after six weeks training. The delta heart rates in the first and final weeks were 54.7 and 51.4 bpm respectively. In addition, the scores of Borg-RPE scales were significant decreased from 16.2 to 12.9 in the final week. The results of the present study revealed that intervention of a stair-climbing exercise program could effectively improve nurses' physical fitness. Nurses need the moderate exercise training to promote nurses' health and enhance quality of life for both nurses and patients. Suggestions of present study were provision of health promotion activities and frequency of exercise at least three times a week.

*Keywords*: health promotion, clinical nurses, exercise, physical fitness

## 1 INTRODUCTION

Night duty rotations are common practice in nursing. It is essential that nurses working in these environments be able to maintain careful and astute observation of their patients. However, shiftwork and work-related stress may affect negatively on nurses' health and safety. This includes cardiovascular disease, gastrointestinal complaints, sleep troubles, mental health problems, fatigue, job dissatisfaction, accidents and injuries at work, reduced vigilance and job performance, absenteeism and turnover (Muecke, 2005). In addition, poor sleep quality may affect normal homeostatic body mechanisms. Insufficient restorative daytime sleep and inadequate recovery time from nightwork may lead to sleep deprivation, which may affect a nurse's ability to provide high standards of care. Problems associated with lack of sleep include disturbance of the circadian rhythm, fatigue, which can lead to physical and psychological problems, and disruption to family life (Barton, 1994; Reid, Roberts & Dawson, 1997).

A survey among nurses revealed that symptoms of musculoskeletal disorders were positively associated with work posture and negatively associated with exercise habits. In addition, a study on physical fitness of nurses at a medical center in Taiwan revealed that the mean physical fitness rating among those nurses was below average (Hwu, Yuan, Yeh & Chang, 2005; Lien, Yu, Lin & Liu, 2009). Therefore, intervention programs aimed at promoting physical fitness might enhance the quality of life for both nurses and patients. The purpose of the present study was to provide a health promotion program for nurses and examine the effectiveness of that intervention program on sleep quality and physical fitness.

## 2 LITERATURE REVIEW

Evidence suggests that sleep disorders and sleep fragmentation are very common among nurses (Alessi & Schnelle, 2000). A variety of factors contribute to sleeping difficulties, including age-related change in sleep, depression, medical illnesses, and sleep-altering medications. Other important factors include common lifestyle characteristics such as inactivity, large amounts of time spent in bed, lack of bright light exposure, poor sleep hygiene, and the disruptive nighttime nursing home environment. Chen, Hwang, Chou, and Tang (2004) explored the relationship between self-reported well-being and diurnal rhythmicity of hormone changes among healthy nightshift nurses. Results of analysis reported that poorer physical condition during nightshift including: easier to feel tired, poorer sleep quality, lack of personal interests, more dehydrated, and poorer attentiveness when compared with their previous experiences of daytime work. The hormone melatonin is very important in relaying environmental information to the brain. Secreted by the pineal gland only during times of environment darkness, melatonin has receptors located in the suprachiasmatic nuclei and when injected into the body, it causes drowsiness and inhibits several endocrine functions (Fox, 1999; Perkins, 2001). Conversely, the cyclical release of the hormone cortisol, a glucocorticoid, helps with day waking. Thus, the pattern of day and night is very important in regulating the circadian clock and sleep-wake mechanisms, as normal sleep-wake patterns rely on the recognition of day and night by the retina.

Stair climbing is a low-cost, inconspicuous, and readily accessible form of exercise and has been shown to be associated with reduced mortality (Paffenbarger, Kampert, Lee, Hyde, Leung & Wing, 1994). Boreham, Wallace, and Nevill (2000) evaluated the effects of a 7-week progressive stair-climbing program on the cardiorespiratory fitness and lipid profiles of previously sedentary young women. The staircase consisted of a 199-step public access staircase with a total vertical displacement of 32.8 meters. The pace of climbing was fixed at approximately 135 sec per ascent, or 88 steps/min. The results showed that individuals in the stair-climbing group displayed a rise in HDL cholesterol and a reduced total HDL ratio over the course of the program. In addition, oxygen uptake, heart rate, and blood lactate levels were reduced during the stair-climbing test. The researchers concluded that a short-term stair-climbing program could confer considerable cardiovascular health benefits on previously sedentary young women, lending credence to the potential public health benefits of this form of exercise.

## 3 METHODS

### 3.1 *Participants*

The study was approved by the Institutional Review Board of the researchers' hospital (No. 97-04-28A). A total of thirty female clinical nurses were recruited from a metropolitan hospital in Taipei. The mean age of the participants was 30.2 years (range, 21–48 years).

### 3.2 *Questionnaires*

The questionnaire was designed to gather background information, work shifts, work hours, and average length of sleep during the past month.

### 3.3 *Health promotion program*

The protocol for the stair-climbing exercise was implemented as reported previously (Boreham, Wallace & Nevill, 2000). Briefly, participants were required to ascend a 199-step public access staircase three times per week for a total of eighteen trials during a six–week period. Nursing stations equipped with a wheelchair, oxygen tube, and BP monitor were set up on the 16th, 19th, and 22nd floors. Each participant received a snack box after the stair-climbing exercise to prevent hypoglycemia. Heart rate, systolic blood pressure, and diastolic blood pressure were measured before and after the stair-climbing exercise by an NIBP monitor (Philip A1) in a nearby nursing station. A resting period of at least five minutes was provided for each participant to allow physiological parameters to reach a steady state before the actual experiment was started. The resting physiological responses served as a baseline measurement. Delta heart rate, delta systolic blood pressure, and delta diastolic pressure were then calculated. Each participant used a stopwatch (Casio, HS 10W) to measure the time it took for completing the stair-climbing activity. After each experimental condition was analyzed, participants reported their psychophysical response with a perceived exertion (Borg-RPE) rating (Borg, 1985). The scales that are constructed 15-points ratings from 6 (no exertion at all), 7, 8 (extremely light), 9 (very light), 10, 11 (light), 12, 13 (somewhat hard), 14, 15 (hard, heavy), 16, 17 (very hard), 18, 19 (extremely hard), and 20 (maximal exertion) are linearly related to heart rate expected for that level of exertion (expected heart rate is 10 times the rating given). All data needed were recorded in the table for further analysis. Researchers followed up the scheduled progress per week.

### 3.4 *Data analysis*

Categorical variables are presented as a percentage and continuous variables are presented as mean ± standard deviation. The paired-sample t-test was used to compare responses before and after the stair-climbing activity. A p value $< 0.05$ was considered to indicate statistical significance. All statistical analyses were performed on a personal computer with the statistical package SPSS for Windows (Version 13.0, SPSS, Chicago).

## 4 RESULTS

### 4.1 *Background information*

A summary of the background data is presented in Table 1. Only 7.7% of nurses participated in sports and less than 4% of nurses exercised regularly, e.g., running, walking, or swimming. About 40% of nurses reported a history of insomnia, but less than 10% of them were taking hypnotic agents. In addition, 34.6% reported that they sometimes consumed caffeinated beverages and 42.3% reported that they regularly consumed caffeinated beverages. Approximately 50% of nurses had a history of constipation.

### 4.2 *Stair-climbing exercise*

The mean values for climbing time and physiological responses after stair climbing at baseline and after the intervention are presented in Table 2. The mean climbing time decreased from 160 s at baseline to 138 sec after participation in the six-week exercise program. Mean waist circumference also decreased after completion of the health promotion program. There was a significant reduction in delta systolic blood pressure after six weeks of participating in the program ($t = -5.67$, $p < .001$). Although other physiological parameters did not differ significantly before and after the intervention, there was a decreasing trend. That finding indicates that a longer intervention period might be more effective in promoting physical fitness.

Table 1. Characteristics of the participants (N = 30).

| Variables | Mean (SD) | Range |
|---|---|---|
| Age (Years) | 30.1 (6.7) | 21–48 |
| Seniority (Years) | 6.4 (6.6) | 0.25–26 |
| Length of Sleep (Hours) | 6.95 (1.1) | 4.5–9 |
| Education | N | % |
| University | 21 | 70 |
| College | 9 | 30 |
| Marital status | N | % |
| Single | 28 | 93.3 |
| Married | 2 | 2.7 |
| Work shifts | N | % |
| Day shift | 13 | 43 |
| Evening shift | 9 | 30 |
| Night shift | 8 | 27 |
| Exercise habit | N | % |
| No | 9 | 30 |
| Sometimes | 19 | 63.3 |
| Yes | 2 | 6.7 |
| Insomnia | N | % |
| No | 18 | 60 |
| Sometimes | 11 | 36.7 |
| Usually | 1 | 3.3 |
| Taking hypnotic agents | N | % |
| No | 27 | 90 |
| Sometimes | 3 | 10 |
| Consumption of caffeinated beverages | N | % |
| No | 4 | 13.3 |
| Sometimes | 6 | 20 |
| Usually | 20 | 66.7 |
| Constipation | N | % |
| No | 14 | 46.7 |
| Sometimes | 11 | 36.7 |
| Usually | 5 | 16.6 |
| Diarrhea | N | % |
| No | 24 | 80 |
| Sometimes | 6 | 20 |

## 5 DISCUSSION

The Institute of Occupational Safety and Health has categorized fitness of laborers into five levels (i.e., poor, slightly poor, median, good, and excellent). Studies on nurses' physical fitness at medical centers in Taiwan revealed that the mean physical fitness rating among those nurses was slightly poor (Hwu et al., 2005; Lien et al., 2009). Among the nurses evaluated in those studies, 56% had a body mass index < X, indicating that they were slightly thin. In addition, the majority of those nurses, especially those who worked in the ICU and medical-surgical units, had slightly poor back muscle tolerance (53.64%) and slightly poor cardio-pulmonary fitness (54.23%) ratings. Huang, Chang, Chen & Chang (2003) also reported that the physical fitness ratings and frequency of exercise among nurses were

Table 2. Comparison of anthropometrical characteristics, climbing time, and physiological responses before and after the intervention.

| Variables | After program | Before program | Paired-difference | t | Sig. |
|---|---|---|---|---|---|
| Stature | 160.32 | 160.30 | 0.02 | .014 | .872 |
| Body mass index | 53.10 | 53.50 | −0.40 | −1.23 | .229 |
| Waist circumference | 67.98 | 70.53 | −2.55 | −2.52 | .017 |
| Climbing time | 138 | 160 | −22 | −7.70 | <.001 |
| Heart rate (beats/min) | 138 | 141 | −3.00 | −0.93 | .362 |
| Systolic blood pressure (mm Hg) | 150.20 | 155.30 | −5.10 | −1.41 | .171 |
| Diastolic blood pressure (mm Hg) | 76.90 | 79.70 | −2.80 | −1.20 | .241 |
| Delta heart rate (beats/min) | 63.10 | 63.70 | −.600 | −.172 | .864 |
| Delta systolic blood pressure (mm Hg) | 30.60 | 47.70 | −17.10 | −5.67 | <.001 |
| Delta diastolic blood pressure (mm Hg) | 4.37 | 7.50 | −3.13 | −1.02 | .351 |

lower than those among individuals in the general population. The researchers, therefore, instructed the nurses to perform calisthenics for 10 minutes per day for eight weeks. That intervention program, however, did not improve the physical fitness or physiological status. Tao (2007) evaluated the relationship between health activities of nurses and nursing quality and found that nursing staff had worked a mean of 8.46 years and that 54% of the nurses never exercised. Among the nurses who exercised, 25% of them jogged for 10 to 30 minutes once or twice a week.

The survey was designed to study the characteristics of sleep and perceived factors promoting and disturbing sleep (Vuori, Urponen, Hasan & Partinen, 1988). The available information suggests that especially light and moderate exercise had mainly positive effects on sleep. O'Connor & Youngstedt (1995) also showed that exercise is perceived as helpful in promoting sleep and suggest that regular physical activity may be useful in improving sleep quality. In addition, Harrington (2001) mentioned the importance of physical fitness and activity in helping workers reduce the problems associated with shiftwork. Exercise, as jogging, and evening walks, was the most frequently reported factor for promoting sleep and improving its quality. About 30% of participants reported exercise as the most important factor for improving sleep (Urponen, Vuori, Hasan & Partinen, 1988). The results of the present study showed that the stair-climbing activity significantly decreased the incidence of sleep disturbance and led to a reduction in waist circumference and delta systolic blood pressure. In summary, moderate physical training has been shown to increase sleep length and night-time alertness (Harma, 1996). Most previous studies examined the association between sleep, physical activity and human health by questionnaires (Huang, 2008; Vuroi et al., 1988; Urponen et al., 1988). The present study was to provide a health promotion program for nurses and examine the effectiveness of that intervention program on sleep quality and physical fitness.

## 6 CONCLUSION

Health promotion is defined as a behavior that is "motivated by desire to increase well-being and actualize human health potential". Indeed, as nurses focus on the health of their patients, families, and communities, practicing health-promoting behaviors for themselves should be concerned, too. Regular exercise is a key factor in promoting health. Many shiftworkers wish to, but cannot, perform leisure activities at the same times as dayworkers. However, stair climbing is a low-cost, inconspicuous, and readily accessible form of exercise. The present study found that stair-climbing effectively improved nurses' sleep quality and health. Thus, we recommend that nurses could perform the stair climbing exercise after a morning or a day shift and if performed after a night shift, some hours before a late evening nap.

## ACKNOWLEDGMENTS

This study was supported by a grant from the researchers' hospital (project no. V97 A-078). The authors wish to thank all of the individuals who participated in the intervention program.

## REFERENCES

Alessi, C.A. & Schnelle, J.F. (2000). Approach to sleep disorders in the nursing home setting. *Sleep medicine Reviews, 4*, 45–56.

Barton, J. (1994). Choosing to work at night: a moderating influence on individual tolerance to shift work. *Journal of Applied Psychology, 79*, 449–454.

Boreham, C.A.G., Wallace, W.F.M. & Nevill, A. (2000). Training effects of accumulated daily stair-climbing exercise in previously sedentary young women, *Preventive Medicine, 30*, 277–281.

Borg, G. (1985). An introduction to Borg's RPE scale. Movement Publications, Ithaca, New York.

Chang, S.F. (2003). Worksite health promotion—the effects of an employee fitness program, *Journal of Nursing Research, 11*(3), 227–230.

Chen, L.K., Hwang, S.J., Chou, S.P. & Tang, Y.F. (2004). Self-reported well-being and circadian hormone changes in nightshift nurses. *Taipei City Medical Journal, 1*, 289–296. (Original work published in Chinese).

Fox, M. (1999). The importance of sleep. *Nursing Standard, 13*, 44–47.

Harrington, J.M. (2001). Health effects of shiftwork and extended hours of work. *Occupational and Environmental Medicine, 58*, 68–72.

Huang, L.S., Chang, M.Y., Chen, S.E. & Chang, L.H. (2003). Evaluation study for promotion of physical fitness exercise in region hospital nurses. *Health Promotion & Health Education Journal, 23*, 67–77. (Original work published in Chinese).

Huang, Y.T. (2008). *The relationship between physical activity and sleep quality among nurses in a medical center*. Unpublished master's thesis, Taipei Medical University, Taipei, Taiwan. (Original work published in Chinese)

Hwu, L.J., Yuan, S.C., Yeh, P.M. & Chang, Y.O. (2005). Physical fitness among nurses at a medical center in central Taiwan. *Chang Gung Nursing, 16*(3), 243–251. (Original work published in Chinese)

Lien, C.W., Yu, C.H., Lin, S.L. & Liu, B.S. (2009, December). *Association of sleep quality and physical fitness comparing shiftwork of nurses*. Paper presented at the 10th Asia Pacific Industrial Engineering and Management System Conference (APIEMS), Kitakyushu, Japan.

Muecke, S. (2005). Effects of rotating night shifts: literature review. *Journal of Advanced Nursing, 50*, 433–439.

O'Connor, P.J. & Youngstedt, S.D. (1995). Influence of exercise on human sleep, *Exercise and Sport Sciences Review, 23*, 105–134.

Paffenbarger, R.S., Kampert, J.B., Lee, I.M., Hyde, R.T., Leung, R.W. & Wing, A.L. (1994). Changes in physical activity and other lifeway patterns influencing longevity. *Medicine and Science in Sports and Exercise, 26*, 857–865.

Perkins, L. (2001). Is the night shift worth the risk? RN, 64, 65–66, 68.

Reid, K., Roberts, T. & Dawson, D. (1997). Shiftwork and health. *The Journal of Occupational Health and Safety, 13*, 439–450.

Tao, Y.H. (2007). *Relationship between health activities and nursing quality of nurses in hospital*. Unpublished master's thesis, Chaoyang University of Technology, Taichung County, Taiwan. (Original work published in Chinese).

Urponen, H., Vuori, I., Hasan, J. & Partinen, M. (1988). Self-evaluations of factors promoting and disturbing sleep: an epidemiological survey in Finland. *Social Science & Medicine, 26*, 443–450.

Vuori, I., Urponen, H., Hasan, J. & Partinen, M. (1988). Epidemiology of exercise effects on sleep. *Acta physiologica Scandinavica Supplementum, 574*, 3–7.

*Ergonomics in Asia – Shih & Liang (eds)*
*© 2012 Taylor & Francis Group, London, ISBN 978-0-415-68414-9*

# Author index

Printed and bound by CPI Group (UK) Ltd, Croydon, CR0 4YY
01/11/2024
01782599-0006